SUPERALLOYS
Analysis and Control of Failure Process

Nataliya V. Kazantseva
Doctor of Sciences in Physics and Mathematics
Chief Researcher of the Institute of Metal Physics
Ural branch of Russian Academe of Sciences
Ekaterinburg, Russia
and
Professor of the Ural State University of Railway Transport
Ekaterinburg, Russia

Natalia N. Stepanova
Doctor of Sciences in Physics and Mathematics
Leading Researcher of the Institute of Metal Physics,
Ural branch of Russian Academe of Sciences
Ekaterinburg, Russia
and
Professor of the Ural Federal University
Ekaterinburg, Russia

Mikhail B. Rigmant, PhD
Senior Researcher of the Institute of Metal Physics
Ural branch of Russian Academe of Sciences
Ekaterinburg, Russia
and
Professor of the Ural Federal University
Ekaterinburg, Russia

Guest Contributor
Yurii S. Vorobiov
Doctor of Technical Sciences
Professor of A.N. Podgorny Institute for Mechanical Engineering Problems
National Academy of Sciences of Ukraine
Kharkiv, Ukraine

CRC Press
Taylor & Francis Group
Boca Raton London New York

CRC Press is an imprint of the
Taylor & Francis Group, an **informa** business
A SCIENCE PUBLISHERS BOOK

Cover illustrations reproduced by kind courtesy of Drs. Nataliya V. Kazantseva, Natalia N. Stepanova and Mikhail B. Rigmant (authors)

Preface

//////////////////////////////////

Superalloys form a class of the structural materials for high-temperature applications. Since long these materials have been used for gas turbine and steam-turbine blades in power engineering and air jet engine turbines. Nickel, Iron, and Cobalt superalloys are extensively used due to their excellent creep, fatigue, and corrosion resistance at elevated temperatures. Superalloys are complex materials which consist of different chemical elements that determine their physical and mechanical properties.

The primary purpose of this book is to present new results in studies of the intermetallic compound Ni_3Al and Ni_3Al-based, austenitic iron-based alloys, and new intermetallic $Co_3(Al, W)$-based alloys. Results of the studies of the working gas-turbine nickel blades are also presented in this book. In this book we have included both our own results and review of the world recent researches. We have tried to describe here the main problems of use of the superalloys in 21st century. In this book the features of the ternary phase diagrams Ni-Al-X (X = {Co, Fe, Nb, Ti, Cr}), effects of the alloying on the long-range order and mechanical properties of the Ni_3Al-based alloys are presented. Description of the strain-induced ferromagnetism in the Ni_3Al-based alloys and magnetic control of the failure of the gas turbine blades are also included. A separate section is devoted to the analysis of the vibration process and strength change in working single-crystal gas turbine blades. We have also included a review of the new intermetallic cobalt superalloys (Chapter 2). This is the first large systematic approach to study and present future applications of these new materials. The structure, crystal lattice parameters, orientation relationships between phases, mechanical and magnetic properties of the $Co_3(Al, W)$-based alloys are described in this book. This book contains a chapter (Chapter 3) devoted to nondestructive magnetic methods of study of Iron and Nickel superalloys. These methods are useful for the detection of the places which have high residual elastic stress that promote the early onset of cracks in the critical parts of industrial products. This book will be useful for material scientists, first-year postgraduate students taking a class in material science and engineering, and engineers developing new alloys for gas turbine technology.

Nataliya Kazantseva

Acknowledgments

The authors would like to thank the Russian Scientific Foundation (Project no. 15-12-00001) for supporting the studies and help in publication of this book. We thank to our colleagues from Institute of Metal Physics (Ekaterinburg, Russia) Nina I. Vinogradova, Denis Davydov, Michael Korkh, and Maria Ogneva for help in studies and preparing this book. Author (N. Kazantseva) thanks to PhD Alexander Menshchikov for help in writing in English.

Contents

Preface   *iii*
Acknowledgments   *v*
Introduction   *ix*

CHAPTER 1 Nickel Superalloys   1
Introduction   **1**
1.1 Structure, mechanical, and magnetic properties of the
     Ni_3Al-based alloys   1
     1.1.1 Effect of alloying elements on Ni_3Al   *1*
     1.1.2 Structure of nickel superalloys   *31*
     1.1.3 Ni_3Al-based intermetallic alloys   *58*
     *1.1.4 Prediction of the destruction processes in nickel
             superalloys by non-destructive magnetic
             methods*   *77*
1.2 Analysis of the strength and vibrations of cooled
     single-crystal blades   103
     1.2.1 Object of the study and calculation model   105
     1.2.2 The temperature condition of the blade   108
     1.2.3 Analysis of the thermal stress of the blade   110
     1.2.4 Analysis of vibrations of a cooled single-crystal blade   117
*Conclusion and future developments of the nickel
heat-resistant intermetallic alloys*   126

CHAPTER 2 Cobalt Superalloys   130
Introduction   **130**
2.1 Structure and mechanical properties of the
     $Co_3(Al,W)$-based alloys   131
     2.1.1 Effects of alloying with Al or W in the cobalt-based alloys   131
     2.1.2 Phase transitions in Co-Al-W system (Co-rich corner)   140
     *2.1.3 Deformation behavior of $Co_3(Al,W)$-based and
             Ni_3Al-based alloys*   155

2.1.4 Anomaly in the temperature dependence of yield stress   162
2.1.5 Dependence of the phase composition of the
*   Co-Al-W alloys on casting conditions*   165
2.1.6 Mechanical properties of the $Co_3(Al,W)$-based alloys   171
2.2 Magnetic properties of the $Co_3(Al,W)$-based alloys   177
2.2.1 Effect of Al and W alloying on magnetic properties
*   of Co-based alloys*   177
2.2.2 Magnetic properties of the phases in $Co_3(Al,W)$-based alloys   180
2.2.3 Effect of heat treatment and plastic deformation
*   on the structure and phase composition of*
* $Co_3(Al,W)$-based alloys*   184
Conclusion and future of the $Co_3(Al,W)$-based superalloys   190

CHAPTER 3 Iron Superalloys   **192**
3.1 Austenitic steels: physical and mechanical properties
     and area of their application   193
3.1.1 Chemical composition of the main grades of
*   corrosion-resistant steels*   193
3.1.2 Effect of ferrite on the operating properties of austenitic steels   195
3.1.3 Application of destructive methods for phase control of
*   austenitic steels*   196
3.1.4 Determination of the phase composition of austenitic
*   steels by methods of nondestructive testing*   198
3.2 Magnetic control of the phase composition of two-phase
     austenitic-ferritic and austenite-martensitic steels   204
3.2.1 Magnetic phase control of three-phase austenitic-ferritic-
*   martensitic steels*   208
3.2.2 Control of products made from heat-resistant
*   austenitic steels and nickel superalloys having*
*   relative magnetic permeability $\mu \leq 1.10$*   227
3.3 Relationship between the specific electrical resistance and the
     phase composition of austenitic steels   230
3.4 Instrumental control of the specific resistance of austenitic steels
     at operated conditions   237
Conclusion   239
Prospects of nondestructive testing products of
high-temperature austenitic steels   241
References   243
Index   273
Authors' Biography   277

Introduction

Nickel, cobalt, and iron-based alloys used for high-temperature application are known as superalloys. Superalloys are widely used for manufacturing of gas turbine and steam-turbine blades in power engineering and air jet engine turbines (Sims et al. 1987). Development of new heat-resistant superalloys covers various areas, such as optimizations of their structure and properties by experiments or computer simulations (Crudden et al. 2014), developments of composite materials, or powder additive technology (Carter et al. 2016, Basak et al. 2016). The method of Direct Metal Laser Sintering (DMLS) is one of the most promising technological manufacturing processes for the aviation industry and other industries. DMLS allows reduction of the phase segregation in heat-resistant alloys, especially in the rhenium-containing nickel superalloys (Zavodov et al. 2017).

There are numerous recent reviews of the structure and properties of nickel superalloys (Sims et al. 1987, Campbell 2006, Kassner 2015, Pollock 2006, Sato et al. 2008, Reed and Rae 2014, Jozwik et al. 2015). Austenitic stainless steels play an important role in the modern industrial world. Reviews of the processing, microstructure, properties and performance of high-temperature iron-based alloys were presented by Plaut et al. (2007) and Cardarelli (2013). We have not found any special review for new $Co_3(Al,W)$-based cobalt superalloys. Short reviews of these cobalt superalloys were conducted by Ishida (2008), Bauer et al. (2012), and Suzuki et al. (2015).

Because of their resistance to creep, oxidation, corrosion, and wear, iron-based superalloys are also used for manufacturing of the high-temperature aircraft bearings and machinery parts subjected to sliding contact (Cardarelli 2013). In the Gas Turbine Equipment (GTE), blades are the most stressed parts. To work in stationary gas turbine, the blades are designed with strength for a long-time operation (tens of thousands of hours). Alloys for aircraft engines have to withstand extremely high temperatures and dynamic stresses. It is difficult to increase the lifetime of turbine blades, because of the cyclical temperature regime of their operation. This leads to an increased demand for the reliability of turbine blades that operate in conditions of multicomponent stresses and contact with aggressive high-speed gas flows.

This book presents results obtained by the authors in studies of the intermetallic compound Ni_3Al and Ni_3Al-based, austenitic iron-based alloys, and new intermetallic $Co_3(Al,W)$-based alloys. Results of the studies of the working gas-turbine nickel blades are also examined in this book. The combination of high-strength Ni_3Al or $Co_3(Al,W)$ intermetallic compounds and a relatively soft nickel solid solution or cobalt solid solution in the structure makes it possible to classify these alloys as natural composites. At present, the Ni_3Al-based alloys are used, first of all, for the production of turbine blades of aircraft engines and stationary facilities in power engineering; the sixth generation of the nickel superalloys is in use. The $Co_3(Al,W)$-based alloys are promising materials for the new generation for high-temperature aero engine applications, currently in intensive development. The Ni_3Al- and $Co_3(Al,W)$-based alloys have a complex composition. The main strengthening phase of the nickel superalloys is $\gamma'(Ni_3Al)$. In the cobalt superalloys, the main strengthening phase is $\gamma'(Co_3(Al,W))$. An optimization of the alloy composition begins with an assessment of the alloying effect on the alloy properties. In this book, the effect of the alloying element on the bonding forces in the Ni_3Al intermetallic compound is shown using ternary model alloys with a third transitional element. The physical and mechanical properties of the doped Ni_3Al are discussed. This information does not lose its relevance and remains interesting for the development of new nickel superalloys.

The authors of this book participated in the development of a number of technological methods included in modern technology for the production of nickel superalloys. One of these was the use of High-Temperature Melt Processing (HTMP) for its homogenization before crystallization. This treatment leads to an optimization of the cast structure and a significant increase of the long-term strength of the alloy. The effect takes place for bulk crystallization. The increase in the long-term strength is especially significant, when homogenization of the melt is conducted before crystallization of the single-crystal aircraft blade. Introduction of Ultra-Disperse Powders of titanium carbonitride (UDP) to the melt before its crystallization has an effect, comparable to the high-temperature melt processing. To increase the long-term strength, this method may be used both independently and together with homogenization of the melt.

For high-temperature nickel materials, the maximum possible operating temperatures have already been reached. However, the problem of increasing the operation time of the product remains urgent. It is essential to ensure stability of the alloy structure during operation. In order to develop new alloys, it is interesting to study the possibility of the formation of new stable and metastable phases in the nickel or cobalt superalloy under crystallization, influence of high temperatures, and loads.

The problem of controlling the structure stability during the operation using non-destructive testing methods is most important. It is especially relevant for work in a forced regime, when the temperature and stress levels are extreme. Magnetic methods of nondestructive testing are not in demand for austenitic (paramagnetic) materials. But the authors have shown that the operation of the nickel gas turbine blade in the forced regime leads to the appearance of a ferromagnetic effect in the paramagnetic nickel superalloy. This effect is associated with structural changes leading to blade destruction. In iron-based austenitic alloys, ferromagnetic effect is associated with the formation of the martensitic ferromagnetic phase. Thus, there is a unique possibility for nondestructive testing to predict fractures before the appearance of a crack. The appearance of ferromagnetic properties in austenitic material under deformation gives a weak magnetic effect; hence sensitive equipment is required for its registration. A new compact portable device has been developed by the authors for magnetic non-destructive testing of products made of low magnetic iron and nickel superalloys.

Acoustic methods of nondestructive testing make it possible to reveal cracks already formed inside a product. Such methods have not been industrially developed for turbine blades from nickel superalloys. This is related to the difficulties in controlling multi-phase composition and dendrite structure of the cast material. In this book, the authors show the principal possibility of using acoustic methods for products made of nickel superalloys, including turbine blades.

Chapter One

Nickel Superalloys

INTRODUCTION

This chapter focuses on casting alloys. Two groups of alloys are considered: single-crystal and polycrystalline state of nickel superalloys and Ni$_3$Al single-crystal intermetallic alloys. All of these alloys are used for manufacturing of gas turbine blades. Results of the study on the influence of severe deformations on the structure and properties of the nickel superalloys are presented in this chapter. Nickel superalloys used in aircrafts operate at the highest levels of stresses and temperatures (up to 1200°C). Also, nickel superalloys used in electric power stations operate at lower temperatures (800-900°C), because of that the nickel gas turbine blades may also be under extreme stresses when the turbine power increases. The main factor for a long trouble-free operation time of the turbine blades is their structural and phase stability at high temperatures. In this chapter, we present the results of the analysis of the gas turbine blade failure under external effects (high temperature, rotation, vibration). We also found that a magnetic non-destructive testing of the structure state of the working nickel gas turbine blade may be used in the industrial monitoring process of its operation, maintenance and failure.

1.1 Structure, Mechanical, and Magnetic Properties of Ni$_3$Al-Based Alloys

1.1.1 Effect of alloying elements on Ni$_3$Al

Ni$_3$Al intermetallic compound (γ'-phase) is the main strengthening phase of the modern heat-resistant nickel superalloys used in the aircraft technology and stationary power gas turbine equipment. Ni_3Al intermetallic compound ($L1_2$, Pm3m) exists within a narrow concentration interval which is close to 75 at.% Ni. The aluminum atoms occupy the corners and the nickel atoms are face-centered in the Ni_3Al crystal lattice.

The main feature of Ni_3Al is its ability to dissolve almost all transition elements. This maintains a high degree of long-range ordering and $L1_2$ ordered structure type with two sublattice (nickel and aluminum). Numerous experimental data, reviewed by Sluiter and Kawazoe (1995), allow us to assert that the atoms of Nb, Ti, V, and W mainly replace the positions of aluminum and Co atoms that are included in the nickel sublattice. Fe and Cr atoms can equally be found in both sublattices.

In this section, we discuss the influence of alloying elements on the properties of the Ni_3Al intermetallic compound.

Lattice parameter. In the literature, the Ni_3Al stoichiometric state has different values for the lattice parameter a, ranging from 0.3566 ± 0.0001 nm (Stoeckinger and Neumann 1970) to 0.3589 nm (Stoloff 1989). It is known (Stoloff 1989, Cahn et al. 1987b) that the value of the lattice parameter of Ni_3Al in cast alloy can be unstable and changes during long-term exposure even at room temperature. In Stepanova et al. (2000), the change of the Ni_3Al lattice parameter on alloying with different chemical elements was studied by X-ray at room temperature and under continuous heating of the sample in a diffractometer chamber. The chamber was evacuated and the homogenized single crystal <001> samples were used for the study. Solution treatment (1578 K-2 hours) and two stage aging processes (1373 K-4 hours, vacuum cooling and 1113 K-20 hours, air cooling) of the samples, as it is shown by Stepanova et al. (2000), eliminates liquation and strain arising during the growth of a single crystal. In this case, the diffraction lines become narrower; for example, for all investigated alloys the (004) diffraction line width at half height was reduced by 28-35% after annealing. Room temperature values of the crystal parameter a (after annealing of the alloys studied) are given in Tables 1.1 and 1.2 (Stepanova et al. 2000). The summary in the study of the lattice parameter change for doped Ni_3Al by Stepanova et al. (2000) coincides with the conclusions obtained in the study of complex-alloying nickel superalloys. These conclusions may be summarized as follows: an introduction of the Nb and Ti atoms into the Ni_3Al crystal lattice leads to increasing values of the crystal parameter a; alloying with V atoms reduces the lattice parameter of Ni_3Al. The Fe, Cr, and Co atoms have a little effect on the values of the Ni_3Al crystal parameter (Mishima et al. 1985).

X-ray studies during continuous heating were done in the temperature range from room temperature to 1200°C under vacuum conditions. The heating rate was 5°C/min. An increase of the lattice parameter a values with increasing temperatures was found in almost all studied ternary alloys, with one exception (Tables 1.1 and 1.2) Savin et al. 2000a, 2000b.

Table 1.1 Parameters of the investigated alloys

Alloy	Composition at.%	a, nm (20°C)	t_a, °C	t_m, °C	S^2 (20°C)
Ni$_3$Al	75.3-24.7	0.35705	1330 ± 5	1925	0.98 ± 0.05
Ni$_3$Al-Nb	75-19-6	0.35956	1330	1925	0.96
Ni$_3$Al-V	75-21-4	0.35668	1260	1750	0.96
Ni$_3$Al-Ti	75-18-7	0.35912	1310	1850	0.94
Ni$_3$Al-W	75-22-3	0.35887	1310	1700	0.96
Ni$_3$Al-Fe	71-21-8	0.35714	1175	1840	0.85
Ni$_3$Al-Co	67-25-8	0.35705	1150	1850	0.85
Ni$_3$Al-Cr	72-24-4	0.35702	1100	1825	0.78

The coefficients α of thermal expansion of alloys based on Ni$_3$Al were experimentally determined by Stepanova et al. (2000). Table 1.2 presents two temperature ranges for the coefficients α.

Table 1.2 Parameters of the series of ternary Ni$_3$Al-based alloys

Alloy	$\alpha \cdot 10^5$, deg^{-1}, (20-1000°C)	$\alpha \cdot 10^5$, deg^{-1}, (1000-1200°C)
Ni$_3$Al	1.51 ± 0.04	1.91 ± 0.04
Ni$_3$Al-Nb	1.46	1.92
Ni$_3$Al-V	1.45	1.83
Ni$_3$Al-Ti	1.30	1.84
Ni$_3$Al-Fe	1.77	2.10
Ni$_3$Al-Co	1.77	1.59
Ni$_3$Al-Cr	1.63	2.41

In the range of 20-1000°C, the α values are similar to each other in alloys Ni$_3$Al-X, where X = {Nb, V, Ti}, as well as for the binary Ni$_3$Al alloy. In the alloys Ni$_3$Al-X (X={Fe, Co, Cr}), the coefficient α increases up to $1.77 \cdot 10^{-5}$ deg^{-1}. In the temperature range of 1000-1200°C, the values of the α coefficient of thermal expansion increase in all alloys except the alloy doped with cobalt, in which the coefficient α has a value of $1.59 \cdot 10^{-5}$ deg^{-1}.

In the Ni$_3$Al-V alloy, the value of the crystal parameter a is smaller than for Ni$_3$Al over the entire investigated temperature range. The changes of the lattice parameter with increasing temperature are shown in Fig. 1.1. The temperature dependence of the α coefficient of thermal expansion may be explained by the atomic substitution process in the Ni$_3$Al crystal lattice.

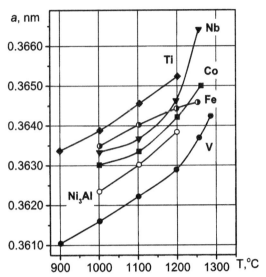

Fig. 1.1 Temperature dependence of the lattice parameter for Ni₃Al and the series of Ni₃Al-based ternary alloys doping with Nb, Ti, Co, Fe, V (Stepanova et al. 2000).

It was theoretically suggested by Enomoto and Harada (1989) that the temperature dependence of replacement type occurs in ternary alloys based on the elements that are placed at room temperature in both sublattices (in this case, chromium and iron). With increasing temperature, the atoms of these alloying elements start to replace mainly the positions of the aluminum atoms and, therefore, the vacancies appear in the nickel sublattice. The number of the nickel vacancies increases with increasing temperature. Work done by Stepanova et al. (2000) experimentally confirms correctness of the Enomoto's calculations. The alloy doped with cobalt (atoms of which are placed in the nickel sublattice) has different values of the α coefficient.

All these results suggest that the alloying elements can be divided into three groups according to the type of substitution: 1) Nb, V, Ti; 2) Fe, Cr; and 3) Co.

Stoloff (1989) gives the value $\alpha = 1.51 \cdot 10^{-5}$ deg^{-1} at 800°C. According to Stoeckinger and Neumann (1970), the coefficient of thermal expansion α in binary Ni₃Al alloy can be estimated as $(1.54 \pm 0.04) \cdot 10^{-5}$ deg^{-1} for temperatures up to 1000°C and $(2.03 \pm 0.04) \cdot 10^{-5}$ deg^{-1} at temperatures above 1000°C. These values are in a good agreement with the above-mentioned results.

Degree of long-range order. From this point of view, the concept of "degree of long-range order" with respect to intermetallic compounds makes no sense, although it is, of course, can be defined formally by

X-ray data. It will always be close to one (Stoeckinger and Neumann 1970) or 0.98 (Pope and Garin 1977).

The nature of the interatomic interactions in the intermetallic compounds is much closer to the pure chemical compounds than to the solid solutions. In addition to the metallic bonds, the covalent (and even possibly ionic) bonds have an important role (Iotova et al. 1996). The Ni_3Al crystal lattice consists of two sublattices with the Ni and Al atoms, respectively (Stoloff 1989). The third element may be placed into these sublattices according to the determined type of atomic substitution.

Stepanova et al. (2000) determined the degree of the long-range order S^2 by an X-ray method using the ratio of the integral intensities of super-structural and structural lines. For Ni_3Al, more accurate values of S^2 are determined from the other pairs, e.g., (110) and (220), not from the intensity ratio of the diffraction lines (100) and (200). The values of S^2 obtained by Stepanova et al. (2000) are given in Table 1.1.

A high degree of the long-range order S^2 in the Ni_3Al intermetallic compound retains up to its melting temperature. Disordering starts at 1330°C and the complete disorder is reached at the melting point (1373°C) (Stepanova et al. 2000). The disordering temperature of Ni_3Al (Cahn et al. 1987a) is defined as 1450°C.

It is known that the γ'-phase super-structural diffraction reflections become wider with increasing temperature in the binary Ni_3Al and in the complex-alloying nickel superalloys (Solly and Winquist 1973). This broadening is much greater than it would be under the influence of temperature only. Solly and Winquist (1973) suggested that the degree of S^2 decreases when the temperature increases to the melting point (or to the temperature of complete dissolution of superalloys). The higher the original degree of the long-range order, the slower this process should be.

The initial changes of the degree of the long-range order before disordering can be imagined as the accumulation of point defects and/or their complexes. Starostenkov et al. (2005) conducted a comparative analysis of the disordering process by modeling the alloy Cu_3Au and intermetallic compound Ni_3Al (both alloys have the $L1_2$ superstructure). It was shown that the disordering process can occur, for example, by means of the association of the point defects in plane (111) by the complexes containing interstitial atom and three vacancies symmetrically located around. The accumulation of such defects along the close-packed directions leads to the disordering in the Cu_3Au alloy in the range of 400-500°C, whereas for Ni_3Al the disordering occurs near the melting point.

It was established in many studies that the degree of the long-range order in Ni_3Al doped with a third element remains close to unity (Goman'kov et al. 2008, Stepanova et al. 2000). However, there are elements, alloying of which leads to the decrease of the long-range order. In the alloy $Ni_{72}Al_{24}Cr_4$, not only the starting point of the order-disorder transition, but also the point of the complete disorder t_c = 1305°C was recorded. In $Ni_{75}Al_{12.5}Fe_{12.5}$ the degree of the long-range order S^2 of 0.85 was obtained (Cahn et al. 1987a).

Stepanova et al. (2000) observed a continuous transition from the temperature dependence $\rho(t)$ characteristic for intermetallic compounds to the dependence $\rho(t)$ characteristic for ordering alloys using a series of alloys Ni_3Al-Fe with different concentrations of the alloying element. As a result, the temperature of the complete disorder is fixed well below the melting temperature. Full disordering in this work was also observed in the $Ni_{72}Al_{24}Cr_4$ alloy (t_a = 1100°C; t_c = 1305°C).

A review of data on the concentration dependence of the temperature of the order-disorder transition is given by Kozubski (1997) for the alloys based on Ni_3Al (with Fe, Mn, Cr).

Measurement of the electrical resistivity as a method of studying the ordered state. This section presents results of the electrical resistivity measurements. Stepanova et al. (1999, 2000) conducted experiments in rotating magnetic field using a contactless method. Previously, Corey and Lisowsky (1967) investigated the temperature dependence of the electrical resistivity $\rho(t)$ for the binary Ni_3Al. Figure 1.2 presents the results of Stepanova et al. (1999). The electrical resistivity $\rho(t)$ linearly increases with temperature up to 1330°C, and then sharply declines toward the melting point. This drop is associated with the onset of the disordering (Corey and Lisowsky 1967).

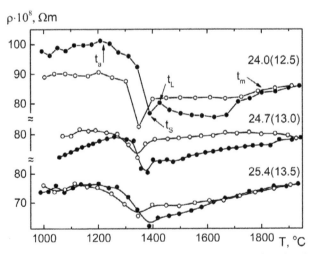

Fig. 1.2 Temperature dependence of electrical resistivity for the three alloys with composition near the stoichiometric Ni_3Al (Stepanova et al. 1999).

As a rule, the decrease in $\rho(t)$ occurs when the ordered state forms; Cu_3Au ($L1_2$) is a typical example of the alloy with such disordering behavior (Rossiter and Bykovec 1978). On the other hand, for Ni_3Al, the sharp drop of the electrical resistivity coincides with the decrease of S^2 (the factor $d\rho(t)/dt$ is negative). This phenomenon is observed in Ni_3Al and Cu_3Au due to the peculiarities of their electronic spectrum.

There are different models explaining the negative coefficient $d\rho/dt$. For example, the decrease in $\rho(t)$ at temperatures above t_a can be explained based on the theoretical considerations developed by Kozlov et al. (2006a), Los et al. (1991), and Nautiyal and Auluck (1992).

According to Kozlov et al. (2006a), an energy gap always appears in the electron spectrum at the ordering when the energy range corresponding to the Brillouin zone boundary of the ordered alloy. For the degree of the long-range order of $S^2 = 1$, the depth of the energy gap is maximal. In this case, the depth of the energy gap does not depend on temperature.

If the degree of filling is such that the Fermi level is far from the energy gap, the resistivity does not change, as demonstrated, for example, by the alloy Cu_3Au (Courths and Löbus 1999). In this case, the change in the electrical resistivity is related to scattering processes, and the decrease in $\rho(t)$ is caused by the ordering. When the degree of filling is such that the Fermi level is near the energy gap, the ordering is accompanied by the increasing resistivity.

The value of the electrical conductivity σ in the alloy is determined by the density of states at the Fermi level $N(E_F)$. In this case, the resistivity increase is related not so much to a decrease of $N(E_F)$ (exciting mainly d-states), but to the decrease of the mobility of the s-electrons (due to the s-d hybridization), which are mainly responsible for the transport phenomenon . After the disordering, an increase in scattering also takes place in the Ni_3Al alloy. In this case, however, the changes in the electronic spectrum have significantly more influence on the resistance, which leads to a sharp drop in the resistivity values.

Measurements of the electrical resistivity ρ are the most rapid method of studying the phenomenon related to the ordering processes. The results obtained by Stepanova et al. (2004) for ternary alloys based on Ni_3Al are given in Table 1.3, where. t_a - temperature of the onset of disordering; t_L - liquidus; t_S - solidus; t_L - t_S is the range of crystallization; t_m - temperature of the melt homogenization. There is a hysteresis loop between the heating and cooling for the temperature dependence $\rho(t)$. In addition, data obtained by the electrical resistivity measurements and DSC are somewhat different. This is because the methods respond to different stages of the melting process and, apparently, the structural state, which is different at the onset of melting. The destruction of the ordered state by melting leads to the formation of the melt state with a short-range order. As a result, there is a hysteresis between high-temperature heating and subsequent cooling, and the critical point t_m

is fixed in the liquid state (Barishev et al. 2010, Tyagunov et al. 2013, 2014, 2015).

The short-range order clusters exist in the dynamic mode (with a relative balance of the processes of disintegration and formation). In the melt, there are also complexes of atoms with a structure similar to the carbide phase.

Table 1.3 Comparison of the phase transition temperatures defined by the measurements of the electrical resistivity $\rho(t)$ and DSC

Alloy, at. %	t_a, °C heating	$t_L - t_S$, °C melting by $\rho(t)$	$t_L - t_S$, °C crystallization by $\rho(t)$	$t_L - t_S$, °C melting by DTA	t_m, °C
Ni₃Al 76.2-23.8	1260	1385-1420	1345-1394	1365-1390	1620
Ni₃Al 75.3-24.7	1330	1380-1405	1345-1390	1370-1390	925
Ni₃Al 74.6-25.4	1275	1385-1420	1345-1390	1365-1410	750
Ni₃Al-Nb 75-19-6	1330	1371-1409	1358-1408	1360-1385	925
Ni₃Al-V 75-21-4	1260	1384-1420	1345-1395	1380-1405	750
Ni₃Al-Ti 75-18-7	1310	1384-1407	1346-1408	1375-1405	850
Ni₃Al-W 75-22-3	1310	1383-1409	1345-1395	1400-1425	700
Ni₃Al-Fe 71-21-8	1175	1371-1409	1357-1407	1370-1395	840
Ni₃Al-Co 67-25-8	1150	1360-1435	1346-1395	1385-1410	850
Ni₃Al-Cr 72-24-4	1100	1384-1408	1370-1408	1375-1405	825

The destruction of clusters completes in the process of heating of the melt to t_m. The temperature of the t_m anomaly corresponds to the transition of the melt into a more homogeneous, equilibrium state. The temperature behavior of the resistivity $\rho(t)$ has a hysteresis during the subsequent cooling after heating above t_m that testifies to the irreversible nature of the changes of the melt structure.

In liquids, the scattering processes provide the main contribution to the resistance; presence of the short-range order reduces its value. Kolotukhin et al. (1995) presented experimental results in an attempt to estimate sizes of such atomic associations (about $5 \cdot 10^{-9}$ m), which decreased in the melt with increasing temperature. The authors studied the temperature dependence of the electrical resistivity and kinematic viscosity in combination with X-ray studies of the melt.

Note, that there is a correlation between the values t_a at the onset of disordering and the homogenization temperatures t_m of the melt. However, the complete correlation between t_a and the value of the melting temperature is not so clear. One can only state that there exists a trend of the increase of the melting temperature with increasing t_a.

Ni-Al phase diagram. Figure 1.3 shows an experimental high-temperature part of the phase diagram (Stepanova et al. 2003a).

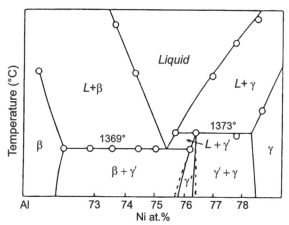

Fig. 1.3 High temperature part of the phase diagram Ni-Al system (Stepanova et al. 2003a).

In the past, there was a debate about the type of the Ni-Al phase diagram (Hunziker and Kurz 1997, Battezzatti et al. 1998, Lee and Verhoeven 1994, Stepanova et al. 2003a, 2004). The Ni-Al phase diagrams were complicated by the metastable states easily formed during the melting process. It is in agreement with the phase diagram obtained by Hilpert et al. (1987). Currently, the main body of the experimental evidence confirms the existence of Ni_3Al in such diagrams.

Metastable phases formed during crystal growth of Ni_3Al by the Bridgman method. According to the Ni-Al phase diagram in the composition range of 75-76 atomic percentage of Ni, three phases form almost at the same temperature (Fig. 1.3): the γ'-phase (Ni_3Al, $L1_2$), the nickel-based γ solid solution (FCC), and the β-phase (NiAl, B2). It is known (Hunziker and Kurz 1997) that stoichiometric single-phase Ni_3Al state is difficult to obtain because of the complex nature of the Ni_3Al crystallization, which involves peritectic and eutectic reactions. Battezzatti et al. (1998) analyzed the influence of the solidification speed on the phase composition in the Ni_3Al-based alloys. These alloys were directionally solidified by the Bridgman method. The character of the growth structure was studied in a series of samples within the homogeneity area of the

γ'-phase (Ni_3Al). $Ni_{76}Al_{24}$ was the only composition among the studied binary alloys, in which a single-phase state was formed directly from the melt.

A single crystal of the Ni_3Al alloys with a chemical composition within the area of homogeneity has a small (about 1%) amount of the residual β-phase in the as-cast state (Stepanova et al. 2003a). This fact is experimentally supported by a directed crystallization with a high temperature gradient.

β-Phase is transformed from $B2$ to the martensitic phase under quenching of the alloys from a high temperature region. It is believed that the type of the crystal structure of the martensitic phase is $L1_0$ (Chakravorty and Wayman 1976). All three phases (γ', β, and $L1_0$) are ordered, therefore, all possible reflexes are observed in the diffraction pattern due to the presence of superstructure. This greatly complicates the phase identification by the TEM studies.

Stepanova et al. (2003a) used a diffraction method of the thermal neutrons to study the phase composition. This method allows us to analyze the entire volume of the single crystal sample directly (due to a high penetrating power), to set its phase composition, and to evaluate the perfection of its crystalline structure. The research was conducted with a longer time of signal registration to increase the sensitivity of the neutron diffraction techniques. Significant improvements in the efficiency of the phase analysis were achieved using the parallel neutron multi-detector (16-channel) system.

Strong Ni_3Al lines were observed in neutron diffraction patterns, along with various low-intensity reflections. The existence of the martensitic phase with the $L1_0$ superstructure should be considered as a disputed fact, because the reflections observed on neutron diffraction patterns were actually banned for $L1_0$. More realistically, the diffraction reflections of the martensitic phase can be described as deformed tetragonal lattice obtained from the $L1_2$ structure (type Ti_3Cu, P4/mmm). The formation of such a phase was observed previously by Sanati et al. (2001).

In addition, another group of diffraction reflections is observed in the neutron diffraction patterns (Kazantseva et al. 2009a). Those reflections do not belong to a cubic crystal system and do not correspond to lines of any known stable phase in the Ni-Al system. In the electron-microscopic images, this phase is visible in the form of dispersed particles, forming a fibrous pattern in the structure of as-cast Ni_3Al single crystals. This metastable phase is formed from non-stoichiometric β-phase by shear. It belongs to the category of the so-called "omega"-phase and has a composition close to Ni_2Al and a crystal lattice ranging from C6 trigonal (P-3m1) to the more equilibrium $B8_2$ (P6$_3$/mmc). High-temperature annealing promotes the conversion of the omega-phase in Ni_3Al. A complete elimination of all non-equilibrium phases requires the prolonged homogenizing annealing (100 hours at 1000°C). A deformation of 38% also contributes to the acceleration of the decay of non-equilibrium structures.

The effect of the third alloying element on the Ni-Al phase diagram. The formation of ternary alloys based on the Ni₃Al intermetallic compound is achieved by adding practically all transition elements (Stoloff 1989, Sims et al. 1987). The incision isothermal ternary diagram Ni-Al-X undoubtedly consolidates the main information source about triple Ni₃Al-X alloys with transition elements at 1100°C (Stoloff 1989).

Stepanova et al. (2004) considered the effect of the third alloying element with a different substitution type on the phase boundaries in the Ni-Al phase diagram. The composition of alloys was selected in the middle of the homogeneity area at 1100°C of the Ni₃Al-X ternary diagram (Stoloff 1989) (Table 1.3). The phase composition scheme of the Ni₃Al-X ternary alloys is shown in Fig. 1.4.

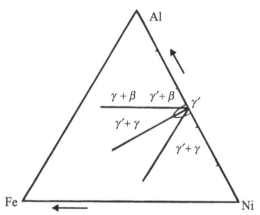

Fig. 1.4 Scheme of the phase composition of ternary alloys based on Ni₃Al with different type of substitution (Savin et al. 1999).

The single crystals obtained by the Bridgman method were used in the study. In the case of the replacement of the Al position or the placement of the alloying atom into both sublattices, the single-phase γ' state forms. A two-phase state $\gamma + \gamma'$ is observed for a higher concentration of the alloying elements.

A cellular growth structure is found in a single crystal; the directional solidification rate is 0.5 mm/min. The β-phase is involved in the crystallization process when the substitution positions are in the nickel sublattice, and a cellular-dendritic growth structure of the single crystal is observed in this case. Temperatures of the phase transformation, defined from the DTA, and the measurements of the temperature dependence of electrical resistivity are given in Table 1.3.

Just as in the case of the crystallization of the Ni_3Al intermetallic compound, which must be formed directly from the liquid according to the equilibrium state diagram, the crystallization of in Ni-Al system occurs in several stages due to the cooling conditions during the directional solidification in the Bridgman method. Metastable $\beta - \gamma$ eutectic forms appear instead of the crystallization and the cellular-dendritic two-phase structures (Stepanova et al. 2003a). The intermetallic γ'-phase is formed by the peritectoid $\gamma \rightarrow \gamma'$ and eutectoid $\beta \rightarrow \gamma'$ phase transitions just in the solid state, which contributes to the appearance of the γ'-phase areas, formed in various ways, with various chemical compositions. The remaining (unconverted) intermetallic β-phase (NiAl) during the ingot cooling can also promote the martensitic transition. Hunziker and Kurz (1997) described this process in detail.

With an example of the Ni_3Al-Fe system alloys, the formation of the metastable states was considered, shown schematically in Fig. 1.5.

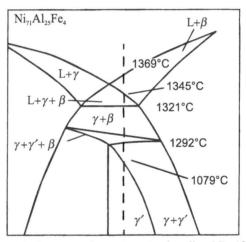

Fig. 1.5 Diagram of phase transformations in the alloy $Ni_{71}Al_{25}Fe_4$ in the case of its crystallization through metastable eutectic $\gamma + \beta$ (Lepikhin et al. 2004).

Elastic moduli for Ni_3Al and the ternary alloys based on Ni_3Al. The values of the elastic moduli for Ni_3Al and for the ternary alloys based on Ni_3Al are given in Tables 1.4 and 1.5 (Rinkevich et al. 2006). The elastic properties are closely related to the interatomic interactions. In the literature, there are data about the modules of elasticity for Ni_3Al alloy (Stoloff 1989, Stoeckinger and Neumann 1970, Iotova et al. 1996, Yoo 1987, Ono and Stern 1969, Dicson and Wachtman 1969, Kayser and Stassis 1981). Rinkevich et al. (2006) conducted an acoustic study of the elastic properties in the $Ni_{75}Al_{19}Nb_6$ and $Ni_{67}Al_{25}Co_8$ single crystal alloys. Table 1.4 compares results for the $Ni_{75}Al_{19}Nb_6$, $Ni_{67}Al_{25}Co_8$, and Ni_3Al alloys.

Doping with cobalt leads to decreasing ultrasonic wave velocities by 8%. Doping with niobium, on the contrary, leads to an increase of the velocities. This trend is evident in all measurements, except one, when the longitudinal wave propagates along the axis <001> of the crystal (Rinkevich et al. 2003, 2006). The measurements of the single crystal doped with cobalt experimentally show an anomalously large value of the anisotropy factor *A*. The values obtained for the *A* factor are confirmed by the results of the attenuation coefficient measurements.

The results for the attenuation coefficient show a significant increase of the attenuation coefficient of the longitudinal waves in Ni_3Al-Co at a frequency of 15 MHz. For the transverse waves and a frequency of 50 MHz, systematic data could not be obtained because only a single echo pulse was often observed. The Debye temperature T_D was calculated from elastic moduli (Rinkevich et al. 2006). There are several methods for calculating T_D from the velocities of elastic waves. Truell et al. (1969) presented an overview of these techniques; one of the simplest methods was developed by Leibfried. In addition to the Leibfried method, T_D was also calculated by the method of the power series expansion. The resulting accuracy of the Debye temperature determination is within 1-2%.

Table 1.4 The elastic moduli c_{ij} for Ni_3Al and the Ni_3Al-based ternary alloys

Alloy	T_D, K	c_{ij}, GN/m^2			Reference
		11	12	44	
$Ni_{75}Al_{19}Nb_6$	464	47	51	35	(Rinkevich et al.2006)
$Ni_{67}Al_{25}Co_8$	379	14	77	03	(Rinkevich et al. 2006)
Ni_3Al	438	24	48	25	(Kayser and Stassis 1981)

Table 1.5 The bulk modulus *B*, the anisotropy factor *A*, and the shear modulus *G* for Ni_3Al and the Ni_3Al-based ternary alloys

Alloy	B, GN/m^2	A	G, GN/m^2	B/G
$Ni_{75}Al_{19}Nb_6$	183	2.80	89	2.06
$Ni_{67}Al_{25}Co_8$	189	5.56	49	3.85
Ni_3Al	179	3.30	73	2.45

There are a few estimates of the Debye temperature for the Ni_3Al alloy in the literature. Stoeckinger and Neumann (1970) measured the value T_D = 360 ± 20 K from the results of X-ray studies. In an earlier work, Ono and Stern (1969) obtained T_D = 462 K from the calculation of the experimental values of the elastic moduli of Ni_3Al. The value is close to T_D = 467 K reported by Rinkevich et al. (2006) for the alloy Ni_3(Al, Nb) (Table 1.6). Whereas niobium doping increases the Debye temperature, alloying with cobalt reduces it. A comparison between the anisotropy factors and the Debye temperatures confirms the trend.

The latter holds true for the nickel-based superalloys: an increase in the anisotropy factor leads to decreasing Debye temperatures.

Table 1.6 The Debye temperature T_D calculated by different methods and the melting temperature T_m for several alloys based on Ni_3Al

Alloy	T_D, K method of Leibfried	T_D, K, method of the power series expansion
$Ni_{75}Al_{19}Nb_6$	464	474
$Ni_{67}Al_{25}Co_8$	379	371
Ni_3Al	438	448

The increase of the elastic modulus and Debye temperature with alloying are characterized by strengthening the bonding forces in the $Ni_{75}Al_{19}Nb_6$ crystal in comparison with the Ni_3Al alloy.

Diffusion in the Ni₃Al intermetallic compound. Numerous experimental data were obtained on nickel self-diffusion in Ni_3Al (Hancock 1971, Hoshino et al. 1988, Minamino et al. 1992, Shi et al. 1995, Zulina et al. 1997), including the results of the diffusion dependence on the alloy composition. *In-situ* measurements of the Al self-diffusion are difficult to carry out because of an insufficient availability of the suitable isotope ^{26}Al. The nickel grain boundary self-diffusion is discussed by Zulina et al. (1997).

The intermetallic compound Ni_3Al is characterized by a high degree of the long-range order. A casual exchange of atom positions in the ordered materials leads to disordering. To preserve the long-range order, an inclusion of more complex self-diffusion mechanisms is required, taking into account such factors as the crystalline structure, degree of the order, concentration, and temperature dependence. In stoichiometric Ni_3Al, the nickel atom has eight Ni atoms and four Al atoms as the nearest neighbors, whereas the aluminum atoms are surrounded only by the Ni atoms. Therefore, the Ni self-diffusion can occur by vacancy motion in the nickel sublattice, without disordering.

The diffusion coefficient D is described by the Arrhenius equation: $D = D_0 \exp(Q/RT)$, where Q is the activation energy of diffusion, R is the gas constant, and T is the absolute temperature.

In contrast to the NiAl intermetallic compounds, the structural vacancies were not found in Ni_3Al by the method of positrons annihilation (Badura-Gergen and Schaefer 1997). Thermal vacancies form predominantly in the Ni sublattice. The absence of the structural vacancies in Ni_3Al is confirmed by the fact that the Arrhenius equation is complied almost perfectly for the Ni self-diffusion (Frank et al. 1995).

The effective enthalpy of the vacancy formation in Ni_3Al is about the same as in the pure FCC metals and it increases with increasing nickel concentrations. In Ni_3Al, a low vacancy concentration has a relatively high mobility, whereas for NiAl, a reversed situation is found: a low vacancy formation enthalpy combines with a high migration enthalpy (Badura-Gergen and Schaefer 1997). Effective enthalpy H_V^F of the vacancy formation in Ni_3Al in (Badura-Gergen and Schaefer 1997) estimated as 1.65 ± 0.08 eV; of 1.81 eV and 2.01 eV for alloys with 74.1; 75.2; 76.5 at.% Ni, respectively.

Wang et al. (1984) estimated the enthalpy of the vacancy formation $H_V^F = 1.6 \pm 0.2$ eV in the Ni position using the method of positron annihilation; the enthalpy of the vacancy migration is $H_v^M = 1.2 \pm 0.2$ eV. Approximately 80% of all vacancies are located in the nickel sublattice. It should be noted, however, that the nickel diffusion coefficient is higher at low temperatures and that at high temperatures, "the fast" element is aluminum. This result was obtained by Shi et al. (1995) in a diffusion study performed with the isotopic method.

The diffusion of the Al atoms is more complicated. Experiments on the positron annihilation showed that the thermal vacancy concentrations in the nickel sublattice exceed those in the Al sublattice by an order of magnitude. It is natural to assume that the Al diffusion occurs by a mechanism that includes the Ni sublattice vacancies. This implies the formation of the antisite defects, at least temporarily. Such a model has been developed and Divinski et al. (1998) found an experimental confirmation.

The Al diffusion rate is determined by the probability of finding a Ni vacancy among its nearest neighbors. Since the vacancy concentration in stoichiometric Ni_3Al at low temperatures is about 10^{-2} (Badura-Gergen et al. 1997), such a mechanism determines the low diffusion mobility of the Al atoms. Experiments have shown that the ratio of diffusion coefficients $D_{Ni}/D_X > 1$ at room temperature, whereas $D_{Ni}/D_X < 1$ at high temperatures (X = {Ge, Ga}). Therefore, the mechanism of the Al diffusion is more difficult to describe than that of the Ni atoms. Another problem is the diffusion of the Nb or Ti atoms substituting the aluminum positions.

Within experimental errors, the diffusion of the atoms of all these elements is described by the Arrhenius equation in the range of temperatures investigated. However, they can be divided into the two groups: Ga, Ge (the elements of the Al homologous series) and Ti, Nb (the elements substituting the Al positions). The diffusion of Ga and Ge has the same temperature dependence, characterized by almost identical values of the activation energy. The activation energy is above of 60 kJ/mol, which is close to the activation energy of Ni-diffusion corresponding to the estimated enthalpies of the antisite defect formation in Ni_3Al (Divinski et al. 1998). In contrast to Ga and Ge, the Ti and Nb atoms have lower diffusion mobility with higher activation energies. Apparently, the diffusion mechanism crucially depends on the effective energy of the formation and migration of the antisite defects.

At 1306°C, the volume diffusion coefficients D_v are $2.45 \cdot 10^{-16}$, $1.12 \cdot 10^{-17}$, and $2.18 \cdot 10^{-18}$ m^2/s for Ge, Ti, and Nb, respectively (the $Ni_{75.6}Al_{24.1}$ sample was coated with a layer of diffusing substance) (Divinski et al. 1998).

Cermak and Rothova (2003) gives the experimental values of the activation energy Q for the diffusion of an alloying element X in the diffusion pairs Ni_3Al/Ni_3Al-X. Kozubski (1997) gives the self-diffusion activation energy of the alloying element X (where X = {Ti, W, V, Mo, Nb}) deposited on the surface of Ni_3Al. These results are consistent with the data presented in Table 1.7. Thus, Ni, as well as Cu and Co atoms substituting the Ni position, is subject to the vacancy diffusion in the nickel sublattice.

Diffusion coefficients for Cu and Co are higher than those for the Ni self-diffusion because of the strong attraction between the vacancy and substitution atoms in the Ni positions. Furthermore, an addition of Cu and Co reduces the degree of the long-range order in the ternary alloy.

Table 1.7 Activation energies of the relaxation process Q_A and diffusion Q in Ni_3Al and Ni_3Al-X ternary alloys

Alloy	Composition, at.%	Q_A, kJ/mol	Q, kJ/mol	Reference
Ni_3Al	75.3-24.7	249 ± 20	269 (Al) 302 (Ni)	(Zulina et al. 1997) (Divinski et al. 1998)
Ni_3Al-Co	67-25-8	374	325	(Minamino et al. 1992)
Ni_3Al-Nb	75-19-6	416	476	(Divinski et al. 1998)
Ni_3Al-V	75-21-4	531	467	(Minamino et al. 1997)
Ni_3Al-W	75-22-3	604	463	(Kozubski 1997)
Ni_3Al-Ti	75-18-7	655	468	(Minamino et al. 1992)

The diffusion mechanism of Ge, Ga, and Si is identical to the Al self-diffusion. The atoms of elements replacing the aluminum position (V, Ti, Nb) form a separate group. Apparently, the Fe or Cr atoms equally occupying the positions in both sublattices have the same diffusion mechanism as Ni: the vacancy diffusion in the nickel sublattice. If the enthalpy of the X atom placement in the antisite positions is higher than it is for the Al atoms, the X element diffusion can be substantially slow down. The relaxation phenomenon may be useful for evaluating the diffusion activation energy (Q_A, Stepanova et al. 2000).

The internal energy of the system is determined by the state parameters, such as stresses, temperatures, fields, etc., as well as a number n of internal parameters v_i. The equilibrium values v_i^e are also determined by the values of the state parameters (Cahn and Haasen 1996). These internal variables may be the order parameters of the alloy, including the degree of its long-range order. If one of the state parameters (for example, temperature) suddenly changes, the other internal parameters (associated with it) relax toward their new equilibrium values. In our case, the relaxation process controls the diffusion mobility D_m. Thus, it is possible to measure the relaxation time τ associated with D_m using the equation: $D_m = ka^2/\tau$, where a is a typical grating spacing; k is a constant determined by the model type.

Changing physical properties during the relaxation process by diffusion typically follow an *S-shaped curve*. We note that the relaxation times for various physical quantities may significantly vary within the same temperature range, because of the association with different types of processes. When the activation energy is determined from the electrical resistivity measurements, the dependence of the resistivity on the relaxation time is exponential: $[R(t) - R(\infty)]/[R(0) - R(\infty)] = \exp(-t/\tau)$, where τ is the relaxation time of the process (Kozubski and Cadeville 1988).

Ordering is a diffusion process, which occurs by a vacancy mechanism (Schoijet and Girifalco 1968). Regarding the measurements of the lattice parameters in the FCC cubic crystals, the relaxation phenomenon and the attainment of the maximum possible degree of the long-range order at a given temperature in the ordered alloys occur in the following way. (1) Rapid heating of the sample. (2) Sharp increase of the number of thermal vacancies during the high-temperature isothermal aging below the temperature of the onset of disordering t_a. At first, this process is accompanied by a decrease of the value of the crystal lattice parameters and, with time, the lattice parameter changes vanish.

The activation energy Q_A is determined from the Arrhenius equation for the relaxation rate: $V = A \exp(Q_A/RT)$, where A is a constant, R is the universal gas constant, and T is the temperature on the Kelvin scale. Results from Stepanova et al. (2000) show that the alloying of the Ni_3Al with all transition elements increase the activation energy Q_A. The highest values of Q_A are typical for the alloys with the tendency to occupy the aluminum positions in the Ni_3Al crystal lattice. Table 1.7 shows the activation energies Q_A for the relaxation process, as well as the diffusion activation energies Q for the Ni and Al atoms in the Ni_3Al intermetallic compound and for the alloying element in the Ni_3Al-X ternary alloys. The main problem of the measurements of the relaxation parameters is that the process is continuous and therefore it is almost impossible to fix its initial moment. Resistivity changes are always somewhat delayed in time. According to Kozubski et al. (1993), the resistivity measurement error cannot be lower than ± 20 kJ per mol.

Factors determining the substitution type for the alloying element in Ni_3A. An addition of the third element to the Ni_3Al crystal lattice leads to changes in its electronic structure. It also changes the character of the interatomic interaction and the level and type of the chemical bonding of the compound. Lawniczak-Jablonska et al. (2000) showed by an X-ray analysis that the electron charge distribution in the $Ni_3(Al,Fe)$ compound is more isotropic than that in the Ni_3Al compound. This increases the metallic component of the bonding and, correspondingly, in the plasticity. Rinkevich et al. (2006) determined the elastic properties of the Ni_3Al-X ternary compounds doped with niobium or cobalt using acoustic measurements. Increasing elastic moduli and Debye temperature were found in the $Ni_{75}Al_{19}Nb_6$ alloy in comparison with Ni_3Al. These facts indicate an increase in the bond strength of the Ni_3Al-Nb crystal lattice. The doping with cobalt strongly affects the normal components of the elasticity modulus and decreases the shear modulus (from 73 GN/m^2 to 49 GN/m^2) and the Debye temperature in comparison with Ni_3Al (Rinkevich et al. 2006). An anomalously large value of the elastic anisotropy factor A was experimentally revealed for the $Ni_{67}Al_{25}Co_8$ single crystal as compared to the Ni_3Al single crystal (5.56 and 2.64, respectively), which was also confirmed by the measurements of the damping coefficient (Rinkevich et al. 2006). This kind of experimental research is ongoing, with the expansion of the range of alloying elements (Bagot et al. 2017).

The review of the experimental results taken from previous studies allows us to subdivide all investigated alloying elements into the three groups: (1) Nb, V, W, Ti; (2) Fe, Cr; (3) Co.

These alloying elements may be characterized as follows:

(1) The elements of the first group are in the atomic positions of the aluminum sublattice. The elements of the second group can be equally placed into both sublattices of Ni_3Al. The elements of the third group are in the nickel sublattice;

(2) The Ni_3Al-X alloys with the chemical elements taken from the above three groups differ by their physical properties, including the temperature dependence of the thermal expansion coefficient;

(3) The elements from the second and third groups can act as plasticizers, whereas the first group elements increase the alloy brittleness in the polycrystalline state;

(4) In the alloys doped with the elements from the second and third groups, the temperature of complete disordering t_C can be obtained in the solid state by increasing the concentration of the alloying element. In the alloys of the first group, the onset of the disordering process happens at temperatures close to the melting point.

There is a point of view that such differences are caused by a variable impact on the ordering energy. Namely, the elements of the first group lead to the increase of the ordering energy, while the elements from the second and third groups reduce the energy.

Positron annihilation study of the doping influence on the 3d-electron states in Ni_3Al. Druzhkov et al. (2010), used positron annihilation spectroscopy to investigate the influence of doping with a third element (Fe, Co, Nb) on the electronic structure of single-crystal Ni_3Al, in particular, on its $3d$ electron states. The positron annihilation spectroscopy (Eldrup and Singh 1997) has two application areas. Firstly, the Angular Correlation of the Annihilation Radiation (ACAR) provides information about the electronic properties of materials: the momentum distribution of the valence and ion-core electrons (Puska and Nieminen 1994). Secondly, positrons have been used as a high-sensitivity probe of the vacancy type defects. A positively charged ion is absent in a vacancy defect, therefore positrons are trapped and annihilate in the defects with surrounding electrons. In this case, the annihilation radiation carries the information about the local electronic structure of the vacancy defect. Apart from the studies of metals (Druzhkov et al. 2008), the positron annihilation spectroscopy is extensively used for diagnosing the structural and electronic properties of the intermetallic compounds (Wurschum et al. 1996, Fu and Painter 1997, Van Pedegem et al. 2004). In particular, Sun and Lin (1994) investigated the influence of doping elements on the tendency of the Ni-Al atomic pairs to form covalent bonds due to the dNi-pAl hybridization in the Ni_3Al compound.

Druzhkov et al. (2010) studied the single crystal samples of the Ni_3Al-X (X = Nb, Co, Fe) ternary alloys; their chemical compositions are listed in Table 1.8. After homogenizing annealing (at 1100°C, for 100 hours), all the samples had a single-phase composition (γ'-phase, Ni_3Al). The ACAR data allowed them to separate the contributions from the annihilation of the positrons with valence electrons (the low-momentum part of the spectrum) and those with ion-core electrons (the high-momentum part of the spectrum) (Siegel 1980).

Table 1.8 Parameters of the ACAR spectra for metals and intermetallic compounds

Material	Parameter S, arb. units	Parameter W, arb. units	N_f, electrons per atom
Ni	0.518 ± 0.001	0.0196 ± 0.0004	1.34
Al	0.683 ± 0.001	0.0061 ± 0.0001	2.98
Ni_3Al	0.527 ± 0.001	0.0189 ± 0.0003	2.06
$Ni_{67}Al_{25}Co_8$	0.538 ± 0.001	0.0171 ± 0.0002	1.78
$Ni_{75}Al_{19}Nb_6$	0.545 ± 0.001	0.0156 ± 0.0001	1.70
$Ni_{71}Al_{21}Fe_8$	0.532 ± 0.001	0.0185 ± 0.0003	1.90

The crystal field affects the strongly bound core electrons only weakly and, therefore, the high-momentum part of the spectrum has information about the type of the atoms surrounding the site of the positron annihilation.

The formation of the intermetallic compounds is accompanied by a change in the d-electron states of the initially pure metals. The contribution of the d-electrons to the positron annihilation may be identified from the curves of the ratio between the ACAR spectra. Druzhkov et al. (2010) obtained the ratios of the pure well-annealed $3d$ (Ti, Fe, Co, Ni) and $4d$ (Nb) metals with respect to the aluminum spectrum. Owing to the Coulomb repulsion from atomic nuclei, the thermalized positrons annihilate predominantly in the interior space. In the transition metals, however, both the conduction electrons and d electrons with a high probability are involved in the annihilation process (Siegel 1980). Since $3d$ electrons are screened from the nucleus by the s and p electrons, the Coulomb repulsion of the positron by the nucleus is insignificant in the region of the $3d$ shell.

The $3d$ electrons are relatively strongly bound to the atom and localized in the r space. Consequently, their momentum distribution is wide and it covers the high-momentum range. Aluminum does not contain d electrons and the annihilation occurs on the $2sp$ electrons characterized by a strong Coulomb repulsion; hence the contribution of the intrinsic core electrons of aluminum (predominantly, in the $2p$ states) to the annihilation is rather small.

The ratio curve shows a maximum in the momentum range $(8–11) \times 10^{-3}\ m_0c$, which is associated with the annihilation of positrons, mostly with the $3d$ electrons of transition metals (Asoka-Kumar et al. 1996). As the number of the $3d$ electrons increases (from 2 in Ti to 8 in Ni), the height of the maximum is increased and it is shifted toward the high-momentum range. Similar results were obtained by numerical calculations. Although niobium has four d electrons, the height of the maximum coincides with that for titanium and the maximum of the peak is slightly shifted toward the low-momentum range. This is associated with the fact that the $4d$ electrons are less localized in the atom (compared to the $3d$ electrons) and, correspondingly, exhibit a narrower momentum distribution.

Druzhkov et al. (2010) compared the ratio curves (with respect to the aluminum spectrum) for pure nickel and the Ni_3Al and Ni_3Al-M (M = Fe, Co, Nb) intermetallic compounds. The curve for the Ni_3Al compound is similar to that for pure nickel. This fact suggests that the positrons annihilate predominantly in the Ni sublattice, as was previously assumed by Wurschum et al. (1996). For Ni_3Al, it was demonstrated that the "free" positron lifetime τ_f (the annihilation from the Bloch delocalized states) is shorter than that expected from the linear interpolation of the τ_f values for pure metals (Ni and Al). This may occur due to the charge transfer from the Al atom to the Ni atoms (Nautiyal and Auluck 1992). The excess of negative charge leads to an increase in the positron density in the transition metal sublattice and, correspondingly, shifts τ_f

toward the value obtained for pure nickel. Therefore, both the ACAR data and the measured lifetimes indicate that the positrons annihilate predominantly in the Ni sublattice of the Ni_3Al crystal lattice.

The height of the maximum for the Ni_3Al compound is lower than that for pure Ni. Several factors may be responsible for the decrease in the peak amplitude:

(1) Some d electrons of the Ni atoms and some p electrons of the Al atoms form covalent chemical bonds due to the strong dNi-pAl hybridization of the nearest-neighbor Ni-Al atom pairs. The electrons involved in the covalent bonds are localized in the interatomic positions. In the momentum space, they are characterized by lower momenta, compared to the electrons localized in atoms. The annihilation probability of positrons with the Ni $3d$ electrons decreases because of the decreasing population of the localized $3d$ states.

(2) The effect of the Ni atom density in the compound. As it is noted above, the positron lifetime in the Ni_3Al compound is close to that in the well-annealed pure nickel. The lifetime characterizes the positron annihilation from the Bloch delocalized state in the crystal lattice (see Table 1 in Wurschum et al. 1996). It indicates that within the sensitivity of the method ($\sim 10^{-6}$ per atom), structural vacancies are absent. This experimental fact is in agreement with the theoretical calculations for the Ni_3Al compound. Accordingly, the deviations from the stoichiometric composition in both directions may be considered as the accommodation by antisites, rather than by the structural vacancies (Fu and Painter 1997).

Pure nickel and the Ni_3Al compound have similar lattice parameters and the same coordination number, which is equal to 12. In the Ni_3Al compound, the nearest environment of each Ni atom includes four Al atoms and eight Ni atoms. Consequently, the density of Ni atoms in the crystal lattice of the intermetallic compound is lower than that in the crystal lattice of the pure nickel. Therefore, we can state that the height of the $3d$ peak in the ratio curve for the Ni_3Al intermetallic compound decreases because of both the lower Ni atom density and the formation of the covalent component of the chemical bond.

Therefore, both the data available in the literature and our results suggest that, in the Ni_3Al crystal lattice, the positrons annihilate from the quasi-free Bloch states and the positron density is predominantly concentrated in the nickel sublattice. The positron concentration in the nickel sublattice allows us to diagnose changes in the $3d$ electrons states due to the doping of the intermetallic compound with a third element.

The ratio curves for the intermetallic compounds doped with a third element do not differ radically from the curve for Ni_3Al. This fact indicates that the doped intermetallic compounds Ni_3Al-X do not contain defects that can trap positrons. Actually, in the case of a ternary compound, the probability of a replacement of a nonstoichiometric vacancy by the doping element is higher than that of the compensation of vacancies by substitutional atoms. Therefore, all changes in the ACAR spectra may be attributed to the modification of the electronic structure of the compound by the doping element.

Let us consider the experimental results for the intermetallic compounds, in which the doping elements (Nb, Fe) predominantly occupy the positions of aluminum atoms. The $3d$ maximum for the compound with niobium (at $\rho_z \approx 10 \cdot 10^{-3}$ m_0c) is lower and narrower than the maxima for the Ni_3Al, Ni_3Al-Fe, and Ni_3Al-Co compounds. Correspondingly, the W and S parameters have the lowest and the highest values, respectively (Table 1.8). The parameter S characterizes the positron annihilation with the valence electrons. In an intermetallic compound, the valence electrons are considered as both the free conduction electrons and the electrons that ensure the covalent bond. It is necessary to elucidate the factor responsible for the increase in the parameter S in the compound doped with niobium. Table 1.8 presents the values of the number of free electrons per atom N_f in the ternary Ni_3Al-based compounds in comparison with the corresponding values for aluminum and nickel.

The value of N_f is almost equal to the number of $3sp$ electrons for aluminum, and it amounts to 1.34 for nickel. Owing to the partial s-p hybridization, this value is considerably larger than the number of s electrons per atom. The value of N_f for the Ni_3Al intermetallic compound is substantially higher than the value of N_f for nickel. For the doped compounds, N_f decreases. The number N_f characterizes the conduction electron density, i.e., the metallic component of the chemical bond. It can be seen from Table 1.8 that the addition of Nb decreases the metallic component of the bond to a greater extent, compared to the addition of Co or Fe.

Let us consider the influence of the doping with niobium on the electronic structure and annihilation parameters for the intermetallic compound. The parameter S for the Ni_3Al-Nb compound indicates that the probability of annihilation of positrons with valence electrons in it is higher than that in the Ni_3Al compound, even though the free-electron density is lower. Consequently, the increase in the parameter S may be associated with the increase in the covalent-electron density due to the strong dNb-dNi hybridization. The low value of the parameter W (Table 1.8) suggests a decrease in the population of the localized $3d$ states in nickel. These specific features are more clearly seen in the curves of the ratio between the ACAR spectra for the compounds doped with the third element and the ACAR spectrum of the Ni_3Al compound (Fig. 1.6). The height of the maximum at $\rho_z = (8\text{-}11) \cdot 10^{-3}$ m_0c for the Ni_3Al-Nb compound that characterizes the localized $3d$ states is much lower than unity, whereas the low-momentum (lower than $5 \cdot 10^{-3}$ m_0c) portion of the curve lies substantially above unity.

For the intermetallic compound doped with iron, all annihilation parameters are similar to those for the Ni_3Al compound (Table 1.8).

This also follows from the ratio curve that does not differ significantly from unity (Fig. 1.6).

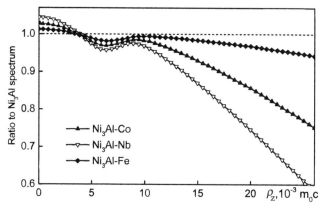

Fig. 1.6 Curves of the ratios between the intensities of the ACAR spectra of the compounds doped with the third element and the intensities of the ACAR spectra of the Ni_3Al binary compound (Druzhkov et al. 2010).

The inference can be made that the iron atoms weakly affect the Ni $3d$ electron states and, correspondingly, the covalent component of the chemical bond. This result does contradict, however, the X-ray spectroscopy obtained by Lawniczak-Jablonska et al. (2000).

For the Ni_3Al-Co compound, the ratio curve (with respect to the Ni_3Al spectrum) exhibits the same features (but less pronounced), as for the compound doped with niobium (Fig. 1.6). The amplitude of the $3d$ peak is close to unity, and the peak itself is smeared in the high-momentum range. This means that the delocalization fraction of the $3d$ states is insignificant, compared to the Ni_3Al compound, although the intensity of the low-momentum component is higher than unity but lower than that for the Ni_3Al-Nb compound. Correspondingly, the free-electron density N_f has an intermediate value (Table 1.8). This means that the degree of the metallic component of the bond is higher than that of the Ni_3Al-Nb compound, but lower than that for the binary and Ni_3Al-Fe compounds. A small enhancement of the covalent bond in the compound doped with cobalt (compared to the Ni_3Al compound) may be associated with a stronger dCo-pAl hybridization.

Wolverton and de Fontaine (1994) demonstrated that, for cobalt replacing the nickel position, the effective pair interaction with aluminum should be stronger than that for the Ni-Al atoms pair. We note that the annihilation characteristics for the Ni_3Al-Co compound can also be affected by the decrease in the Ni atom density in the nickel sublattice, where the positrons are concentrated. As was shown by Druzhkov et al. (2010), the $3d$ maximum for cobalt is lower than that for nickel due to the smaller number of $3d$ electrons in comparison to cobalt. Consequently, the decrease in the height of the $3d$ maximum for the Ni_3Al-Co compound may be associated with both the stronger interaction of the Co-Al pair and with the specific features of the Co $3d$ shells.

The increase in the covalent component of the chemical bond because of the d-d hybridization is in agreement with the results of the calculations of the energy of the effective pair interaction by the tight binding method (Mitrokhin et al. 2005).

Thus, the niobium atoms increase the covalent component of the chemical bond (compared to the binary compound) due to the dNb–dNi hybridization. The doping with cobalt atoms also enhances the tendency toward the formation of the covalent bond. The iron atoms have a weak effect on the electronic structure of the compound.

Calculation of the pair interaction energy in the ternary alloy Ni_3Al-Nb. A relatively simple crystal structure of Ni_3Al is a convenient model for theoretical studies including the conditions of alloying with any transition element. The effect of doping in such works is considered, as a rule, in terms of a site preference (Wu and Li 2012, Raju et al. 1996, Wolverton and de Fontaine 1994, Kozlov et al. 2011, Mitrokhin et al. 2017) or a simulation of Ni_3Al-X properties (Kumar et al. 2015, Bai Lin Lü et al. 2013, Mitrokhin et al. 2008a,b).

The problem of the substitution type for a given alloying element is directly connected with the stability of the ordered state. As a rule, calculations are based on the ratio of the interatomic interactions in the alloy. Mitrokhin et al. (2005) performed the calculations using the OXON software package, developed by A.P. Horsfield at the Center of Materials Science at the University of Oxford. In this package, the method *ab initio* molecular dynamics (CPMD) is implemented (Horsfield et al. 1996, Colombo 1998) in the form of the bond-order potential.

In this (semi-empirical) method, the computational costs grow in proportion to the number N of particles in the system, not N^3 (N is a measure of the system size). Mitrokhin et al. (2005) determined the adjustable parameters from the moduli of elasticity obtained experimentally by Rinkevich et al. (2003). All calculations were performed at room temperature (298 K).

Let us return to the analysis of the factors leading to the increase of the ordering energy in the intermetallic compound Ni_3Al doped with Nb. In the case of the ordered alloy, the addition of a third element X would lead to disordering. Ni_3Al is an intermetallic compound, therefore the configuration disordering does not occur in the alloy. This is because the intermetallic compound is close to a chemical compound in one sense; inter-atomic bonds are not only metallic, but they also have a significant covalent component. It is believed that the ionic bonds are present along with the covalent and metallic ones (Iotova et al. 1996).

The calculations of Mitrokhin et al. (2005) show that, when Ni_3Al is doped with niobium (two atoms of the alloying element for a supercell of 32 atoms), the energy of the chemical bond for a Ni-N pair of the nearest-neighbor atoms is down by more than 50%, while for a Ni-Al pair of nearest-neighbor atoms this value is almost unchanged (Table 1.9, Mitrokhin et al. 2005). It can be explained by the strong influence of the d-states of Nb on the d-states of Ni and a weak influence of Nb on the s- and p-states of the matrix. The niobium atom interaction with the nearest-neighbor atoms is twice as strong than it is in the case of the interaction between Ni-Al. Therefore, the niobium increases the ordering energy of the Ni_3Al alloy.

Table 1.9 Pair interaction energy in the Ni_3Al alloy doped with niobium

Alloy	J_{Ni-Al}, eV	J_{Ni-Ni}, eV	J_{Ni-Nb}, eV
Ni_3Al	−1.092	−1.091	—
$Ni_{75}Al_{19}Nb_6$	−1.008 −0.994	−0.461	−2.033 −1.959

Mitrokhin et al. (2005) calculated the free energy E_i for the substitutions of various types (Table 1.10). The calculations did not take into account changes in the lattice parameter after alloying; the temperature of 298 K was constant. Aiming to estimate the ordering energy of the Ni_3Al, Mitrokhin et al. (2005) calculated the total energy of all atoms in the supercell. For both configurations ordered and for a statistical arrangement of atoms, the calculations resulted in E_{order} = −165.928 eV and $E_{disorder}$ = −159.057 eV per cell. The ordering energy is the difference between these two values: ΔE = 6.87 eV per cell or 0.21 eV per atom, which is 15.7 mRy/atom. The values are consistent with the results obtained by other authors. For example, Sluiter and Kawazoe (1995) reported ΔE = 14 mRy/atom using TB-LMTO-ASA and Lu et al. (1991) measured ΔE = 16 mRy/atom. Various methods of the calculations present the same trend.

Table 1.10 The total energy E and the Fermi energy E_F calculated for a supercell consisting of 32 atoms

Types of substitution	E, eV	E_F, eV
Ni_3Al	−165.9	5.294
Nb (2 at) → Al	−178.6	5.396
Nb (2 at) → Ni	−174.4	5.812
Co (2 at) → Al	−165.4	4.970
Co (2 at) → Ni	−168.4	5.274

The optimal substitution type corresponds to the minimum total energy of the system; for Ni_3Al, the total energy amounts to -206 eV. As a rule, the free energy of the ternary alloys is higher than that of the Ni_3Al binary alloy (Table 1.11).

Thus, the description given above is based on the analysis of pair interaction.

Table 1.11   The total energy E (eV), calculated for different substitution types in Ni_3Al

Element	At position of Al	At position of Ni	Substitution type
Nb	−219.8	−213.9	Al
Co	−207.9	−208.8	Ni
Cr	−220.1	−213.2	Al
Fe	−213.3	−211.1	Al
Pd	−201.8	−201.4	in both sub-lattices
Rh	−212.5	−211.7	Al

Due to the high symmetry of the ordered phases, there are additional factors of their stabilization, related to the electronic structure of such alloys (Kozlov et al. 2006a). Stability of the ordered state is found to be dependent on the quasi-localized and collectivized states in the electronic spectrum of the alloy. Mitrokhin et al. (2017) calculated, from the first principles, the total density of states $N(E)$ for two different types of substitution in the Ni_3Al crystal lattice. A comparison of these results with the experimental data of an XPS study suggests that the preferred type of Nb substitution is the aluminum position and that cobalt atoms prefer the nickel positions.

Thus, the Ni_3Al crystal containing transition elements have two competing energies: (1) the energy U associated with the interaction between the d-states of the transition element and the states of the nearest-neighbor atoms and (2) the Coulomb interaction energy Q of the two d-electrons in the same atom. In the one-electron approximation, Q is included in a periodic potential and never considered explicitly. One could calculate an average value of the density of states $N(E)$. However, many important properties for an atom of the alloying element are washed out and become less noticeable in that quantity. For $U < Q$, it is better to assume that initially each atom has the same number of electrons and then the activation energy is needed to transfer an electron from one atom to another. This is called the approximation of localized electrons (Harrison 1989).

Mitrokhin et al. (2005) performed an analysis of the Local Density Of States (LDOS) for an atom of the alloying element in different substitution positions of the Ni_3Al crystal lattice (Al or Ni). Local $N(E)$

values characterize the electronic spectrum of an individual atom, depending on its nearest neighbors.

The local density of states at the Fermi level is practically zero in Ni_3Al for the atoms of nickel and aluminum.

From the zone calculations (Harrison 1989), it is well known that, when the Fermi level falls in the main peak of $N(E)$, such a state is unstable. Accordingly, this type of substitutions is not implemented. In the optimal position for substitution of the atom of the alloying element, the Fermi level is close to zero of the local density. A classic example is the study the stability of the ferromagnetic state for iron: the state is stable, when the main peak of $N(E)$ in the paramagnetic state is split into two peaks that are separated by $\Delta E = 2.2$ eV, and the Fermi level lies between the peaks, in a minimum of the density of states.

In the total density of states (TDOS), these details also appear, though are largely smoothed, because density of states at the Fermi level is mainly formed by d-electrons of nickel. Apparently, the same criterion applies for the intermetallic compound Ni_3Al. The system is unstable when the niobium atoms stay on the nickel positions and the cobalt atoms stay in the positions of aluminum. For example, the Fermi level of the $Ni_{67}Al_{25}Co_8$ alloy corresponds to the main peak of the d-states when the Co atom substitutes the Al atom. When replacing Ni atoms, the Fermi level is shifted by 0.304 Ry, because the charge q flows from the nickel nearest-neighbor atoms to the cobalt atom ($q_{Co \to Al} = 8.471$ e; $q_{Co \to Ni} = 8.617$ e).

Deformation of Ni_3Al in different temperature ranges. The temperature dependence of the mechanical properties of Ni_3Al causes the industrial interest to this compound.

The temperature anomalies of the Ni_3Al deformation characteristics and the associated effect of thermal hardening are interesting features of this intermetallic compound (Stoloff 1989). With increasing temperatures, the yield strength $\sigma_{0.2}$, the resistance to deformation, and the rate of hardening of the alloy grow and pass through a maximum, reaching values significantly exceeding those found at low temperatures. This is abnormal, if we compare them to the change rates observed in pure metals or solid solutions. For example, the temperature dependence of $\sigma_{0.2}$ for Ni_3Al with different Al contents, given by Stoloff (1989) and Greenberg and Ivanov (2002), illustrate the anomalous temperature dependence of the yield stress.

It was found that the maximum yield strength and the shape of the $\sigma_{0.2}$ temperature dependence are strongly related to the Al content. With the increasing Al content, the anomalous temperature dependence shifts, whereas the position of the temperature maximum remains practically unchanged.

The nature of the positive temperature dependence of the yield stress is related to the structure of the sliding superdislocations and planar defects in the $L1_2$ superstructure. The observed temperature anomalies are accompanied by changes in the active dislocation slip systems. The most significant experimental fact is the replacement of the octahedral (111) dislocation slip to the cubic (100) one at temperatures close to those of the yield stress peak (Demura et al. 2007a, Keshavarz and Ghosh 2015).

The practical application of the intermetallic compound Ni_3Al is limited by its low ductility in the polycrystalline state. At room temperature, Ni_3Al samples can be tested only by compression. Tensile deformations with loading of less than 1% lead to brittle intergranular fractures of the Ni_3Al crystal. The main reason for the lack of ductility of the polycrystalline Ni_3Al may be the existence of the impurity segregation at grain boundaries. It should be noted, however, that ductility is low even if there are no impurities at grain boundaries. Another reason may be the small number of active dislocation slip systems.

Improving the Ni_3Al ductility may be achieved with micro- and macro-alloying (Koneva et al. 2012). The failure time and the long-term strength are largely increased by a small addition of boron. Such an effect is associated with the boron segregation at grain boundaries, which slows down the grain boundary diffusion. Alloying with cobalt, iron, or chromium may also increase the Ni_3Al ductility.

Despite the complexity of Ni_3Al deformation under tension, different deformation allows one to deform this intermetallic compound. For example, deformations up to 40% can be achieved by cold rolling of Ni_3Al polycrystalline samples (Aoki et al. 1995, Kobayashi et al. 2005, Li et al. 2008). In contrast to the polycrystalline state, single crystals of Ni_3Al have high plasticity (Demura et al. 2007b, Zhang and Wang 1998, Golberg et al. 1997).

Li et al. (2008) studied the temperature dependence of the fracture resistance σ_B. Although the yield stress shows anomalous of positive temperature dependence between room temperature and 600°C, σ_B slowly decreases in this interval and sharply falls above 600°C. On the other hand, elongation of Ni_3Al single crystals is almost 100% at room temperature. However, with increasing temperatures, the elongation decreases and reaches a minimum of only a few percent at 600°C. Thus, the yield strength and elongation show opposite temperature dependences.

The yield strength and elongation strongly depend on temperature and the crystallographic orientation of the crystal. The elongations from 45 to 100% can be obtained by the tensile tests at room temperature with the crystal orientations (<110> or <100>) (Thomton et al. 1970).

The yield strength of the samples with the [100] orientation is higher than that in the samples with the [110] orientation at the same temperature. The samples with the [100] orientation show a higher elongation at room temperature compared to the [110] samples. Elongation of the samples with different orientations decreases and the difference between the elongations becomes smaller with increasing temperatures. Sliding of partial dislocations on the {100} planes can also be seen at temperatures above 400°C and it becomes dominant at 700°C. The samples with the [110] orientation are more plastic at 700°C because of the development of additional slip systems with increasing temperatures.

Superplastic flows with elongations of up to 650% were observed in the Ni_3Al polycrystalline samples (with grain sizes of 1.6-6 μm) at temperatures above 800°C (Zhang and Wang 1998, Semiatin et al. 2004, Mukhopadhyay et al. 1990). Development of super-plasticity is caused by dynamic recrystallization. These results were partly obtained using nickel superalloys that have a two-phase structure (Zhang and Wang 1998). The effect has pronounced temperature and strain-rate dependences. The elongation δ of the Ni_3Al polycrystalline sample is decreased from 440 to 100%, with the strain rate increasing from $8 \cdot 10^{-4}$ to $8 \cdot 10^{-3}$ s^{-1}. According to the literature data, the temperature interval of the superplasticity development for the nickel superalloys is 950-1100°C on the deformation rate from 10^{-3} to 10^{-5} ms^{-1}.

Another example of the Ni_3Al superplasticity is related to the high-temperature deformation of single crystals, as shown by Stepanova et al. (2011a) in their studies of the mechanical properties of Ni_3Al at high temperatures (1100-1250°C) under active loading.

All samples studied by Stepanova et al. (2011a) exhibited superplasticity; Fig. 1.7 displays one of their stress-strain curves. Behavior of the curves (appearance of serrations) is typical for the stress relaxation conditions

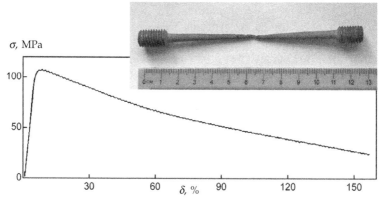

Fig. 1.7   Stress–strain curve of a single crystal of an as-grown Ni_3Al alloy at 1200°C; the inset: form of the sample after deformation.

of the testing process. This raises a question of which process caused the relaxation in the alloys. Results of the mechanical tests at different temperatures are given in the Table 1.12. Firstly, there is a difference in the mechanical properties of the as-cast and homogenized samples.

The composition of the alloy completely corresponds to the homogeneity range of Ni_3Al. Single crystals are crystallized through the eutectic reaction in the form of a lamellar eutectic consisting of wide γ'-phase (Ni_3Al) plates with thin layers of the β-phase (NiAl, $B2$); the alloys have a higher nickel content relative to the stoichiometric composition. β-phase is metastable and must be completely dissolved under crystallization.

Table 1.12 Mechanical properties of the Ni_3Al single crystals with the <001> orientation in as-cast and homogenized (HS) states

State	t, °C	σ_u, MPa	$\sigma_{0,2}$, MPa	δ, %
as-cast	150	200	190	5
as-cast	200	102	93	55
as-cast bicrystal	200	103	94	2
HS	150	143	125	40
HS	200	101	98	25
HS	250	66	60	10

In practice, after the cooling of the single crystal ingot, a small amount of NiAl is preserved. During the cooling, the β phase is transformed to a martensitic $L1_0$ type tetragonal phase (Hunziker and Kurz 1997). Stepanova et al. (2003a) found 1% of the residual NiAl in the as-cast single crystal of Ni_3Al by the method of the neutron diffraction analysis. Table 1.12 shows that β-phase strengthens the as-cast alloys. A sharp increase in plasticity with increasing deformation temperatures from 1150 to 1200°C is apparently related to β-phase dissolution.

We also observe the recrystallization development in some local regions with high levels of stresses in the structure of the studied samples in the entire range of the deformation temperatures. Small recrystallization grains are observed in the fracture zone (the neck zone) in the structure of all tested samples; recrystallization is not observed in the major part of the crystal. The samples are single-phase after homogenization at 1100°C for 100 hours, according to the TEM and X-rays diffraction analysis.

The mechanical tests at 1150°C reveal lower values of σ and $\sigma_{0.2}$ for homogenized samples in comparison with the as-cast samples, caused by the strengthening effect of the dispersed metastable β phase. The plasticity of the as-cast alloys is lower than those of the homogenized one; the elongation-at-break values are 55 and 140%, respectively. The strength properties of the as-cast and homogenized samples are practically identical after a deformation at 1200°C. This fact confirms once again the above conclusion about the dissolution of the metastable β-phase in the as-cast samples at such temperatures.

Large quantities of the rectilinear crystallographic surface cracks were observed in both the as-cast and homogenized samples. Directions of the cracks in the neck regions were close to $\langle 110 \rangle$; which are two simultaneously acting directions of easy glide in the $L1_2$ superlattice. Under superplasticity conditions, the formation of the crystallographically oriented surface cracks appears to be an additional mechanism of relaxation, along with the dynamic recovery and the dynamic recrystallization, which ensures high elongation in the samples.

After a deformation at 1250°C, the microstructure of the alloy is analogous to that observed at 1200°C in many respects. High densities of dislocations are observed inside the subgrains. The dislocation lines intersect each other in two directions at angles of about 110°. An analysis of the bright field micrographs and selected-area electron-diffraction patterns made it possible to determine these directions as [112] and [11$\bar{2}$] The misorientation of the nearby subgrains is 2°-3°. Single coarse twins are revealed in the fracture zone.

Twining represents an additional method of stress relaxation under the deformation process of the sample. The twinning plane, calculated from TEM, corresponds to the [$\bar{1}$11] plane. The appearance of twins indicates switching of the additional slip systems that are necessary to guarantee the relaxation process. Recrystallization and dynamic recovery are the alternative mechanisms of relaxation. In practice, all the mechanisms of stress relaxation are observed, because of the inhomogeneity of the high-temperature deformation process. Dynamic recovery is the predominant mechanism of relaxation.

1.1.2 Structure of the nickel superalloys

Heat-resistant nickel alloys are multicomponent systems. The complex chemical composition of these alloys is necessary for the formation of the structure and phase composition and stability under long-time exposure under stresses during heating in aggressive environments (Sims et al. 1987).

Intermetallic γ'-phase based on Ni_3Al (FCC, superstructure $L1_2$) and solid solution (γ-phase, FCC) are the main phases of the nickel superalloy. Cobalt, chromium, molybdenum, and tungsten are the strengthening elements of the nickel solid solution. In addition, chromium plays an active role in the protection of the alloys against oxidation. Tungsten reduces the diffusion processes in the alloy. Titanium, aluminum, niobium, and tantalum are the alloying elements of the Ni_3Al intermetallic compound

(γ'-phase). As a result, the chemical composition of the intermetallic γ'-phase can be complex; for example, $(Ni, Co)_3(Al, Ti, Nb, Ta)$.

The alloys with less than 30% of the γ' strengthening phase are called deformable alloys. When the content of γ'-phase increases beyond that value, they are called casting alloys. The development of the casting heat-resistant nickel superalloys is primarily related to the greater strengthening effect due to the increase the γ'-phase content and a higher structural stability of the alloys.

Turbine blades of the stationary power equipment (600-900°C) are in the polycrystalline state. Diffusion coefficient on the grain boundary, as a rule, is two or three orders of magnitude higher than that inside the grain. As a result, the presence of grain boundaries in the sample becomes the major factor influencing the high temperature creep (Sims et al. 1987). Therefore, for a stable operation at the highest temperatures and stresses (in the aviation), the turbine blades are made by the Bridgman method of the directional solidification from the single-crystal alloys with a high content of the strengthening γ'-phase.

Let us consider the structural state of superalloys on the example of a modern Russian carbon-free rhenium-containing nickel superalloy ZhS-36VI. This alloy is an analog of the high-temperature CMSX-4 alloy (USA), which is also used as a material for single-crystal blades; it is a carbonless single-crystal rhenium-containing alloy. The main problem of such alloys is the phase instability associated with the release of the Topologically Closed Packed (TCP) phases.

One of the successful methods for obtaining high-rhenium nickel superalloys is the High-Gradient Directional Crystallization (HGDC). Crystallization with high gradient is used to obtain the blades with the directional and single-crystal structure (Kablov and Petrushin 2008). The temperature gradient G at the solidification front in the HGDC method is changed from 60 to 220°C/cm, while the traditionally used radiation cooler gives the gradients only in the range of 20-30°C/cm. A high level of the blade operation characteristics generated by the HGDC technology ensures the absence of surface carbides and defects like freckles, reduced micro-porosity, and the crystallographic orientation in the best direction of the crystal plasticity.

The HGDC single crystal superalloy has the cellular-dendritic growth structure with the main axis of growth of [001]. Increasing of the temperature gradient at the solidification front to 200°C/cm allows one to obtain castings with a fine structure and high degree of homogeneity and with decreased inter-dendrite distance to twice its original value. A series of the melting equipment for HGDC is developed to implement the process of directional solidification with the melting temperatures of up to 1700°C (Kablov and Petrushin 2008).

To reduce the micro-porosity of the nickel superalloy single-crystal blades, the Hot gas-Isostatic Pressing (HIP) is used (Carona and Ramusat 2014). To eliminate the small micro-pores, the compressive stresses must be greater in order than those in the healing of the large pores.

For the ZhS36-VI [001] superalloy, the technological parameters of this process are worked out and the greatest effect of healing pores and homogenization of the chemical composition of the alloy is achieved by heating in the single-phase γ-region at the temperature of 1320°C and pressing of 180 MPa (Morozova et al. 2009, Petrushin et al. 2016).

The volume fraction of the strengthening γ'-phase in the modern aircraft alloys comes to about 60-70 vol.%. The modern nickel-based superalloys work at temperatures of up to 1100°C. The operation temperature can be above the beginning of γ'-phase dissolution (approximately 850°C); for example, the operation temperatures of carbon-free nickel superalloys are in the range of 950-1050°C. Thus, the question about the structural stability in the region of operating temperatures becomes important.

γ'-phase is released in the form of dispersed particles, with the sizes controlled by the cooling rate. The particles have spherical or cubic shapes, depending primarily on the volume fraction of the strengthening phase in the alloy. The spherical γ'-phase particles form at low volume fraction in the alloy. With increasing intermetallic γ'-phase content to approximately 40% and above, the elastic fields of particles begin to interact, promoting the cuboid shapes of the particles (Fig. 1.8).

The effect of particle sizes on the long-term strength can be seen in Fig. 1.9 (Kuznetsov et al. 2008). The optimal size and shape of the γ'-phase particles are ensured by the step-wise annealing.

0.2 μm

Fig. 1.8 Particles of intermetallic γ'-phase in the single crystal of superalloy ZhS-36VI, dark-field image in the reflex (001) of γ'-phase, TEM (Baryshev et al. 2010).

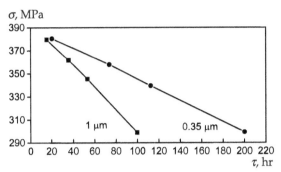

Fig. 1.9 Long-term strength of the alloy depending on the particle size of the strengthening phase (Kuznetsov et al. 2008, reprinted with the kind permission of the author).

In the carbon-containing alloy, MC-type carbides are formed – for example, NbC or TiC. During blade operation (with heating), the carbide reactions take place, with a participation of MC and γ'-phase promoting the formation of the $M_{23}C_6$-type carbides. A further stage of the carbide reactions during a long-term operation leads to the formation of the M_6C carbide that has undesirable needle-like morphology resulting in the embrittlement of the alloy. The composition of these carbides is complex.

Significant amounts of the alloying elements are diverted from the solid solution for their formation. For example, the $M_{23}C_6$ composition may be $Cr_{21}(Mo,W)_2C_6$ and M_6C may be $(Ni,Co)_3Mo_3C$ or $(Ni,Co)_2W_4C$. In the ZhS-32 alloy containing 0.15% of carbon, the carbide M_6C emerges after 50 hours at 1050°C. At a temperature of 1150°C, M_6C is formed within 10 hours, whereas at 850°C it takes 1500 hours (Zhivushkin et al. 2011). The temperature values at the beginning and the end of γ'-phase dissolution are often used to characterize the stability of the superalloy structure (Sims et al. 1987). At the operating temperature the long-term strength of the alloy increases with the increase of the γ'-phase volume fraction.

Another parameter that largely determines the high-temperature stability of the nickel-based superalloys is the ratio of the lattice parameters of the γ solid solution and strengthening intermetallic γ'-phase. These parameters are close but typically not equal to each other, therefore a lattice parameters mismatch (misfit) is introduced as $\Delta = (a_{\gamma'} - a_\gamma)/a_\gamma$ (Protasova et al. 2011, Samoilov et al. 2011). The sign and the magnitude of the misfit mainly determine the morphology of the γ'-phase particles and their coagulation rate under stress at high-temperature conditions.

Lattice coherence of the γ'-phase and γ-matrix remains constant up to high temperatures and leads to considerable elastic stresses impeding the movement of the dislocations and delaying the particles coagulation.

The specific lamellar structure (the raft structure) is formed during the high-temperature loading in the single crystal superalloys with high volume fractions of γ'-phase (about 70%) (Kuznetsov et al. 2008). The raft structure is formed at the first stage of creep by splicing of separate γ'-phase cuboids due to the redistribution of alloying elements in the matrix (γ-solid solution) under stresses. The raft structure with a high thermal stability determines operation durability of turbine blades. Morphology of the raft structure is largely determined by the misfit sign. A negative value of the γ'/γ misfit leads to the formation of raft structures with the lamellas arranged perpendicular to the axis of the applied stress. This is the most favorable arrangement of the γ'-phase lamellas for long-term operation. Coarsening of raft structures takes place at the third (accelerated) stage of creep, when the boundaries of the plates become zigzag-shaped and their location becomes chaotic.

Fractures of superalloys often begin with the formation of micro pores arising in the process of operation (Morozova et al. 2009, Petrushin et al. 2016). At the initial stage, the micro pores appear at small-angle boundaries of the grain blocks. The fracture zones are formed by the micro cracks originating from the pores.

An important aspect of phase stability is the stability of the superalloy against the formation of excess intermetallic phases. The so-called Topologically Close-Packed phases (TCP-phases), such as the σ-phase or μ-phase, are brittle and have unfavorable lamellar morphology. The problem of TCP formation is greatly exacerbated with rhenium alloying (Kablov et al. 2006b).

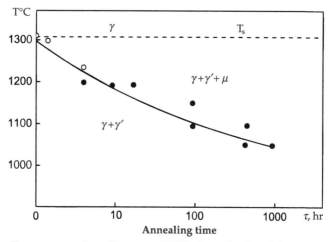

Fig. 1.10 Temperature-time diagram of TCP-phase (μ-phase) formation in single crystal blade made on superalloy ZhS-36VI in the process of high-temperature heating (T_s γ'-temperature of complete γ'-phase dissolution) (Kuznetsov et al. 2008).

Formation of TCP-phase in the blades made of the ZhS-36VI superalloy during the high-temperature exposure was described by Kuznetsov et al. (2012). Pre-exposure of the single crystal samples for more than 10 hours at the temperatures of 1200°C and above leads to additional γ'-phase precipitations and increases their mechanical properties at room temperature (Fig.1.10). High temperature tests show degradation of the long-term strength because of the instability of the additional γ'-phase precipitations, with numerous instances of the formation of TCP-phase (μ-phase). μ-phase has a plate morphology and the A_7B_6 chemical composition like $(Ni,Co)_7(Cr,W,Re,Mo)_6$.

According to the temperature-time-transformation phase diagram, μ-phase is formed at 1050°C after an exposure for 1000 hours and at 1100°C after 100 hours. At 1200°C, this renium-contained phase appears after an exposure for 10 hours and at 1300°C is formed within 1 hour. Kuznetsov et al. (2015) concluded that the degradation of the long-term strength after pre-exposure at high temperatures is caused primarily by the formation of the hardened γ'-phase layers around them, not by the plates of TCP-phase.

Modules of elasticity. Rinkevich et al. (2003, 2008) determined the elasticity modules in single-crystal nickel-based superalloys using the acoustic method. Compositions of the studied superalloys are given in Table 1.13. The studies were performed on samples of the carbon-free superalloys, in order to exclude the presence of a carbide phase. The EI-437B alloy is a Russian analog of NCF80A (Japan); the ZhS-36 alloy is a Russian analog of CSMX-4 (USA).

Table 1.13 Chemical composition of the superalloys, wt. %. Nickel-base

Grade	Cr	Ti	Mo	W	Re	Al	Co	Nb	C	γ', vol. %
EI-437 B	1.0	2.5	-	-	-	1.0	-	-	0.07	7
CNK-8MP	2.3	4.5	-	6.7	-	4.1	8.75	1.5	≤0.02	49
ZhS-36	3.9	1.1	1.2	1.5	2.0	5.9	8.9	0.9	≤0.02	64

Particles of the strengthening γ'-phase in the superalloys are too small and the phase's presence is manifested in acoustic experiments (this requires a significantly higher frequency of the acoustic signal). Therefore, the research did not reveal any correlation between the elastic moduli and the volume fraction of the strengthening phase.

The elasticity moduli for the polycrystalline ZhS-36 superalloy are defined by $E = 246$ GPa and $G = 95$ GPa (Rinkevich et al. 2008). For the technologically important metals and a group of alloys based on them, the modules vary slightly with a spread of no more than 10-15% (Truell et al. 1969); the above results confirm this trend. The elastic moduli of high-temperature superalloys are similar to those for pure nickel; for example, $B_{Ni} = 187$ GPa, $G_{Ni} = 78$ GPa (Truell et al. 1969).

The elastic modulus in single crystals depends on the crystallographic orientation. At 1000°C, the modulus value E in the alloy ZhS-32U varies between 85 GPa for <001>, 162 GPa for <011>, and 235 GPa for <111> (Petrushin et al. 2016). This means that the crystals with the <001> orientation have advantages over the samples with the <111> orientation in the long-term strength tests. The crystals with the <001> orientation are more resistant to cracking.

The results of the measurements of the elasticity moduli (c_{ij}), the density (ρ), the Debye temperature (T_D), the volume fraction of γ'-phase are presented in Table 1.14.

The anisotropy factor A, the bulk modulus B, and the shear modulus G of the alloys are presented in Table 1.15.

Table 1.14 Components of the γ'-phase in the superalloys

Grade	γ'-phase vol. %	T_D, K	c_{ij}, GPa 11	12	44	ρ, kg/m^3
EI-437 B	7	465	34	49	16	7.87
CNK-8MP	49	472	54	52	31	8.46
ZhS-36	64	447	49	52	30	8.65

Table 1.15 Components of the studied alloys

Grade	A	B, GPa	G, GPa
EI-437 B	2.73	177	78
CNK-8MP	2.57	186	90
ZhS-36	2.68	84	88

Golubovskiy et al. (2015) obtained the experimental data at low-cycles fatigue tests in a rigid cycle for $R(\varepsilon) = \{0, -1\}$ and high-cycles fatigue for $R(\sigma) = \{0, 1\}$ in the rhenium-containing nickel superalloys of the VZhM type used for the manufacturing of nozzle blades. The tests were conducted at 850° and 1050°C for the single-crystal samples with the <001> orientation. The nucleation of the fatigue cracks at 850°C was related to the stress concentrators in the volume of the ingot, such as the separate pores or their clusters. At 1050°C, the surface corrosion cracks were the main sites of failure. The anisotropy of the growth rate of the fatigue cracks was detected. It was found that cracks grow faster along the [001] direction than along the [011] direction.

Concluding, the heat resistance of the single crystal alloy is determined by the complex interactions of several factors, such as:

- Growth dispersion of the crystalline structure and its perfection (small misorientation between the grain blocks),
- Volume fraction of the strengthening intermetallic γ'-phase, morphology and dispersion of the latter;

- Lattice mismatch between the γ-solid solution and intermetallic γ'-phase;
- Content of the intermetallic TCP-phases;
- Volume fraction and morphology of the carbide particles.

For heat resistance, the most important factor is the thermal stability of the strengthening γ'-phase. The diffusion processes have an important role because they define the kinetics of the phase formation, coagulation, and dissolution. In addition, in multicomponent alloys, the chemical composition of the phase components may change, depending on the crystallization conditions, leading to changes in the mechanical properties of the alloys. Researchers are faced with the challenging problem of developing new alloys, improving the production technology and choices of the optimal operation modes, and developing methods for monitoring of the structural state of alloys in the manufacturing process and operation of the turbine blades.

Treatment of melts to modify the physical and technological properties of nickel-based superalloys. At present, active attempts are made to change the physical and technological properties of the heat-resistant nickel alloys in the solid state affecting the melt before crystallization. Two kinds of the melt treatment are considered here. They are the high-temperature treatment of melt (HTMT), which has wide industry application (Barishev et al. 2010, Tyagunov et al. 2013, 2014, 2015), and the introduction of ultrafine titanium carbonitride powder to the melt (Baryshev et al. 1998, Stepanova et al. 2000). Both methods of the melt treatment are discussed in this chapter on the example of modern rhenium containing the industrial nickel superalloys like ZhS-32VI, ZhS-36VI, ZhS-26VI.

The method of high-temperature X-ray analysis is used to study *in-situ* the processes of γ'-phase dissolution in single-crystal samples of the ZhS-type nickel superalloys (Stepanova 1999, 2003b).

There are also other methods of the determination of the γ'-phase volume fraction at a given temperature. Previously, a method for determining the relative volumetric content of the strengthening γ'-phase in the nickel-based superalloys was proposed, based on the analysis of the temperature dependence of the electrical resistivity and on the data from differential thermal analysis (Petrushin et al. 1977). An equation relating the volume fraction of the γ'-phase at a given temperature with the volume fraction at room temperature was found using the temperature ratio at the onset of the dissolution and the completely dissolved γ'-phase.

High-temperature treatment of the melt (HTMT) for the nickel-based superalloys. High-temperature treatment of melts significantly affects alloy properties in the solid state (Barishev et al. 2010, Tyagunov et al. 2013, 2014, 2015). It increases the content of the strengthening γ'-phase and the uniformity of its distribution, improves the γ'-phase and carbide phase morphology. With HTMT, it is possible to achieve a more uniform distribution of alloying elements in micro-volumes of the alloy. An application of such melt treatment leads to a significant increase in the long-term strength of single-crystal samples.

Stability of the strengthening intermetallic γ'-phase is an important factor determining the level of the long-term strength of such single crystals. HTMT has no practical effect on the temperature of the γ'-phase complete dissolution (Petrushin and Cherkasova 1993). X-ray experimental data of an *in situ* γ'-phase dissolution process by HTMT are given by Stepanova et al. (2003b).

The chemical compositions of the investigated alloys are presented in Table 1.16.

Table 1.16 Chemical compositions of the studied nickel superalloys, wt.%

Alloy/element	Cr	Ti	Mo	W	Re	Ta	Al	Co	Nb	Zr	C
ZhS-26 58 vol. % γ'	5.0	1.0	1.1	1.7	-	-	5.8	9.0	1.6	0.05	0.15
ZhS-32 67 vol. % γ'	5.0	-	1.0	8.3	4.0	4.0	6.0	9.0	-	0.05	≤ 0.02

The alloy ZhS-32 has a composition with a carbon content of 0.16 wt.%. The cast ingots were obtained by vacuum induction melting. Single crystals were grown by the Bridgman method. The metal melting and heating up to 1600°C was carried out in the vacuum. Further heating and HTMT were carried out in the argon atmosphere and then the melt was poured into a mold. The mold fell down in a bath with a liquid metal using a Liquid Metal Cooling technology (LMC); the cooling rate was 10 mm/min for the alloys ZhS-26VI and ZhS-32VI.

For the alloy ZhS-36VI, the cooling was slower (5 mm/min). The metal temperature in the crystallization bath at the beginning of the process did not exceed 740°C. The temperature gradient in all cases was 100°C/cm. We used the various temperatures of the melt treatment: 1650, 1700, 1740, 1830°C.

Figure 1.11 shows the temperature dependence of the electrical resistivity $\rho(t)$ for the alloy ZhS-32VI (Stepanova et al. 2003b).

During the heating of the melt, an increase of the electrical resistivity starts immediately above the melting point. In this temperature interval, the melt state is preserved until crystallization, following the traditional technology (without HTMT); the melt with a temperature of 1600°C is poured into a mold for the crystallization of the ingot. An anomalous behavior of the temperature dependence of $\rho(t)$ between 1630°C and 1700°C is observed in Fig. 1.11.

The electrical resistivity ρ remains almost constant despite the fact that the temperature increases. The abnormal behavior of $\rho(t)$ disappears at $t_{k1} = 1740°C$, when the resistance increases again with increasing

temperatures. The anomaly is revealed in the same temperature range
in studies of other physical properties of the melt, such as its viscosity.

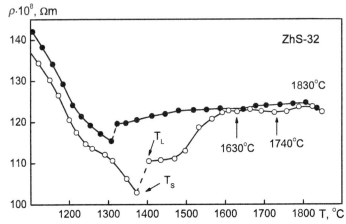

Fig. 1.11 Temperature dependence of the electrical resistivity $\rho(t)$ of the heat-resistant nickel superalloy ZhS-32 in solid and liquid state (Stepanova et al. 2003b): T_S - solidus; T_L - liquidus.

According to modern concepts (Barishev et al. 2010, T'sepelev et al. 2008), the nickel superalloys have atomic clusters with the middle range order of Ni_3Al-type after melting. Such clusters exist in the dynamic mode, with a relative balance of the disintegration and formation processes. The disintegration of the atomic Ni_3Al-type clusters occurs in the temperature interval of the anomaly, which corresponds to the transition of the melt into the homogeneous equilibrium state. In Fig. 1.11, the anomaly temperature is denoted as t_m and the disintegration temperature of the Ni_3Al-type atomic clusters is indicated as t_{k2}. Atomic complexes with structures similar to carbide phase are also present in the melt. The destruction of all types of clusters completes when the melt is heated to t_{k2}; for the ZhS-32 alloy, this temperature is 1830°C. The temperature behavior of $\rho(t)$ in the melt during a subsequent cooling after heating above t_{k2} has a hysteresis as can be seen in Fig. 1.11.

The results of mechanical tests allow us to obtain the optimal regimes of crystallization as follow: 1740°C for the ZhS-26VI and ZhS-32VI alloys, and 1760°C for the ZhS-36VI alloy. Application of such melt treatment leads to a significant increase in the long-term strength of the single-crystal samples in the solid state.

Influence of HTMT on the structure and properties of the ZhS-26VI superalloy. The chemical composition of the Russian ZhS-26VI superalloy is given in Table 1.14; it is similar to the MAR-M247 (USA) superalloy. The observed structures in the ZhS-26VI superalloy after different regimes of HTMT as well as the morphology and size of the γ'-phase particles are similar (Table 1.17).

Table 1.17 Parameters of the single-crystal growth structure of the ZhS-26VI superalloy, depending on the HTMT regime:

HTMT	D, µm	d, µm	θ, grad.	$m_{\gamma'}$, µm
1600°C	320 ± 20	32 ± 3	12.0 ± 1.0	0.40 ± 0.05
1650°C	310	38	8.0	0.40
1740°C	300	40	6.0	0.45
1830°C	290	40	6.0	0.60

In Table 1.17, D is the dendritic cell size, d is the distance between the secondary branches of dendrites, θ is total misorientation of blocks in single crystals, m_γ is size of γ'-particles (Stepanova et al. 1999, 2003b).

An application of HTMT with 1740°C improves quality of the crystalline structure of single crystals (growth rate 10 mm/min), increases the volume fraction of the strengthening γ'-phase (Table 1.18), as well as increases resistance of γ'-phase to dissolution. Apparently, these factors provide improved mechanical properties of the superalloy in the solid state (Table 1.19).

Table 1.18 Volume fraction of γ'-phase at room temperature in single crystals of the ZhS-26VI superalloy (growth rate 10 mm/min) obtained by different methods

Method of melting	Volume fraction of γ'-phase, vol. %		
	method of investigation		
	Metallography	Dilatometry	X-ray analysis
Without HTMT	54	-	-
HTMT at 1740°C	62	61	62

An X-ray study establishes that the γ'-phase dissolution in the ZhS-26VI superalloy occurs at the same speed for all HTMT regimes, except that at 1740°C. In the latter case, the γ'-phase resistance to dissolution increased. This leads to the advantage of the samples crystallized with HTMT at 1740°C in the long durability tests.

Table 1.19 Failure times of the ZhS-26VI superalloy in the long-term strength testing at 1000°C and applied stress $\sigma = 260$ MPa

Grade	Treatment of the melt	h
ZhS-26VI	without HTMT	58
	HTMT 1740°C	105

Influence of HTMT on the structure and properties of the ZhS-32VI superalloy. The ZhS-32Vi superalloy was manufactured by the directional solidification method in the Russian Research Institute of Aviation Industry (RIAI). This superalloy belongs to the class of rhenium-containing nickel superalloys and is the most heat-resistant within this class (Petrushin et al. 2016, Stepanova et al. 2003b). The alloy can be used for the production of the single-crystal blades for the gas turbine engine with a long-time operation at temperature above 1000°C. The alloys exists in two versions, with carbon and carbon-free. The alloys with carbon have high heat-resistance, improved casting, technological plasticity, and high corrosion resistance. The carbon-free alloy has high levels of heat-resistance and a lesser tendency to form topologically close-packed phases (TCP). The chemical of the alloys are given in Table 1.16. The ingots are melted with an induction furnace under vacuum. High temperature structure refining is provided by the introduction of Ca and Re to the melt; the melt filtration through a ceramic foam filter is used. The ultra-low content of impurities is obtained, namely $O_2 \leq 0.001$ wt.%, $N_2 \leq 0.001$ wt.%, and $S \leq 0.0015$ wt.% using the melt treatment at 1740°C (HTMT). The structure of the single crystal alloy ZhS-32Vi is cellular-dendritic and it consists of the γ solid solution, strengthening γ'-phase ($Ni_3(Al,Ti)$), carbides MC type, and the $\gamma + \gamma'$ eutectic. Temperatures of the phase transformations are given in Table 1.20, where t_{cd}- temperature of the γ'-phase complete dissolution, t_{end} -point of the onset of melting of the eutectic γ'-phase, t_s -solidus, t_L -liquidus, t_{MC} -the temperature at the beginning of carbide formation. Stable strength characteristics of the ZhS-32VI superalloy at 20°C and at the range of 1000-1100°C are obtained after homogenization at 1285°C for 1 hour with a cooling temperature gradient of 70-80°C/min (Stepanova et al. 2003b).

Table 1.20 Temperatures of the phase transformations of the ZhS-32VI superalloy without carbon, °C

Grade	t_{cd}	t_{eut}	t_S	t_L	t_{MC}
ZhS-32VI	1282	1305	1316	1394	1345

Structure of the homogenized ZhS-32VI superalloy has γ'-phase particles within dendritic cells that are uniform in size and shape. Shapes of the γ'-particles are similar to cuboidal with sizes of about 300-400 nm; the eutectic γ'-phase with the particles sizes of about 1 μm is found in the inter-dendritic spaces. The application of the HTMT does not change the parameters of the growth of structure too much. Sizes of the dendrite cell (D) are somewhat reduced and the perfection of the structure of the single crystal is increased (Table 1.21, where: D - dendritic cell size; d - distance between the secondary branches of dendrites; θ - total misorientation of crystal blocks in the single crystal; m_γ - size of γ'-particles, Baryshev et al. 2010, Stepanova et al. 2003b).

Table 1.21 Growth structure of ZhS-32VI single crystals depending on the HTMT regime

HTMT	D, μm	d, μm	θ, grad.	$m_{\gamma'}$, μm
1600°C	280 ± 20	32 ± 3	15.0 ± 1.0	0.42 ± 0.05
1650°C	240	33	10.0	0.40
1700°C	240	35	10.0	0.43
1740°C	260	40	8.0	0.40
1830°C	240	40	8.0	0.45

The γ'-phase dissolution occurs at the same speed in the ZhS-32VI samples after different HTMT regimes (the initial volume fraction of γ'-phase before HTMT is 67 vol. %), with an exception of HTMT at 1740°C. In the latter case, the γ'-phase resistance to dissolution becomes increased (Fig. 1.12). The operation temperatures of the details manufactured from this alloy are 1000-1100°C. At such temperatures, there are differences in the γ'-phase dissolution for the ZhS-26VI and ZhS-32VI superalloys. However, the deformation processes accelerate the γ'-phase dissolution; obtained results allow one to estimate the general trend.

The HTMT treatment at 1740°C leads to the destruction of the eutectic carbide colonies with the Chinese script morphology. Their content decreases and they lose their solidity. This is important because the carbide colonies are stress concentrators in the alloy and impair mechanical characteristics of the alloy due to their suboptimal form and low thermal stability. X-ray studies show that HTMT leads to increased homogeneity of γ-solid solutions. The degree of coherence between the γ and γ'-phases increases and the mismatch of the parameters of their lattices reduces (Table 1.22, where: $a_{\gamma'}, a_{\gamma}$ - lattice parameters for γ'-phase and γ phase, respectively; $\Delta = (a_{\gamma'} - a_{\gamma})/ a_{\gamma}$ - mismatch of the lattice parameters, Baryshev et al. 2010).

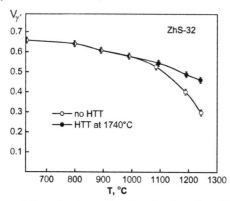

Fig. 1.12 Decrease of the γ' volume fraction in the alloy ZhS-32VI during the heating at a speed of 5°C/min, depending on the HTMT application (Stepanova et al. 2003b).

Table 1.22 Lattice parameters of γ and γ'-phase in the ZhS-32VI superalloy obtained by HTMT in different regimes

Regime of melting	a_γ, nm	$a_{\gamma'}$, nm	Δ, %
1600	0.35869	0.35816	0.189
1650	0.35846	0.35810	0.131
1700	0.35846	0.35812	0.134
1740	0.35858	0.35818	0.125
1830	0.35845	0.35805	0.137

The long-term strength values obtained from the various tests for the ZhS-32VI superalloy (with carbon) are given in Table 1.23, (Baryshev et al. 2010). Characteristics of the long-term strength were determined by testing of the 15 samples using in each test mode.

Table 1.23 Long-term strength of the single-crystal samples of the ZhS-32VI superalloy (HTMT at 1740°C) in tests at the range from 10 to 10,000 hours

Temperature	Crystal orientation	σ_{10}, MPa	σ_{100}, MPa	σ_{500}, MPa	σ_{1000}, MPa	σ_{5000}, MPa	σ_{10000}, MPa
	[001]	743/680	512/463	83/344	36/300	44/216	10/187
900°C	[011]	530/464	437/371	72/306	44/278	79/213	51/185
	[111]	850/681	543/434	96/317	46/276	52/200	19/175
	[001]	82/362	58/42	90/78	66/155	19/11	02/95
1000°C	[011]	69/31	49/19	80/56	155/133	06/90	89/75
	[111]	92/57	85/256	22/98	98/175	49/31	131/115
	[001]	11/171	29/03	0/71	6/60	2/1	44/35
1100°C	[011]	93/74	10/7	7/57	52/ 44	8/3	20/17
	[111]	23/83	34/08	2/4	8/3	3/3	45/36

Anisotropy of the static elasticity modulus of single crystals is maintained within a temperature range from 20 to 1000°C (Table 1.24).

Table 1.24 Elastic moduli of the single crystals of ZhS32U with different orientations

Alloy	Orientation	Values of the elasticity moduli, E, GPa, at different temperature				
		20°C	700°C	800°C	900°C	1000°C
	001	125	104	98	92	85
ZhS32U	011	225	188	180	171	162
	111	312	265	256	246	235

The single crystals with the [111] orientation have maximal moduli and those with the [001] orientation show minimal moduli. It should be noted that the orientation dependence of the mechanical properties is not so clear. For example, in conditions of hard loading cycles, the largest resistance for the low cycle fatigue was found for single crystals with the [001] orientation and the minimal values were observed for the [111] orientation. These results were obtained at a temperature of 850°C on the basis of 100 cycles (Baryshev et al. 2010).

Stability and strength properties of the carbon-free ZhS-36VI alloy for single-crystal blades. At present, the carbon-free ZhS-36VI superalloy is one of the best Russian nickel superalloys (with respect to rational alloying and costs) for the production of the single crystal blades. This is true not only for the aircraft engines, but also for marine and new stationary industrial gas turbines (Kuznetsov et al. 2011b, 2012, 2015, Baryshev et al. 2010, Kozlov et al. 2006b). Sets of the single-crystal blades made from the ZhS-36VI superalloy are tested in the composition of the modern aircraft engines PS-90A and PS-90A2 with a gas temperature at the turbine inlet of 1850K (1580°C).

The alloy is suitable for casting of blades with a single-crystal structure and predominantly the [111] crystallographic orientation. Alloys with this orientation show high levels of heat resistance (σ_{100}^{1000} = 310 MPa). For the blades with an internal cavity cooling, [001] is the preferred crystallographic orientation, although the heat resistance with [001] orientation is slightly lower (σ_{100}^{1000} = 250 MPa). The chemical composition of the ZhS-36VI superalloy is given in Table 1.25.

Table 1.25 The ZhS-36VI carbon-free alloy chemical composition, wt.%

Ni	Cr	Co	Al	Ti	W	Re	Nb	Mo	Ta
Bal.	4.5	9.0	6.0	1.0	2.0	2.2	1.0	1.0	0.5

The structure of the ZhS-36VI superalloy is discussed earlier in the section "Structure of Nickel-based superalloys". The temperatures of phase transformations in this alloy are given in Table 1.26.

Table 1.26 Temperatures of phase transitions in the ZhS-36VI superalloy

Phase transition		t, °C
heat	beginning of γ'-phase dissolution	1050
	maximum of γ'-phase dissolution	1260
	complete γ'-phase dissolution	1310
cooling	beginning of γ'-phase allocation	1257
	maximum of γ'-phase allocation	1225

The carbon-free composition significantly increases the temperature of the γ'-phase dissolution up to 1050°C (in the ZhS-32VI superalloy, this temperature is approximately 850°C). It should be noted that the anomalous behavior of the temperature dependence of the electrical resistivity $\rho(t)$ for the ZhS-36VI melt vanishes at t_{f1} = 1760°C. This temperature is chosen for a high-temperature treatment of the melt (HTMT).

To eliminate harmful effects of impurity elements (O, S, P, N, C) on the mechanical properties of the ZhS-36VI superalloy, alloying of rare earth elements (yttrium and lanthanum) was introduced to the chemical composition of the alloy. The ZhS-36VI single-crystal blades with the [001] crystallographic orientation were manufactured in a furnace for directional solidification UVNK-8P without a liquid cooler. High-temperature treatment of the melt (HTMT) at 1760°C was used. The melting seed from the Ni₃W alloy is used to ensure accurate crystallographic orientation of the single crystals. The step-wise heat treatment starts with the homogenizing annealing as:

(1) 1320 ± 10°C, exposure 4 hours, quenched at a speed of ≥100 degrees/min;

(2) 1030 ± 10°C, exposure 4 hours, quenched at a speed of ≥50 degrees/min;

(3) 870 ± 10°C, 32 hours.

Such a complex treatment is aimed not only at the dissolution of the strengthening γ'-phase, but also at the complete dissolution of large globules of the eutectic γ'-phase segregations, as well as at the maximum reduction of dendritic segregation. As a result of this heat treatment, a two-phase (γ + γ') structure is formed with high volume fraction of the intermetallic phase (about 75%). The γ'-phase particles within the dendrites are orderly arranged along [001] directions and have dimensions of 300-400 nm. In the interdendritic areas, sizes of the γ'-phase particles are larger (close to 1 μm). Carbides or TCP phases were not found. The high temperature melt treatment (HTMT) also affects the γ'-phase stability to the dissolution under heating (Fig. 1.13).

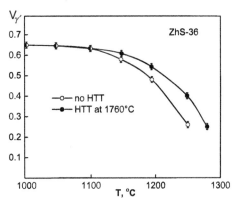

Fig. 1.13 The volume fraction of γ'-phase for alloys ZhS-36VI in the heating process at a speed of 5 degrees per minute, depending on the application of high temperature melt treatment (HTMT) (Stepanova et al. 2003b).

Single crystals of carbon-containing ZhS-36VI superalloys usually consist of some crystal blocks. The total angular misorientation θ of the crystal blocks can reach 15°. The angular misorientation between the nearby blocks is within 1°. For the carbon-free ZhS-36VI superalloy, the total misorientation of the crystal blocks is significantly lower.

The crystallographic orientation of the ZhS-36VI single crystal along the axis of tension is preserved until its destruction. The local azimuthal reversals of the blocks are within 5°. In the process of testing of the long-term strength, raft structures form perpendicular to the axis of loading. The structural parameters of the cast carbon-free single-crystal ZhS-36VI superalloy are given in Table 1.27, where D - dendritic cell size; d - distance between the secondary branches of dendrites; θ - total misorientation of blocks in the single crystal, m_γ - size of the γ'-particles.

Table 1.27 Cast structure of the ZhS-36VI carbon-free single crystals, depending on the regime of HTMT

HTMT	D, μm	d, μm	θ, grad.	$m_{\gamma'}$, nm
without HTMT	340 ± 30	67 ± 5	8 ± 1	420 ± 50
1650°C	340	65	7	420
1700°C	350	67	4	430
1760°C	310	53	4	400
1830°C	375	70	4	450

HTMT leads to an increasing long-term strength. Reduction of the sulfur content in the alloy is another factor increasing the long-term strength. Melting in a vacuum-induction furnace, vacuum refining, and filtering the melt through a foam-ceramic filter leads to a decreasing content of impurities.

Kuznetsov et al. (2008), (reprinted with the permission of the author), determined characteristics of short-term strength using the standard ten-fold cylindrical specimens with a diameter of 5 mm. Nine single crystalline samples for the temperature measurement were used to study the short-term strength, whereas to study the long-term strength, 15 samples in each test regime were used. The results of the mechanical tests are given in Table 1.28.

Table 1.28 Average values of the tensile mechanical characteristics of the [001] ZhS-36VI single crystal after crystallization under HTMT at 1760°C

Temperature, °C	σ_u, MPa	$\sigma_{0,2}$, MPa	δ, %	ψ, %
20	1060	1000	7	3
975	920	900	0	3

All these technological methods reduce the sulfur content and further allow one to reduce the probability of the spurious grains in the structure of a single-crystal ingot. This permits an increase in the production and improves the operation properties of the products. Reducing the sulfur content from 0.0085 to 0.0005% by mass, it was possible to increase the long-term strength of the ZhS-36VI samples at 975°C (without protective coating) from 75 to 110 hours at a stress level of 340 MPa and from 40 to 75 hours at 360 MPa; the optimal particle size of the strengthening γ'-phase is 0.35 µm (Kuznetsov et al. 2008, reprinted with the permission of the author).

The reduction of the sulfur content from 0.0085 to 0.0005 wt.% corresponds to an increase in the failure time from 270 to 420 hours at 1000°C. It should be noted that the gradient on the crystallization front changes from 70 to 150°C/mm. In the low cycle fatigue tests with an amplitude of 950 MPa at 900°C, the decrease of the sulfur content from 0.0085 to 0.0005 wt.% leads to an increase of the number of cycles to failure from 1900 to 3300 (Kuznetsov et al. 2008).

Average values of the long-term strength of the ZhS36VI superalloy at different temperatures are given in Table 1.29 (Kuznetsov et al. 2008).

Table 1.29. Long-term strength of the single-crystal ZhS36VI superalloy with the [001] crystallographic orientation

Alloy	σ_{100}^{900}	σ_{1000}^{900}	σ_{100}^{1000}	σ_{1000}^{1000}	σ_{100}^{1100}	σ_{1000}^{1100}	σ_{100}^{1200}
ZhS-36VI	450	360	320	240	200	160	130

Introduction of the ultra-dispersed powder of titanium carbonitride to the melt of the nickel-based superalloys. This section presents the study of the structural stability of the single-crystal samples obtained by introduction of the Ultra-Dispersed Powder of titanium carbonitride (UDP) to the melt. Introduction of the UDP to the melt before crystallization was proposed by V.P. Saburov to improve the long-term strength of the carbon-containing nickel-based superalloys as a method to affect the carbide subsystem of the alloy (Saburov et al. 1989).

The solidification of the nickel-based superalloys begins with the primary dendrite formation of a γ-solid solution. When the temperature decreases, carbides (C) appear from the eutectic in the interdendritic space. The morphology and orientation of the carbides are largely determined by the orientation of the γ-phase primary dendrites. The eutectic $\gamma + \gamma'$ or $\gamma + \gamma' + C$ are formed at the last stages of crystallization.

Addition of the dispersed powders of refractory compounds to the melt brings crystallization centers leading to changes in the order of phase formations. The UDP introduction also reduced the carbide concentration in the overcooling melt, because some of the components of the melt are taken away for the carbide phase formation at the early stages of crystallization. For example, in the ZhS-32VI superalloy, the carbides formation temperature increases from 1355 to 1364°C. All the results described below were obtained with the samples crystallized without HTMT.

Below, we consider the effects of crystallization with the UDP with respect to the γ'-phase stability and mechanical properties of single crystal ZhS-32VI superalloy with different carbon content. The chemical composition of the ZhS-32VI superalloy is given in Table 1.16.

Ultra-dispersed powder of titanium carbonitride TiCN (approximately 0.02%) is introduced into the melt in the form of tablets sintered with a metal activator. Synthetic TiCN refractory particles have sizes of 10^{-7}-10^{-6} m. The particles of refractory carbonitride can exist in the superalloy melt up to 1650°C without sintering with a metal activator. The type of the metal activator has a significant influence on the processes of particles interaction with the melt (Stepanova et al. 2000, Baryshev et al. 2010).

This fact can be explained by the formation of the diffusion layer at the boundary between the carbonitride particles and a metal activator during sintering, which promotes their interaction with the melt. Stepanova et al. (2000) used nickel, titanium, and chromium as the metal activators (the additives were designated as Ni-TiCN, Ti-TiCN, Cr-TiCN).

Without the UDP, the ZhS-32VI superalloy consists of the intermetallic compound Ni_3Al (γ'-phase) with the volume fraction of approximately 67% (about 10% are in eutectic) and γ'-phase (nickel solid solution). In the carbon-containing alloy, carbides are present in a range of no more than 2% by volume. Introduction of the UDP into the melt of the alloy with carbon changes the growth structure of the single-crystal ingot. The UDP increases the dendrite cell size D, while the distance between the dendrite branches d (second kind) remains virtually unchanged. The UDP promotes growth of the third dendrite branches. The growth parameters of the structures are given in Table 1.30 (Stepanova et al 2000). The long-term strength of the sample significantly depends on the structure perfection of a single crystal. Table 1.30 shows the angular misorientation between the neighboring crystal blocks in the single crystal ($\Delta\theta$) and the total misorientation $\Delta\theta_{tot}$ (between any two points of the crystal). With the UDP, the melt is close to an equilibrium state before crystallization. As a result, the angular misorientation of the blocks decreases from 7° to 1-2°. Addition of the UDP can be used for the production of single crystals with high growth rates. Introduction of the refractory disperse particles of titanium carbonitride with the diffusion layer based on metal activator leads to a redistribution of the diffusion flows in the melt ahead of the solidification front. This significantly reduces the influence of side heat gradients and the solidification front becomes flat-shaped.

Table 1.30 Parameters of the growth structure of the ZhS-32VI single crystals (0.16 mass % C)

UDP addition	D, µm	d, µm	Δθ, grad.	Δθ$_{tot}$, grad.	m, nm
no UDP	260±30	40±10	1.8±0.2	15.0±1.0	480±50
Ni-TiCN	350	40	0.8/1.5	7.0/9.0	460
Cr-TiCN	410	37	0.7/0.3	6.4/8.0	420
Ti-TiCN	370	50	0.6/1.5	7.0/12.0	450

During crystallization without the UDP, the Chinese-character-type carbides are oriented according to the matrix dendrites, because they are formed when the proportion of the solid parts is 75-80%. Introduction of the UDP leads to the carbide formation at the early stages of crystallization and changes of carbide orientation along the axes of the dendrites. The carbide morphology is also changed because large numbers of dispersed carbides with sizes from 20 to 300 nm appear. All of this should have a positive influence on mechanical properties of the single-crystal ingot, which becomes less anisotropic. Under certain input modes of the UDP, fully dispersed carbides can be obtained. The upper limit of the angular misorientation corresponds to the longitudinal cross-section, whereas the lower limit is the transverse cross-section. The chemical composition of the carbides can be defined as (Ta, Nb, W, Mo)C (Table 1.31).

Table 1.31 Chemical elements (at.%) found in the MC carbides of the carbon-containing ZhS-32VI superalloy

UDP addition after crystallization	Ta	Nb	W	Mo	Ti
No UDP	46	38	11	3	-
Cr-TiCN	38	23	30	5	4
Ti-TiCN	34	22	27	6	12
Ni-TiCN	36	22	30	10	4
UDP addition after heat treatment (1285°C, 1 hour)	Ta	Nb	W	Mo	Ti
No UDP	45	40	12	3	-
Cr-TiCN	34	46	12	4	-
Ti-TiCN	31	38	9	3	-
Ni-TiCN	37	45	10	3	-

Composition of the carbides was studied with the method of anodic dissolution and chemical analysis of the carbide precipitation. By an X-ray diffraction analysis, it was found that carbides belong to the MC type. The lattice parameter in the case of the UDP addition increased slightly: 0.4410 nm without additives and 0.4420 nm, 0.4430 nm, and 0.4450 nm after the introduction of Cr-TiCN, Ti-TiCN, and Ni-TiCN, respectively.

By an X-ray diffraction analysis, it was found that carbides belong to the MC type. The lattice parameter in the case of the UDP addition increased slightly: 0.4410 nm without additives and 0.4420 nm, 0.4430 nm, and 0.4450 nm after the introduction of Cr-TiCN, Ti-TiCN, and Ni-TiCN, respectively.

The technological parameters of UDP input in the melt are given in Tables 1.32 and 1.33 (Baryshev et al. 2010).

Table 1.32 Effects of the quantity of UDP (Ni-TiCN) added to the melt on the volume fraction of carbides in single crystals of the ZhS-32VI superalloy with the orientation [001]; exposure time is 10 minutes

UDP quantity, wt.%	0.02	0.03	0.05	0.1	0.14	0.2
volume fraction of carbides, %	2.2	2.2	2.3	2.6	2.75	2.9

Table 1.33 Effects of exposure of the melt at a temperature after the UDP input (Ni-TiCN) on the volume fraction of carbides in the ZhS-32VI single crystals with the orientation [001]; the UDP quantity is 0.03% by weight

Exposure, minutes	10	15	20	25	30
volume fraction of carbides, %	2.2	2.6	2.85	3.1	3.35

The γ'-phase particles in the ZhS-32VI superalloy have a cuboidal regular morphology. Introduction of UDP does not change the volume fraction and particle sizes of the strengthening phase in the cast alloy. Short-term annealing at 1285°C for 1 hour leads to the dissolution of the hardening γ'-phase and (partially) carbides. The amount of the eutectic γ'-phase is reduced and that of the strengthening γ'-phase in the alloys with UDP increases by 2% after annealing. Apparently, the increase in the amount of the strengthening phase is the main result of crystallization of carbon-containing alloy with the UDP. The impact of the UDP on the γ'-phase dissolution rate is more moderate.

A significant result was obtained in case of the carbon-free alloy, especially with the addition of Cr-TiCN (Fig. 1.14).

Introduction Ni-TiCN into the melt dramatically reduces the stability of the intermetallic phase to dissolution. Apparently, structure of the [001] single crystals of the carbon-free ZhS-32VI superalloy obtained by UDP introduction into the melt changes slightly. It should be noted that carbon is not specifically introduced in the carbon-free alloy. Carbon is present as an impurity introduced with the furnace charge with a concentration of 0.02% by weight. Volume fraction of the eutectic phase in the alloy is small (nearly 6%).

Introduction of the UDP during crystallization does not change the growth structure, whereas sizes of the γ'-phase particles are somewhat reduced. The total misorientation between the crystal blocks is small and never exceeds 3°, which allows us to speculate about high perfection of the single-crystal structure. Carbides are not observed in the alloy structure during TEM or SEM studies.

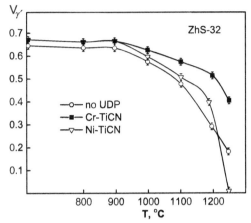

Fig. 1.14 Dissolution γ'-phase with temperature of ZhS-32VI single crystals, crystallized with the UDP additions, carbon-free alloy (0.02 mass percentage of carbon) (Baryshev et al. 2010).

Crystallization with the UDP in the carbon-containing alloy affects the temperature of the γ'-phase complete dissolution t_{cd} less than in the carbon-free alloy (Table 1.34, Baryshev et al. 2010). In the alloy with 0.02% by weight, addition of Cr-TiCN increases the temperature of complete dissolution, whereas Ni-TiCN significantly reduces it ($t_{cd} = 1257°C$). Tensile testing along the growth axis were performed using the single crystal samples of the ZhS-32VI superalloy with the [001] orientation and different carbon contents.

Table 1.34 Temperature t_{cd} of the complete dissolution of the γ'-phase according to the DTA in the carbon-containing and carbon-free ZhS-32VI superalloys

UDP addition	t_{cd}, °C	
	0.16% mass.	0.02% mass.
No UDP	1275 ± 3	1280
Cr-TiCN	1273	1295
Ni-TiCN	1268	1257

The maximum value of $\sigma_{0.2}$ in the carbon-containing alloy was found in the alloy with the Cr-TiCN addition (Table 1.35, where τ - the time to failure relative to that for the sample without UDP, Baryshev et al. 2010). The high σ_u values can also be obtained in the alloy with the addition of Ni-TiCN. The most significant effect of crystallization with the UDP was found in the long-term strength tests. The long-term strength tests were conducted at 1000°C at two stress levels.

The previous section was dedicated to the studies of the single-crystal nickel-based superalloys produced under high temperature treatment of the melt (HTMT). It was noted that the HTMT can provide the formation of the optimal growth structure of the single crystal and increase γ'-phase stability. Addition of the ultra-dispersed powder of titanium carbonitride (UDP) into the melt is similar to the HTMT influence on the structure of the single crystal. A combination of these treatments of the melt makes it possible to widely vary the parameters of the growth structure and the properties of the nickel-based superalloys.

Table 1.35 Average values of the short-time tensile characteristics and results of the long-term strength tests for the ZhS-32VI single crystal (0.16 wt.% C) with the [001] orientation and with different additions

UDP addition	σ_u, MPa	$\sigma_{0.2}$, MPa	δ, %	ψ, %	τ, h 185 MPa 1000°C	τ, h 300 MPa 1000°C
No UDP	1095	900	9.4	9.2	1.00	1.00
Cr-TiCN	1168	953	8.4	9.0	1.32	1.12
Ti-TiCN	1107	847	8.8	9.3	1.09	0.92
Ni-TiCN	1210	905	8.2	9.5	1.19	1.20

Testing at low stress showed benefits in the samples crystallized with addition of Cr-TiCN due to the increasing γ'-phase stability to dissolution at high temperatures (Table 1.36, where τ - the time to failure relative to that for sample without UDP, Baryshev et al. 2010). At high stress levels, high values of time to failure are obtained in the alloy with addition of Ni-TiCN due to the increase of the volume fraction of the γ'-phase.

Table 1.36 Average values of the short-time tensile characteristics and results of the long-term strength tests for the ZhS-32VI single crystal (0.02 wt.% C) with the [001] orientation and with different additions

UDP addition	σ_u, MPa	$\sigma_{0.2}$, MPa	δ, %	ψ, %	τ, h 140 MPa 1000°C	τ, h 280 MPa 1000°C
No UDP	1143	893	3.8	13.3	1.00	1.00
Cr-TiCN	1215	893	3.8	12.5	1.37	1.34
Ni-TiCN	1105	927	5.8	7.8	1.18	1.74

Contact interaction of the alloying titanium carbonitrides with a molten nickel-based alloy. Zhilyaev and Patrakov (2016a, b) modeled mechanisms and kinetics of the solid-phase and liquid-phase interactions of titanium carbide and titanium carbonitride with the melts of pure nickel or nickel-molybdenum. The data allowed understanding of the formation of composition of the alloys with the titanium carbonitride and metal-activator by a solid-phase mechanism and of the interaction of UDP addition with the nickel melt.

It was established that molybdenum is dissolved in TiCx ($x \geq 0.9$) with the formation of molybdenum carbide by the solid-phase mechanism:

$$TiC_x + Mo \rightarrow TiC_{x-m} + Mo_2C \rightarrow (Ti,Mo)C.$$

This process is accompanied by an initial decrease in the lattice parameter TiCx (formation of Mo$_2$C) with its subsequent increase (formation of a layer of solid solution). An interaction of the highly defective carbide TiCx ($x \leq 0.8$) with molybdenum is characterized by mutual solubility of the chemical components. A study of the interaction of the titanium carbide with small amounts of molybdenum (1, 2, 3, and 5%) suggests that the mechanism is based on the diffusion of molybdenum in TiC$_x$ that passes through the vacant sites in the carbon sublattice, forming a layer around carbide particles. According to an X-ray analysis, this layer consists of a complex carbide (Ti,Mo)C$_x$ with a NaCl type crystal lattice, which is an isostructural solid solution of two carbides, such as TiC and MoC$_{1-x}$. The presence of nickel in these systems (Ni:Mo = 3:1) greatly accelerates the process of the solid-phase interaction at temperatures of 600-1200°C. This increasing rate is due to the dissolution of carbide or the carbide component of titanium carbonitride in pure nickel at lower temperatures. The carbide Mo$_2$C formed in these conditions is not detected in the TiC$_{0.96}$-Ni-Mo system after sintering at 1200°C (1 hour), because it is dissolved in nickel and titanium carbide. This carbide remains in the TiC$_x$N$_z$-Ni-Mo system because it is only partially dissolved in nickel. These data show that formation of such layers by the solid-phase mechanism is fundamentally possible in the TiC$_x$-Mo-(Mo$_2$C)-Ni system. The presence of nitrogen in the titanium carbide prevents the layer formation and deviations in titanium carbide composition from stoichiometry content significantly complicate this process.

In the presence of a liquid phase surrounding the carbide particles, a layer of solid solution between the cubic carbide α-MoC$_{1-x}$ and the titanium carbide is formed by the dissolution and deposition mechanism; the layer is radially inhomogeneous. The layer composition depends on the temperature, time of sintering, ratio of the components, and amount of carbon in the system. Diffusion of molybdenum from the melt in the refractory alloys does not occur (temperature range is 1300-1500°C; exposure time is 0.5-25 hours).

Under the interaction between the titanium carbonitride with different compositions and nickel or Ni-Mo melt, the carbide component of the carbonitride is mainly dissolved. Nitrogen barely participates in this process due to its weak solubility in the Ni-Mo melt. Components of the titanium carbide are dissolved in the Ni-Mo melt in a non-equiatomic way and the amount of dissolved titanium has a complex dependence on the composition of TiC_x and it is at minimum in $TiC_{0.80}$. For alloys with $x \geq 0.9$, carbon is predominantly dissolved, while for alloys with $x \leq 0.8$, titanium is fully dissolved. In the latter case, the intermetallic compound Ni_3Ti with dissolved molybdenum (up to 10% by weight) is always present in the system, together with a layer of the carbide solid solution (Fig. 1.15).

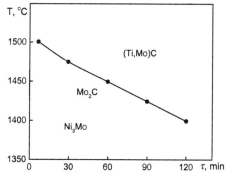

Fig. 1.15 Temperature-time diagram of the chemial compaunds formation in the interaction of $TiC_{0.5}N_{0.5}$ with melt (Ni + 25% of Mo) (Zhilyaev and Patrakov 2016a, reprint with the kind permission of the authors)

The presence of nitrogen in the titanium carbide prevents the formation of the solid solution layer. This fundamental difference between the alloys based on the carbide and titanium carbonitride defines the technological benefits of the latter in cases when it is desirable to avoid complete dissolution of the refractory particles, preserving them as crystallization centers.

The contact interaction of nickel melts with the hot-pressed powders of the titanium carbonitrides, doped with transition elements of the V-VI groups is characterized by the incongruent dissolution of the solid-phase components (mostly carbon and alloying metal) into a liquid. The main causes of this dissolution are:

- Low solubility of nitrogen in molten nickel compared to carbon;
- Greater affinity of titanium for nitrogen, compared to the transition elements of V-VI groups;
- Destabilizing contribution of the elastic deformation energy;
- Contribution of the atomic energy of the transition metals of the groups IV-VI in the free energy of the formation of the solid solution $Ti_{1-n}Me_nC_xN_z$ (and $Ti_{1-n}Me_nC_x$); the free energy is at maximum with adding of the metals of the IV group.

As a result, the peripheral parts of the carbonitride grains are enriched with nitrogen and titanium. After the solidification of the melt, the areas near the carbonitride particles have a structure that forms before eutectic, such as the primary nickel-based crystals and the triple carbonitride eutectic produced by the interaction of the titanium carbonitrides with the nickel melt. The introduction of molybdenum in the nickel melt accelerates the processes.

Ion-plasma coatings to protect superalloys from oxidation. Nickel superalloys have some protection from oxidation. Their resistance to oxidation at high temperatures is caused by the formation of Cr_2O_3 at the surface of superallloy ingot. In the Ni-Cr system alloys, the NiO oxide dominates at the surface, when Cr content in the alloy is less than 10% by weight. Close to the alloy-oxide interface areas, Cr_2O_3 particles are surrounded by pure nickel. An increasing chromium content leads to the disappearance of the mixed Ni-Cr_2O_3 layer.

A minimum oxidation rate is found at the optimal ratio of the NiO and Cr_2O_3 oxides at the surface. During long exposures at high temperatures (above 1000-1050°C), these oxides turn into $NiCr_2O_4$ spinel. A superalloy must contain 20-25% of Cr for real protection from oxidation. High chromium content negatively affects the properties of the intermetallic γ'-phase and significantly reduces the long-term strength of the alloy. Therefore, as a rule, superalloys with protective coatings are used. In operation, the coatings should not flake off or become a source of cracks that might spread from the surface into the material (Caron and Lavigne 2011).

The NiAl (β-phase)-based coatings on the intermetallic compounds are often used to protect the turbine blades from oxidation. This nickel aluminide meets the requirements for the high-temperature operation. The Al_2O_3 oxide formed on the surface of aluminide coatings has high protective properties under the conditions of high-temperature oxidation.

A significant positive effect is achieved by micro alloying of the coating with some elements, such as Zr, Y, and Rare Earth Elements (REE) (Ren et al. 2015, Cao et al. 2017, Lin et al 2017, Liu et al. 2016, Zhong et al. 2016).

In practice, platinum is often included into the coating composition (Caron and Lavigne 2011, Parlikar et al. 2017, Xu et al. 2016). This element is not used in the Russian coatings. The coatings containing platinum are not absolute guarantee against cracking or delamination of the coating (Caron and Lavigne 2011).

Coating methods can be different (Kablov and Muboyadzhyan 2012, Wang et al. 2017). Below, we will focus on the Russian experience in the coating by ion-plasma application (Kuznetsov et al. 2011b, Budinovskii et al. 2011). This coating process ensures high coating density due to ion bombardment, good adhesion, high precision and reproducibility of the deposition process, sub-microcrystalline structure of the coating, in combination with the simplicity and versatility of the industrial equipment using a vacuum-arc evaporation method.

Transition layers between the coating and the main alloy are primary interdiffusion and secondary reaction zones. In superalloys doped with rhenium, the composite condensation-diffusion coatings supplemented by a barrier layer are used.

Here we consider their application on an example of the ZhS-36VI superalloy (Budinovskii et al. 2011, 2014). Carbides of refractory metals can be formed in the carbon-containing heat-resisting alloys doped with rhenium. Although the ZhS-36VI superalloy does not contain carbon, the use of barrier layers with the chromium carbides does not exclude the possibility of the barrier layer destruction by its interaction with the blade material under high-temperature loading. The barrier layer containing metal oxides and nitrides is more resistant to high temperatures. The chemical composition (by weight) of the alloys used for coating: 20% Cr, 12% Al, 0.3% Y, Ni - balance (SDP-2 coating); 1.5% Y, 5% Ni, Al - balance (VSDP-16 coating).

The composite barrier layers are formed by the vacuum arc discharge (with a current of 700 A and a flow of the reaction gas (acetylene, nitrogen, or oxygen) at pressures of 0.03-0.1 Pa. The thickness of the barrier layers in the coatings is 10 microns. After the application of the barrier layer, the reaction gas supply is interrupted. Then the heat-resistant layer of the SDP-2 coating alloy with a thickness of 60 mm and a layer of the VSDP-16 aluminum alloy with a thickness of 20 μm are successively applied by the ion-plasma spraying in the vacuum. The final formation of the multilayer ion-plasma coatings is observed in the process of the vacuum heat treatment at a temperature of 1050°C for 3 hours.

The coating formed in an oxygen atmosphere shows the highest microhardness (twice as high, compared with the coating without the barrier layer) and it consists of different phases, such as the γ-phase, γ'-phase, Al_2O_3, and $YCrO_4$ (Budinovskii et al. 2011).

The metallographic study of the coatings is conducted to detect the TCP-phases on the boundary between the alloy and coating. In the oxide barrier case, the layer deposition speed is lower, but the microhardness of the coating is much higher than in the case of SDP-2 without barrier layer. The oxide coating has also high heat resistance after exposure of 1000 hours at 1100°C, as shown by the gravimetric control of the specific weight change of the samples.

The coating layers are formed during annealing by the aluminum diffusion from the VSDP-16 alloy into the substrate and the nickel diffusion from SDP-2 in the opposite direction. The result is the outer layer of the coatings based on NiAl (β-phase) doped with chromium and yttrium in the range of solubility of these elements in β-phase at

1050°C. This layer does not contain refractory elements which could get into the coating from the substrate.

Two-phase layers consisting of NiAl (β-phase) and inclusions of α-chrome solid solution are between the main material and outer layer. Part of the SDP-2 remains on the boundary between the coating and alloy. A zone of diffusion interaction with a depth of 10 µm also starts to form. In all cases, these layers do not contain the refractory elements taken from the basic alloy. After a long high temperature exposure, the coating structure significantly changes. The aluminum and chromium contents become reduced. Diffusion of these elements occurs on the surface, as well as in the substrate. In the first case, these processes occur due to the consumption of these elements on the formation of protective oxide films, when oxide protection forms again after cracking and chipping. In the second case, the diffusion processes occur due to the presence of a concentration gradient, because the aluminum and chromium contents in the protective layer are twice as high than in the base alloy. Then the zone of the diffusion interaction can be observed and the secondary reaction zone containing TCP-phases with a different morphology behind it. After 500 hours of testing at 1100°C, the formation of the oxide inclusions (mainly oxides of chromium and nickel) begins in the outer zone. The introduction of the composite barrier layers to the coating design provides a higher aluminum content after 500 hours of testing in all layers of the coating than in the case of the pure metal coatings. It is possible to avoid the appearance of micro cracks running into the internal volume of the alloy under testing for the long-term strength.

When the test duration is increased up to 1000 hours, the amount of the oxide inclusions are increased and the thickness of the aluminum oxide layer becomes decreased. In the samples coated without the barrier layer, TCP-phases with needle morphology are formed due to the diffusion of the refractory elements in the base alloy directly near the coating. As a result, the coating loses its protective properties and it can be destroyed. The barrier layers reduce the TCP-phase distribution depth by a factor of two. It should be noted that after 1000 hours of aging, the TCP-phase amounts become lower than after 500 hours. This happens due to the depletion of the solid solution surrounding TCP phases by alloying elements diffused into the coating, which leads to a partial dissolution of TCP phases.

1.1.3 *Ni₃Al-based intermetallic alloys*

The development of new intermetallic alloys is dictated by the fact that the operation temperatures of nickel superalloys are limited by 1050-1150°C. The limit is determined by the alloy softening due to reduction of the volume fraction of γ'-phase (Ni$_3$Al) at such temperatures. It should be noted that the alloying of the γ + γ'-phases nickel alloys with refractory metals improves their mechanical properties but does not eliminate the γ'-phase dissolution.

The intermetallic materials based on nickel aluminate (90% of γ'-phase, Ni$_3$Al) are the most promising ones to replace the nickel-based superalloys with a low content of γ'-phase in some designs. This is because such intermetallic materials have a high melting point (t_M = 1373°C), low density (ρ = 7.5 g/cm^3), and they do not need protection from oxidation due to their higher Al content. In addition, the production of complicated details from intermetallic alloys is possible with a well-worked technique of processing developed for the nickel superalloys. An overview of the composition and properties of the modern alloys based on the Ni$_3$Al intermetallic compound was given by Jozwik et al. (2015).

The Russian high-temperature Ni$_3$Al-based intermetallic alloys are those of the VKNA series developed at the *Russian* Scientific Research *Institute of Aviation Materials* (*RIAM*) by V.P. Buntushkin and co-workers (Buntushkin et al. 2004). The VKNA-type alloys have favorable combination of ductility at low and medium temperatures and heat resistance at temperatures above 1000°C (Povarova et al. 2011a,b 2015).

The VKNA-type alloys are tested in the promising aircraft engine PD-14, the next generation turbofan engine. This engine is supposed to be one of the alternative power sources for the Ilyushin Il-76 and Irkut MS-21 passenger aircrafts. The VKNA-4 alloy has passed the testing as the material for the turbine blades of the AL-31F engine and MD-120 engine (Bazyleva et al. 2014, Jozwik et al. 2015).

Short-time mechanical properties of single-crystals of the VKNA-type alloys in the temperature range of 1100-1250°C. This section presents results of a comparative study of the structure and mechanical properties of the single-crystal VKNA-1 and VKNA-4U superalloys with the ⟨001⟩ orientation (both cast and after heat treatment) using tensile tests in the temperature range of 1100-1250°C (Rodionov et al. 2010a,b). For the VKNA-4U alloy, different crystallographic orientations were tested. At the melting temperature of the nickel superalloys, it is recommended to add separately (stepwise) the basic and alloying elements, according to their reactivity, difference in the melting and boiling points, and density of the components (Bazyeva et al. 2016). This is especially important for alloys doped with refractory metals. At the first stage of reactions, the inactive elements Ni, W, and Mo are melted using the prepared W$_x$Ni$_y$ ligature to eliminate the melt overheating and prevent the depletion of heavy elements at the bottom of the ingot. At the second stage of reactions, Cr and Ti are added to the melt and, at the third stage, the chemically active alloying elements, such as Al, are added. This melting method is accompanied by melt refinement, which leads to the minimal content of C, N_2, O_2 and provides high mechanical properties of the alloys. The nickel superalloys discussed below were melted by Yu.N. Akshentsev at the Department of Precision Metallurgy at the Institute of Metal Physics of the Ural Branch of the Russian Academy of Science (IMP UB RAS). He used an induction vacuum furnace applying all three stages of the addition of the alloying elements into the melt to synthesize the alloys. The single crystals of the VKNA-1 and VKNA-4U superalloys in the form of cylinders with a diameter of 18 mm were grown using the Bridgman method. The growth rate was 1 mm/min and the temperature gradient was 80°C/cm. The chemical composition of the alloys, according to a chemical analysis, is given in Table 1.37 (Rodionov et al. 2010a, b).

Table 1.37   Chemical composition of the intermetallic nickel superalloys, Ni- balance, wt.%

Grade	Composition, wt. %							
	Al	Cr	Mo	W	Ti	Co	C	other
VKNA-1V	8.8	5.3	3.4	3.14	1.03	-	0.02	0.3 Hf, 0.6 Fe, 0.4 Si
VKNA-4U	8.4	5.04	5.05	2.52	0.96	3.75	0.02	1.2 Zr

The single crystals were used to prepare tensile specimens. The high-temperature tensile tests were carried out at 1100, 1200, and 1250°C. The loading rate was 1.32 mm/min (2×10^{-5} m/s). The samples were investigated in the as-cast state and after a heat treatments at 1280°C (for 5 hours) and at 900°C (for 27 hours).

The structure of the VKNA superalloys is close to the $\gamma' + \gamma$ eutectic one (Fig. 1.16) after crystallization.

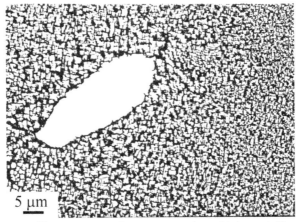

5 μm

Fig. 1.16   The structure of as-cast samples of intermetallic alloy (Rodionov et al. 2010a,b).

In the as-cast alloy, massive layers of the eutectic (primary) γ'-phase were interspersed with areas of the γ solid solution.

On cooling, the secondary γ'-phase precipitates, with sizes of 200-1000 nm, are formed inside the γ solid solution layers. After the heat treatments at 1280°C (for 5 hours) and at 900°C (for 27 hours) to obtain maximum possible amounts of the γ' phase, small γ'-phase particles with sizes of 30-50 nm are additionally precipitated within the interlayers of the solid solution. Simultaneously, the γ'-phase particles coalesce in the central regions of dendritic cells and a bimodal structure of the intermetallic phase with different particle sizes (1-2 µm and 100 nm) is formed. After the heat treatment, the amount of the γ'-phase in the alloy increases up to 95%.

Tension tests were conducted at different temperatures. According to our data, three temperature ranges can be identified in all the samples of the VKNA type alloys, corresponding to different mechanisms of stress relaxation in the process of high-temperature tension.

In the samples tested for short-term strength at 1100°C, a sharp increase of the dislocation density near the fracture surface was observed. The dislocations were concentrated in the γ-phase areas. The tests at 1200°C revealed large areas, in which dissolution of the γ' phase occurred on heating; during a subsequent cooling, the γ' phase precipitated again. In this case, the rate of cooling controlled sizes of the γ'-phase nanoparticles of about 120 nm; the samples tested at 1250°C showed particles of 80 nm in size. In the deformation process at 1200°C, the dynamic recovery is the basic mechanism determining stress relaxation of the samples. The structures of both the VKNA-1 and VKNA-4U alloys demonstrated an increase of the dislocation density. The dislocations were concentrated on the low-angle boundaries penetrating the entire sample. Such boundaries spread around large particles of the intermetallic compound and intersect the regions of the solid solution, where smaller γ' phase particles were again precipitated on cooling. The boundaries were very wide, with small misorientation (1-2°) between the adjacent regions. The resulting SAED pattern taken from the region of the boundary between sub-grains looked similar to the one taken from a single crystal. However, the diffraction reflexes, located far from the central one, were fragmented, because they were formed by the crystal areas with a comparatively small rotation relative to each other. The regions inside the sub-grains were virtually free from the dislocations. In this case, the structural equilibrium is supported by a continuous generation and annihilation of the dislocations and by the formation and destruction of sub-boundaries. As a rule, the larger particles of the intermetallic phase were free of defects.

The detailed structural study did not reveal any recrystallization in the sample. The critical conditions for dynamic recrystallization in the alloys depend substantially on the temperature, orientation of single crystals, rate, and degree of deformation. At 1200°C, the amount of the dissolved γ'-phase increased; dislocation density in the intermetallic particles increased. Thus, additional ways of stress relaxation appeared to exist, and therefore conditions for the recrystallization process were not fulfilled. The VKNA-type alloys were specially designed as the high-temperature, high-strength alloys; they possess an increased thermal resistance to recrystallization. After the tensile tests at 1250°C, the optical metallography revealed the initial stages of directional coalescence of the γ' phase (raft structure) in the neck part of the specimens. Directly near the fracture surface, the appearance of cracks was observed, whose propagation was retarded because of the high fracture toughness of the material. The TEM study revealed the areas with a high dislocation density and large quantity of low-angle boundaries. The relaxation occurs due to the dynamic recovery and recrystallization does not develop. A characteristic property of the microstructural state of the alloy after deformation at 1250°C is the fact that the deformation involves both the spaces between large γ'-phase particles and the particles themselves. The γ'-phase particles participate in the deformation in any event. On the deformation at 1200°C, a few dislocations were also visible in some γ'-phase large particles. At 1250°C, fragmentation of large particles of the γ' phase began with an increase in the dislocation density; the particles were divided by low-angle boundaries. A characteristic contrast appears in the dark-field images taken with the superlattice reflex, such as the particle does not "shine" as a whole; the different areas of the particle "lights up" consecutively when the foil is inclined in the column of the microscope. Such a structural state of the alloy is a consequence of the active mode of loading of the sample.

It should be noted that no γ'-phase fragmentation is observed during the long-term strength tests of the VKNA-type alloys, in which the structural changes are mainly reduced to the formation of a raft-type structure by the directional coalescence of particles under external stresses. At the same time, fragmentation is possible in the samples subjected to an accidental fracture at a high temperature.

The development of the relaxation process during the high-temperature tension tests can clearly be seen in the changing shapes of the stress–strain curves (Figs. 1.17 and 1.18), typical for the high-temperature deformation.

It may be noted that the sample has a very small uniform elongation. During its deformation, a neck appears almost immediately, concentrating the entire plastic deformation. These results, like the data in Figs. 1.17 and 1.18, reveal specific features of the alloy behavior, depending on the type of alloying.

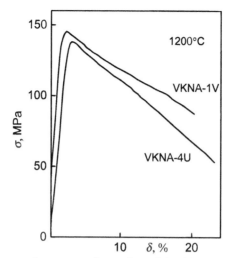

Fig. 1.17 Stress–strain curves of samples of the VKNA-1 and VKNA-4U
alloys at 1200°C (Rodionov et al. 2010a).

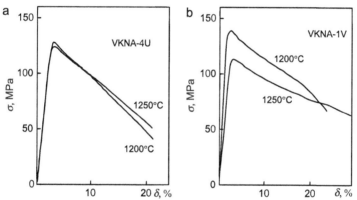

Fig. 1.18 Stress–strain curves taken at 1200°C and 1250°C for the samples of
the alloys: (a) VKNA-4U and (b) VKNA-1 (Rodionov et al. 2010a).

At 1200°C, the mechanical properties of the VKNA-1 and VKNA-4U
superalloys are similar to each other, with some superiority of VKNA-1
(Fig. 1.17). At 1200°C, VKNA-1 has a somewhat larger (within 5%)
modulus of the normal elasticity E, in comparison with VKNA-4U,
which ensures its advantage in the mechanical tests. However, when
temperatures increase to 1250°C, the VKNA-4U superalloy preserves
the modulus value (Fig. 1.18a), whereas VKNA-1 reveals a decreasing
modulus of E (Fig. 1.18b). This difference in the temperature dependence
of the elasticity moduli can be attributed to the differences in the
chemical compositions of the alloys, which influence the binding forces
in the Ni_3Al compound.

Rinkevich et al. (2006) modeled the ternary alloy $Ni_{67}Al_{25}Co_8$, showing that alloying with cobalt leads to a considerable reduction of the elasticity moduli for a number of crystallographic orientations of a single-crystal sample. The chemical compositions of the VKNA-4U and VKNA-1 superalloys differ in that the VKNA-4U superalloy contains cobalt. In the intermetallic alloy, cobalt enters the γ'-phase crystal lattice. Therefore, the VKNA-4U superalloy, whose properties are mainly determined by the intermetallic phase, has inferior properties at 1200°C than VKNA-1. At 1250°C, when the γ'-phase dissolution is more pronounced and the contribution of the solid solution is greater, the mechanical properties of the VKNA-4U superalloy become better, since cobalt significantly strengthens the solid solution. In addition, higher temperatures (up to 1250°C) lead to a significant increase in the plasticity of the VKNA-1 superalloy, which is related to the dissolution of a larger amount of the γ' phase as compared to the VKNA-4U superalloy.

The results of the short-term strength tests are given in Table 1.38 (Rodionov et al. 2010a).

Table 1.38 Mechanical properties of $\langle 001 \rangle$ single crystalline samples of the VKNA-4U superalloy: as-cast and HT (heat-treated at 1280°C for 5 hours and at 900°C for 27 hours)

Temperature, °C	Alloy state	$\sigma_{0.2}$, MPa	σ_B, MPa	δ, %
1100	as-cast	408	414	22.0
	HT	420	430	22.0
1200	as-cast	130	136	20.8
	HT	118	127	22.4

According to the X-ray diffraction data for the superalloy both in the as-cast and heat-treated states, the γ' phase dissolution rate remains the same upon heating to 1100°C. In this case, the higher content of the strengthening phase is advantageous for the heat-treated samples over the as-cast ones in mechanical tests. Higher temperatures of up to 1200°C reveal the instability of the γ' phase in the heat-treated sample with respect to dissolution, so that the as-cast sample has better mechanical properties.

Table 1.39 (Rodionov et al. 2010a) illustrates the influence of Heat Treatment (HT) on the mechanical properties of the VKNA-4U samples.

Table 1.39 Mechanical properties of $\langle 001 \rangle$ single crystalline samples of the VKNA–type superalloys

Alloy	Temperature, °C	$\sigma_{0.2}$, MPa	σ_u, MPa	δ, %
VKNA-4U	1200	130	136	20.8
	1250	123	131	24.6
VKNA-1	1200	138	142	17.8
	1250	115	122	29.0

The elasticity moduli of the VKNA-4U superalloy, determined by Rinkevich et al. (2003), are given in Table 1.40 (Rinkevich et al. 2003).

Table 1.40 Moduli of elasticity c_{ij}, density ρ and Debye temperature T_D, the bulk modulus B and shear modulus G for the VKNA-4U superalloy

Alloy	T_D, K	c_{ij}, GPa			B, GPa	G, GPa	ρ, kg/m³
		11	12	44			
VKNA-4U	382	215	145	117	168	74	8.35

All of above results were obtained for the ⟨001⟩ crystallographic orientation in the samples. Single crystals of the VKNA-4U superalloy with different crystallographic orientations were also grown by the Bridgman method at the Institute of Metal Physics (Russia), at a rate of 10 mm/min using a temperature gradient of 100°C/cm. Figure 1.19 shows the influence of the orientation of the single-crystal sample of VKNA-4U on its mechanical properties.

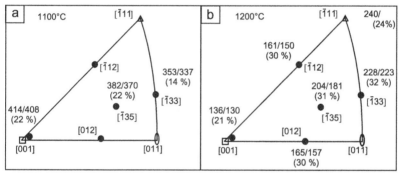

Fig. 1.19 Orientation dependence of the mechanical properties for single crystal samples of the VKNA-4 alloy: (a) 1100°C and (b) 1200°C (Rodionov et al. 2010a).

The results are presented at the stereographic triangles for two tested temperatures. At 1100°C, a change in the orientation from ⟨001⟩ to ⟨133⟩ leads to a certain reduction in the strength properties. It should be noted that the best properties have the samples with ⟨111⟩ orientation. This conclusion is applicable to the samples tested at 1200°C. The values of σ_u = 230 MPa and δ = 24% for the ⟨111⟩ sample given by Kablov et al. (2006a) agree well with the results of tests at 1200°C presented in Fig. 1.19b.

Metastable phases in the Ni₃Al-based intermetallic alloys under directional solidification. The method of directional solidification (Bridgman method) is widely used to obtain single crystals. As shown by experimental data, directional crystallization in the Ni₃Al alloy leads to the formation of different phases and microstructures, including the metastable ones, which differ in chemical composition. In this method, the microstructure and phase composition of the alloy depend on the crystallization rate, and therefore transformations take place according to the diagram of metastable states (Huziker and Kuz 1997). According to the equilibrium phase diagram, the Ni_3Al intermetallic compounds (γ'-phase, $L1_2$ superstructure, structure type Cu_3Au) should be formed directly from the liquid. However, in the Bridgman method, this process occurs in several stages: metastable γ-β eutectic form first, followed by the cellular two-phase dendritic structures (Stepanova et al. 2003a). Kazantseva et al. (2009a) provided an analysis of the structural changes and formation of metastable phase states in the Ni-9.6 wt.% Al-6.7 wt.% Fe-1 wt.% Cr alloy, obtained by crystallization with a Bridgman method. To get the maximum amount of metastable states during annealing or under cooling after annealing we performed a short-term annealing of the crystal at 1000°C for 5 hours.

According to an X-ray analysis after the short-term annealing, the cast alloy has a single-phase (γ'-phase) state; the structure of the alloy is inhomogeneous. In the TEM images, areas with high dislocation densities and those free of dislocations can be seen. In addition, one can see a very specific structure with a fiber texture. The structure is formed by the elongated cells whose boundaries are decorated with the dispersed particles. The texture can be seen more clearly near the boundaries of the eutectic γ/β.

Widths of the texture fibers are dissimilar in different parts of the foil, varying between 0.02 and 0.1 μm. This fact allows us to exclude the dislocation nature of the fiber formation, i.e., that they are not caused by the sub-grain boundaries. The dislocations easily pass through the fibers. The texture is clearly visible in the structural reflexes, which also excludes the idea about the domain origin of the fibers. On the edge of the foil texture, the fibers look like the thin teeth fringes, which also excludes their interpretation as a specific contrast or moiré. Perhaps the nature of such structures is similar to spinodal decomposition and is associated with the heterogeneity of the alloying elements distribution. Fiber texture cells have a characteristic orientation along the [100] direction of the γ'-phase crystal lattice. Thus, we may assume that the formation of such structures occurred during directional solidification. Another feature of the structure after the short-term annealing was the presence of diffraction rings in the SAED patterns, in addition to the point diffraction reflexes of γ'-phase (Table 1.41). The dispersed particles "shined" at these ring positions in the dark-field images. According to the TEM results, the presence of the weak superstructure reflections from the {110}γ' planes in the absence of strong structural reflections from the {220}γ' planes suggests that these rings cannot belong to γ'-phase; they do not belong to β-phase either. The relationship $Q_i = \sin^2\theta_i/\sin^2\theta_1$ is used for indexing reflections from an unknown phase. Different crystal lattices have various relations for Q_i: 2:3:4 in BCC and primitive cubic lattices; 1.33:2.66:3.66 for HCC; 2:3:4 for Ni_3Al and NiAl. In our case, Q_i is 1.43:2.85:4.75 for the observed rings. Therefore, the observed reflections cannot belong to NiAl nor Ni_3Al. Moreover, they do not belong to the cubic system at all and do not correspond to reflections of any known stable phases in Ni-Al, Ni-Al-Fe, Ni-Al-Cr, Ni-Al-Fe-Cr (JCPDS-ICDD PDF-4 Database 2007). However, the emergence of a metastable phase formed from $B2$ with non-stoichiometric composition by the shear is possible in all these systems. This phase belongs to omega phases and has a composition similar to Ni_2Al and crystal lattice varying from the trigonal C6 (P-3m1) to the $B8_2$ ($P6_3/mmc$), with the transition to a more equilibrium $B8_2$ (Muto et al. 1993). As shown by Sadi and Servant (2000), the high- temperature annealing promotes the conversion of omega phase to Ni_5Al_3 or Ni_3Al in this case. To identify the crystal structure of a new phase, the neutron diffraction analysis was used, because the amount of metastable phase is small. The neutron analysis shows that the position and intensity ratio of the observable diffraction lines do not coincide with reflections for neither a long-period tetragonal $D0_{22}$ phase nor superstructure $L1_0$.

Table 1.41 The interplanar distances measured from the ring-like SAED patterns after a short-term annealing at 1000°C for 5 hours

Ring number	1	2	3	4
Interplanar distance	0.2511	0.205	0.156	-
Intensity of ring (conv. units)	10	100	20	-
Interplanar distance for γ'-phase (Ni_3Al, a = 0.355 nm), nm	0.2510	0.2051	0.159	0.1255
Intensity of γ'-phase diffraction lines (Ni_3Al, a = 0.355 nm)	(110) 8.5	(111) 100	(210) 3.3	(220) 22.1
Intensity of β-phase diffraction lines (NiAl, a = 0.289 nm)	0.289	0.2049	0.1449	0.1292
Intensity of β-phase diffraction lines (NiAl, a = 0.289 nm)	(100) 25.2	(110) 100	(200) 13.3	(210) 5.3

In our case, it is not possible to describe all of the observed reflexes as the FCC reflexes or tetragonal structure. A better agreement may be achieved in both line positions and intensity ratios by calculating the diffraction lines as those of a trigonal superstructure $B8_2$. The adopted orientation relationships between the lattices of the new phase and γ'-phase are (Muto et al. 1993, Sadi and Servant 2000):

$$(001)_\omega \parallel (111)_{B2}, \quad <11\text{-}20>_\omega \parallel <1\text{-}10>_{B2};$$
$$\{111\}_{\gamma'} \parallel \{110\}_{B2}, \quad <10\text{-}1]_{\gamma'} \parallel <111>_{B2}.$$

The sample was also studied after the short-term annealing and additional compressional deformation of up to 38% at room temperature.

A large numbers of twins are observed in the structure of the deformed samples. The plane {111} of twinning corresponds to the FCC twinning of γ'-phase. Reflections of the metastable phase are still observed and remain more pronounced (Table 1.42).

Table 1.42 The interplanar distances obtained from the ring-like SAED patterns after deformation by 38%

Ring number	1	2	3	4	5	6
Interplanar distance, nm	0.245	0.205	0.149	0.118	0.108	0.090
Ring intensity (conv. units)	10	40	100	3	2	2

The deformation process increases the amount of omega phase precipitations in the alloy, indicating a possibility for this phase to form by shift. It should be noted that this phase transition can be both from the γ'-phase depleted with nickel $\gamma' \to \omega$ and from the residual beta phase $\beta \to \omega$ without intermediate transitions. The latter variant of the phase transition is preferable, since in reality, the lattice parameters of this metastable phase are not as similar to those of Ni_5Al_3 or Ni_3Al as the lattice parameters of the hexagonal Ni_2Al_3 phase (Kazantseva et al. 2009a). As is known, the temperature range of existence of the omega phases is 300-500°C and, apparently, the omega phase is formed by cooling in our alloy.

A prolonged annealing of the alloy was carried out at 1000°C for 100 hours, in order to determine the possibility of homogenization of the alloy. This temperature is higher than the temperature region of the omega phase existence. Indeed, the long-term annealing at high temperature eliminates the fibrous textures and omega precipitations.

Metastable phases in the intermetallic Ni₃Al-based alloys under severe plastic deformation. The main working phases in the nickel heat-resistant alloys are γ (nickel solid solution), γ' (Ni_3Al, $L1_2$), γ'' ($Ni_3(Nb,Al,Ti)$ $D0_{22}$), carbides, topologically closely packed σ, μ, G and Laves phases (Sims et al. 1987). The phase γ'' (gamma double-prime) is the main strengthening phase for the alloys used to manufacture turbine disks (Inconel 718, Inconel 706, Rene 95, Udimet 630, etc.) operating at temperatures lower than 780°C. This dispersed phase is released as round disks or plates and appears in alloys containing niobium or tantalum of 3-5% by weight (Shoemaker et al. 1990). In the crystal lattice γ'-phase (Ni_3Al) niobium replaces the aluminum positions, actually forming the seeds of the long-period γ''-phase Ni_3Nb (Sims et al. 1987, Cozar R. and Pineau A. 1973).

In complex alloys, molybdenum, titanium, and tungsten can also be included into the composition of the γ''-phase, in addition to niobium and tantalum. It is known that this phase is easily formed in the presence of iron and chromium, in addition to niobium, whose electrons are believed to contribute to the necessary electron concentration. The orientation relationships between the γ matrix and γ''-phase are $(001)_{\gamma''} \parallel \{001\}_{\gamma}$ and $[100]_{\gamma''} \parallel <100>_{\gamma}$. A coherent γ''-phase (an unstable phase) forms from the γ solid solution under annealing at medium temperatures (up to 700°C). Long time exploitation at 650-760°C of the alloy with γ''-phase can arise the $\gamma'' \rightarrow \gamma'(L1_2)$ or $\gamma'' \rightarrow \delta$ (D0a) transitions. The transition $\gamma' \rightarrow \gamma'' \rightarrow \delta$ also exists. The phase transitions from γ'' to γ' or δ are accompanied by a significant decrease in the alloy strength.

The γ''-phase containing alloys, with high strength, satisfactory plasticity, and a good creep resistance, are successfully used in welded joints. The crystal lattice of γ''-phase, called Kurnakov phase (tetragonal, $D0_{22}$, Al_3Ti-type), has ordered atomic positions of nickel and niobium. Crystal parameters of γ''-phase in nickel alloys are $a = 0.362$ nm and $c = 0.741$ nm (Sims et al 1987). This phase with the $D0_{22}$ superstructure can also form in nickel alloys without niobium or tantalum (Ramesh et al. 1990. 1992, Bhattacharya and Ray 2000a,b, Chowdhury et al. 1995., Kazantseva at al. 2005).

According to the X-ray results, under annealing of the intermetallic alloys Ni-25at.%Al; Ni-25at.%Al-0.1at.%B, Ni-24at.%Al; Ni-24at.%Al-0.1at.%B, it was found that the cubic crystal lattice of γ'-phase is stable up to 600°C. At higher temperatures (near 800°C), splitting of the (220)-type diffraction lines was found in the X-ray diffraction patterns, indicating a tetragonal distortion of the initially cubic lattice. With annealing at 1000°C, splitting of the (200)- and (110)-type diffraction lines and distortion of the (111) line were observed. In this case, the degree of the γ'-phase ($L1_2$, Ni_3Al) long-range order corresponds to the maximum value $S^2 = 1$. According to Ramesh et al. (1990), the alloy becomes two-phased at these temperatures, with the second phase having a lower symmetry than the cubic one $L1_2$.

The existence of the two-phase structure was confirmed in this work with a cyclic exposure of alloys in the temperature range between 600 and 1000°C. Formation of γ''-phase with the $D0_{22}$ superstructure was discovered by Bhattacharya and Ray (2000a) by cold rolling $Ni_3Al(B)$ and $Ni_3(Al,Zr,B)$ alloys. Bhattacharya and Ray (2000a, 2000b) and Chowdhury et al. (1995) found that an increasing deformation degree promotes strongly decreasing intensities of the (100) diffraction line, whereas those of the (200) diffraction line are preserved. Such an effect is usually associated with a decrease in the degree of the long-range order. In this case, however, the intensity ratio for another pair of diffraction lines, (110) and (220) must be practically unchanged. This fact is attributed to the appearance of a tetragonal γ''-phase, the lattice parameter of which is similar to that of the cubic $L1_2$ structure. However, the intensity of the {100} diffraction line is decreased, because the {100} plane has a smaller probability of its repetition in the tetragonal lattice. A distinct change in the intensity of the diffraction lines in the $Ni_3Al(B)$ alloy begins at a deformation of 35%.

Changes of the degree of the long-range order were observed after shock wave loading in the single-crystal VKNA-4U superalloy (90% of γ'-phase) which did not contain niobium nor tantalum (Kazantseva et al. 2005). In the studies, the single-crystal VKNA-4U superalloy with the [001] orientation was used.

The crystals were grown by the Bridgman method. The VKNA-4U superalloy consisted of 90% of γ'-phase (intermetallic compound Ni_3Al, LI_2 type) and 10% of γ phase (FCC, nickel solid solution).

Deformation of the samples was done by shock wave loading by an impact with a steel plate (maximum pressure on the sample surface was 100 GPa, pulse duration of 1 µs). A low degree of the long-range order in the sample volume was found. The lattice parameter decreased in both the center and one-half the radius of the deformed surface and the dislocation density of the sample after the impact was about 10^{12} cm^{-2}. In addition, the dislocation density of the deformed sample changed from $9 \cdot 10^{12}$ cm^{-2} at the loading surface to $4 \cdot 10^{11}$ cm^{-2} at a distance of 4 mm from it, because of the shock front damping in the material. The smallest curvature radius of the crystal lattice (i.e., the greatest distortions) were found at the center of the contact surface of the VKNA-4U sample after shock wave loading. A high curvature of the crystal lattice of the deformed sample was found by the X-ray analysis too.

A TEM study of the sample after shock wave loading did not show the disordered γ-phase regions that would explain the low degree of the long-range order. The structure of the deformed sample contained misorientation bands; the band direction coincided with the {111} reflexes in the SAED pattern. Two band directions with alternating contrasts and contours of the initial γ'-phase cuboids could be seen simultaneously at inclinations of the foil in the microscope. Misorientation angles between the bands with different contrasts were about 2-3°. Brittle cracks were observed at the boundaries of the misorientation bands; near the cracks, regions with thin γ''-phase twins were found. Neutronography supported the TEM results: the neutron diffraction patters of the single-phase sample after the shock-wave loading showed the additional diffraction lines of γ''-phase.

Formation of a metastable γ''-phase in nickel alloys during deformation and in the absence of such alloying elements as niobium or tantalum (that form the Ni_3Nb and Ni_3Ta intermetallic compounds with a $D0_{22}$ superstructure) is a more complex process than it may seem at the first glance.

The long-period tetragonal superlattices ($c/a = 2M$, where M is a period of the (001) plane alternation) with the ordered arrangement of stacking faults like antiphase boundaries (APB) can be obtained by introducing parallel antiphase boundaries with the displacement vector ½ a<110> into each M cubic plane along the cube axis of the $L1_2$ superstructure. The modulated $D0_{22}$ structure is formed at $M = 1$ (Yamaguchi and Umakoshi 1990). Formation of the stacking faults significantly increases the total crystal energy and, therefore, formation of APBs in crystals is a thermodynamically disadvantageous process. The equilibrium crystal state corresponds to the formation of the ordered APBs consisting of long-period structures. The long-period modulated structures are usually observed in alloys with a low energy of the APBs formation. Ni_3Al intermetallic compounds are materials with high energies of the APB formation. Mishin (2004) did a theoretical calculation of the energy change in a Ni_3Al crystal lattice on an introduction of an APB. An APB energy of 80 mJ/m^2 in the (001) cube plane was adopted. The minimum APB energy, experimentally calculated from the APB width, is 43 mJ/m^2. The development of periodic APBs causes a significant decrease in the crystal energy. According to the calculations, the formation energy of the structures with ordered APBs is similar to that of the $L1_2$ structure (4.486 eV). For example, the formation energy of the long-period $D0_{22}$ structure (period $M = 1$) is 4.3 eV, and that of the long-period $D0_{23}$ structure ($M = 2$) is 4.4 eV (Mishin 2004). Thus, considering the energy gain, it is favorable for the Ni_3Al crystal to acquire some new ordered state under certain conditions (e.g., when deviating from stoichiometry or under deformations).

Another possibility for the formation of the modulated structure may be opened by the presence of a large number of vacancies in the material and their migration. This allows the dislocations to move actively. This would lead to a partial annihilation of a large number of dislocations, as well as the rearrangement of the remaining dislocations and defects that arise as a result of the dislocation interactions and transitions to energetically more favorable configurations.

It is known that both APB and APB tubes are easily formed as a result of an interaction of dislocations during deformations of a Ni_3Al crystal. As shown by Makoto and Yasumasa (2006), APBs are the places for the long-period modulated phase nucleation during annealing. Rentenberger et al. (2003) observed APB tubes formed under a 4% deformation of Ni_3Al in both superstructural and structural diffraction reflexes of γ'-phase. This amazing fact was explained by the authors invoking some additional atomic displacements inside the APB tubes.

Thus, if we assume that formation of long-period structures under deformation involves the process of nucleation and growth, then the APB (or APB tubes) can serve as the seeds for the new phase nucleation. However, formation of the regions having long-period $D0_{22}$

superstructures as a result of deformations cannot be explained only by the dislocation displacements. For the formation of a long-period $D0_{22}$ crystal lattice with the period M of the (001) plane alternation $M = 1$ these shifts must have a minimum resolved wavelength in the crystal $\lambda = 2d$. It is known that the minimum distance between the dislocations of different signs is about $10^4\,b$ (where b is the Burgers vector). This distance decreases when the dislocation density is 10^{12} cm^{-2}, however it still remains quite large, of the order of $10^2\,b$, and it cannot explain the formation of the long-period structure regions.

Under deformation, a sample can be considered as a system of a large number of atoms in a highly non-equilibrium condition caused by the supply of significant amounts of energy (mechanical, impact). It is known that ordered dissipative structures can appear in such systems. The dissipative structures arising in solids interacting with the environment are formed as a result of the coordinated, interdependent motions of individual particles. In the non-equilibrium, dissipative systems spontaneous structures are formed, whose symmetry may differ qualitatively from that of the initial state. Such a system behavior is known as self-organization. Under strongly non-equilibrium conditions, the systems can obtain a new ordered state with localization of deformations and conservation of weakly deformed fragments. Therefore, it is possible to consider formation of the regions of a new metastable phase with long-period structures in Ni_3Al and nickel alloys under deformation as a relaxation process of the crystal self-organization, occurring due to compensatory periodic shifts of a non-dislocation origin. It is not completely clear, whether the dislocations take part in the generation of the concentration fluctuations, necessary for the formation of this phase. As the experiment shows, the size increase of the regions of the phase with a long-period structure directly depends on the deformation degree (dislocation density).

Metastable carbide transformation in nickel superalloys under severe deformation. The implementation phases, including carbides, nitrides, hydrides, oxides, silicides, as well as the more complex crystal systems, such as carboxides, etc., play a very important role in metallurgy and in the strengthening of steels and alloys. Such crystal systems are formed as a result of the introduction of non-metallic atoms with relatively small sizes in the interstices of the crystal lattice formed by the atoms of transition metals. The small sizes of the introduced atoms lead to their high mobility, even at relatively low temperatures. Because of this, the phase transitions between compounds with different crystal lattices and chemical compositions occur easily (Maslenkov 1983).

Chemical compositions of the iron and nickel superalloys used to manufacture the high-temperature details include a large number of elements improving the alloys resistance against hot corrosion and their strength at high temperatures. Carbon is the essential element and formation of fine carbides inside the grains and at their boundaries significantly reinforces the materials. Chromium is the main alloying element enhancing corrosion resistance of the materials. In the iron superalloy, chromium can dissolve in either alpha or gamma iron and can form carbides by replacing the iron atoms in the crystal lattice of cementite ($FeCr)_3C$. Iron may also enter into the crystal lattice of complex chromium carbides by replacing the chromium atoms (Maslenkov 1983). In the iron superalloys, chromium carbides exist in two variants, namely Cr_7C_3 and $Cr_{23}C_6$. The stoichiometric carbide $Cr_{23}C_6$ is not found in the Cr-C system; some of the metal atoms in this system must be replaced by other elements (Fe, Mn, etc.).

Chromium carbides have high melting points: 1250°C ($Cr_{23}C_6$), 1665°C (Cr_7C_3), 1890°C (Cr_3C_2); they are refractory compounds with the covalent type of chemical bonds. The crystal lattice of chromium carbides contains a lot of vacancies inside the metal sublattice. This fact explains the ease of dissolution in the carbide lattice of the alloying elements, such as iron, molybdenum, tungsten, etc. (Maslenkov 1983). A notable feature of chromium carbides is also a variable carbon concentration within the volume of carbide particles, which leads to changes of the chromium content along the direction from the center of the carbide particles to their edges. Thus, with exposure of the alloy in a certain temperature range, the carbon depletion of the carbide particles may occur, promoting the carbide phase transition to a different crystal lattice.

The nickel-based superalloys have complex chemical compositions that include up to 12 components, carefully balanced to obtain the desired mechanical properties of the alloys. The carbon percentage is relatively small, approximately 0.05-0.2% by weight, which is sufficient for the formation of complex carbides MC with Cr, Mo, W, Nb, Ta, and Ti. Dispersed primary carbides MC have a correct cubic shape and cubic orientation relationship with the matrix. During heat treatment and creep testing, the MC carbides undergo transformations with the formation of the secondary globular complex carbides of chromium, molybdenum, and tungsten (Sims et al. 1987).

Kazantseva et al. (2009b) discovered the formation of modulated structures in the chromium carbides $M_{23}C_6$ and M_7C_3 after severe plastic deformations of the nickel superalloys. A comparative study of planar defects formed in the chromium carbides $M_{23}C_6$ and M_7C_3 was done using samples of the iron superalloy after quenching and annealing, as well as samples of the white cast iron after crystallization. The chemical compositions of the studied samples are given in Table 1.43.

Table 1.43 Chemical compositions of the white cast iron and iron superalloy

Alloy	Concentration, wt. %						
	Ni	Fe	Cr	Mo	V	Mn	C
chromium-nickel steel	17.5	base	18.3	1.50	1.58	-	0.46
white chromium cast iron	-	base	17.3	-	-	4.7	2.96

Compositions of the EP-800 and ZhS-32 alloys are presented in Tables 1.16 and 1.44 (Maslenkov1983). Samples of the iron superalloy were subjected to many steps of the heat treatments: water quenching from temperatures of 1150-1175°C and annealing at temperatures of 600, 650, 700, and 750°C for 1, 5, 10, and 20 hours, respectively.

Table 1.44 Chemical composition of the EP-800 superalloy, wt.%

Cr	Mo	W	Al	Co	Nb	C	Fe	Ni
12.2	6.0	7.4	4.5	9.5	2.2	0.05	≤ 1	balance

Undissolved primary carbides (3-4% by volume) with sizes of 100-200 nm were found in the iron superalloy samples after quenching. As shown by an X-ray spectral analysis of the samples, carbon is allocated in two carbide types, MC and $M_{23}C_6$, after quenching.

The $M_{23}C_6$ carbides have complex variable chemical compositions $(Cr_{59}Fe_{22}V_{18}Mo_1)_{23}C_6$; the MC carbides contain V and Mo up to 8% (atomic). According to the TEM studies, particles of the $M_{23}C_6$ primary carbides in the iron superalloy have an internal defect structure consisting of alternating dark and light thin plates after annealing at 650°C for 20 hours. Twinned reflections and satellites are observed in the SAED pattern taken from the defect area. The twinned reflections probably arise due to the formation of the axial domains of the modulated structure in three <100> directions. These domains look like C-domains in Cu_3Pd or axial domains that are typical for the implementation phases (Kositsyna et al. 1997, de Novion and Landesman 1985). The satellite diffraction reflexes are typical for the long-period modulated structures.

Samples of the EP-800 superalloy were cut from the destruction zone of a gas turbine blade subjected to an impact by fragments of a neighboring blade destroyed during its operation. Carbides with different chemical compositions were found inside the grains and near their boundaries, with different shapes. The latter are cubic in the primary MC carbides formed inside the grains and elongated (more rounded) in the $M_{23}C_6$ and M_6C carbides at the grain boundaries. In the nickel superalloys, $M_{23}C_6$ and M_6C are the secondary carbides. The M_6C carbide, a complex η-carbide found in steels and nickel superalloys, is stable only in the presence of two or more metals; in our case, its composition is $(Mo, W, Nb)_6C$. Carbides with very irregular shapes and specific striping are detected at grain boundaries. This banding is similar to the modulated structure of the M_7C_3 carbides in the chromium white cast iron (Fig. 1.20a). According to the optical metallography and X-ray diffraction analysis in the white cast irons, volume content of the primary eutectic chromium carbides M_7C_3 is 30–33%. Minor amounts of the secondary carbides M_7C_3 with the composition $(Cr,Fe)_7C_3$ are also present. The secondary carbides with sizes of 20-50 nm arise directly from austenite during temperature heating. Shapes, sizes, and mutual arrangements of the primary carbides are not changed by the heat treatment.

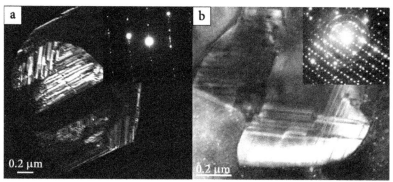

Fig. 1.20 The modulated structure of the carbides: (a) the $M_{23}C_6$ carbide in the steel, aging 650°C-20 hours, the dark-field image in $(\bar{1}31)$ $M_{23}C_6$; (b) the M_7C_3 carbide in the ZhS-32 nickel superalloy, the bright-field image.

A TEM study reveals large primary carbides $(Cr,Fe)_7C_3$ with a complex internal structure. In a dark-field image taken with the (05.0) carbide reflex, a lamellar contrast consisting of the alternating light and dark stripes is observed; the satellite reflexes are also observed in the SAED pattern. Such diffraction behavior is typical for long-period structures. The structure periodicity in this case is about of 5 nm ($a = 10a_0$, where a_0 is the lattice period of the $(Cr,Fe)_7C_3$ carbide) and it appears to be associated with the formation of a layered ordered structure consisting of the alternating APB planar defects in the (100) planes. Similar results were obtained in the samples of the EP-800 nickel superalloy. The satellite reflexes are in the SAED patterns taken from the striped regions. In this case, it can be excluded that these striped regions belong to the $M_{23}C_6$ or M_6C carbides, because the latter are also present in the alloy structure and have another type of the diffraction reflex positions in the SAED patterns. The SAED pattern taken from such striped areas can be identified most closely as that of the metastable long-period carbide M_7C_3 with a twice as large lattice parameter $a = 2a_0$. It should be noted that the M_7C_3 carbide in the white cast irons has the modulation period $a = 5a_0$ (Fig. 1.20b).

ZhS-32 superalloy samples were subjected to intense deformation by shear under high pressure ($\varphi = 360°$, $e = 4.6$) and subsequent annealing at a temperature of $1000°C$ for 15 minutes (Bachteeva et al. 1997, Kazantseva et al. 2009b). The severe plastic deformation and subsequent annealing change the structure of the nickel superalloy by forming sub-microcrystalline grains. It also affects the chemical composition of the carbides providing the phase transitions. The $M_{23}C_6$ and M_7C_3 carbides with different shapes and chemical compositions were observed in the structure after the deformation and subsequent annealing. Planar defects were also detected inside the $M_{23}C_6$ carbide particles. These planar defects could form during the annealing process due to the stress relaxation after severe plastic deformation. The M_7C_3 carbides also have the modulated structure with two directions of domain formation and the modulation period of $a = 2a_0$ (Fig. 1.20b).

In the intermetallic compounds and ordered alloys, the formation of long-period modulated structures is caused by ordered arrangements of the planar lattice defects, such as antiphase boundaries (APB) arising either during deformation or by atomic diffusion during annealing. In the implementation phases, formation of such defects differs from that in the intermetallic compounds. The long-range ordering of the carbon vacancies in the implementation phases is directly related to the atomic diffusion of the nonmetal elements, such as carbon, hydrogen, nitrogen, etc., and composition changes leading to distortions of the metal lattice.

In the implementation phases, non-metal atoms are located between the atomic layers of the metal atoms. Stoichiometry deviations in carbon, hydrogen, and nitrogen lead to formation of vacancies inside the layers of metal atoms, resulting in an acceleration of diffusion. In addition to the vacancies, clusters distort the arrangement of metal atoms. Depending on the composition and heat treatment, such vacancies can also be ordered. They can form a new long-period modulated lattice, built on the basis of the cubic or hexagonal initial crystal lattices. A long-period crystal lattice in the carbide Ti_2C with parameter $b = 2a_0$ (initially cubic crystal lattice of the NaCl-type structure) is interpreted as a consequence of the $1/4$ and $3/4$ alternating fillings with carbon atoms of the $\{111\}$ planes.

Two long-period structures were found in the vanadium carbides V_8C_7 and V_6C_5. The structure of V_8C_7 is formed from a cubic lattice with parameter $b = 2a_0$ and it consists of a mixture of the antiphase domains $P4_132$ and $P4_332$. The structure of V_6C_5 is more complex, it consists of the periodic alternating (111) planes filled fully or at $2/3$ of carbon. At the departure from stoichiometry, the modulation period in the carbide crystal lattice may be increased up to 4 nm (de Novion and Landesman 1985). The long-period modulated lattices were also detected in V_2C, Nb_2C, Ta_2C. The ordered domain structure of these carbides may consist of antiphase domain boundaries and axial domain walls. In the case of the formation of long-period structures based on cubic lattices, axial C-domains can be observed in three possible <100> orientations, formed from thin plates separated by antiphase boundaries (de Novion and Landesman 1985). Hiraga and Hirabayashi (1979) found the formation of disordered planar defects in M_3C carbides in pearlite.

Chromium carbides are characterized by complex crystal lattices. The chromium carbide $Cr_{23}C_6$ has a Fm3m cubic crystal lattice with 116 atoms in the unit cell. Data on the type of the crystal lattice for the chromium carbide Cr_7C_3 vary in the literature (JCPDS-ICDD PDF-4 Database 2007). In various publications, the crystal lattice of this carbide was reported as (1) orthorhombic crystal lattice Pmna, (2) orthorhombic crystal lattice Op40 (prototype Cr_7C_3) with 82 atoms in the unit cell and the lattice parameters $a = 0.45$ nm, $b = 0.701$ nm, $c = 1.214$ nm, or (3) P31c trigonal or pseudo-hexagonal lattice with the parameters $a = 1.398$ nm and $c = 0.4523$ nm. Such disagreement is probably due to the sensitivity of this carbide to a departure from stoichiometry, similarly to what occurs in the titanium carbide Ti_2C, which also depends on the chemical composition and is indicated in different types of crystal lattices, cubic or trigonal (de Novion and Landesman 1985).

In our study, a good agreement for the M_7C_3 carbide was achieved for an orthorhombic crystal lattice, with a modulation along axis *a*.

In the nickel superalloys, the carbide Cr_7C_3 is the metastable one. In the nickel superalloys presented in this study, the carbide appears probably due to carbon depletion of $M_{23}C_6$ due to the deformation and subsequent annealing of the ZhS-32 superalloy and to the long-time operation of the EP-800 superalloy. Modulation of Cr_7C_3 observed in the nickel superalloys can also be related to the applied external stresses (an impact in EP-800 and a shift under high pressure in ZhS-32). They are able to accelerate the diffusion of carbon and stress relaxation by ordering of the layers with carbon deficit. In the white cast iron, the carbide M_7C_3 is stable and its modulated structure forms during annealing. In the investigated alloys, the composition of chromium carbides was complex. The M_7C_3 carbides contain iron in the iron superalloys and white cast iron, whereas they contain molybdenum and tungsten in the nickel superalloys. The differences of the modulation periods observed in the Cr_7C_3 carbides in the white cast irons, austenitic steels, and nickel superalloys shows that the composition changes of the chromium carbide can influence the vacancy formation in the carbon layers.

1.1.4 Prediction of the destruction processes in the nickel superalloys by non-destructive magnetic methods

Recently, technological developments have been intensively conducted to enhance the thermal efficiency and output capability of the power generation Gas Turbines (GTs) (Scobie et al. 2015, Tsukagoshi et al. 2007). The development is associated with the improvement of engineering designs of the turbine, but the main way of increasing the gas turbine power is to increase the operating temperatures and rotation speeds. In this case, the turbine blade material operates in extreme conditions due to high temperatures and stress levels. Selection of an optimal time of operation of turbine blades within the forced regime requires a detailed structural analysis and evaluation of the degradation degree of the blades material. Recently, several structural studies were performed to predict the fatigue life of the blades made from the nickel superalloys (Jiang et al 2015, Doremus et al. 2015); usually, such works were presented results of the laboratory experiments. However studies of the turbine blade structure after industrial operation are of particular importance (Jahangiri and Abedini 2014, Sun et al. 2015, Kumari et al. 2014).

This section presents the results of the structure research of the gas turbine blades at an industrial gas turbine plant, operated with increasing operation temperatures and rotation speeds. The blades were made of the Ch-70V and EP-800 superalloys, widely used in Russian power industry (Maslenkov 1983). These superalloys consist of nickel solid solution, strengthening γ'-phase (40% by weight, intermetallic Ni_3Al compound with $L1_2$ type superstructure), and a small amount of carbides (2% by weight of MC и $M_{23}C_6$). Although the magnetic methods of nondestructive testing are widely used in the industry to evaluate the operability of details, they have never been used before for the heat-resistant nickel-based superalloys. The magnetic properties of the intermetallic compound Ni_3Al are sensitive to its chemical composition. $Ni_{74}Al_{26}$ is paramagnetic at low temperatures down to 4 K, and $Ni_{75}Al_{25}$ is a weakly-itinerant ferromagnetic with a Curie temperature $T_C = 41$ K (De Boer et al. 1969, Min et al. 1988, Rhee et al. 2003).

It is believed that the ferromagnetic-ordered regions can exist in the structure of weakly-itinerant Ferro magnets deviating from the stoichiometric composition even if the temperatures are much higher than T_C. Sun et al. (2015) concluded that the detection of two T_C values (of 87.5 and 252 K) in Ni_3Al with non-stoichiometric composition (depleted in aluminum) cannot be explained using the theory of spin fluctuations. Although alloying can increase Curie temperature, it is well known (Sims et al. 1987) that all phases of the heat-resistant nickel superalloys are paramagnetic at room temperature in the initial state and they retain the paramagnetic state under operation during the warranty period. Nickel superalloys may be related to austenitic materials, such as the austenitic steels. However, experimental studies have revealed a dramatic effect of deformations on the magnetic behavior of nickel superalloys. For example, cyclic deformations were shown to result in a super-paramagnetic behavior of a nickel superalloy (Umakoshi and Yasuda 2006, Umakoshi et al. 2004). An increase of the magnetic susceptibility in the VKNA-4U superalloy was observed after different types of deformation, such as cold rolling up to 40% (Stepanova et al. 2011b), shock-wave loading (Kazantseva et al. 2006, 2014b), and high-temperature deformation (Stepanova et al. 2011c).

It is important to note the existence of the phenomenon of strain-induced magnetism in some intermetallic compounds: a paramagnetic intermetallic compound may become partly ferromagnetic upon severe plastic deformations (Zeng and Baker 2007, Baker and Wu 2005). In the intermetallic compounds, the deformation-induced magnetism after high degrees of plastic deformation (exceeding 40%) was observed for the first time in the Fe_3Al alloy. This phenomenon is now observed in a wide range of the intermetallic compounds, such as FeAl, CoAl, CoAl, CoGa, Ni_3Sn, Fe_3Ge_2, Pt_3Fe, and Co_3Ti. At room temperature, these compounds are the single-phase and paramagnetic alloys; a super-paramagnetic state is observed in them after deformations.

The deformation-induced magnetism has the following features (Zeng and Baker 2007):

- The magnetization of the sample increases quite slowly, whereas beyond some threshold magnitude, magnetization grows faster;
- For a given stress, the effect becomes stronger, when composition deviates from the stoichiometric composition of the intermetallic compound;
- When temperatures decrease, the effect becomes more pronounced; annealing returns samples to their paramagnetic state;
- In practice, one observes a super-paramagnetic condition, rather than a ferromagnetic one;
- The magnetic field dependence of magnetization demonstrates a small hysteresis, and even if the saturation is not achieved, the magnetic susceptibility (the slope of the magnetization curve) is higher than that in the well-annealed paramagnetic state.

A problem is that the detection of this phenomenon by magnetic methods, as a rule, is not accomplished by structural studies. As a result, at present it is unclear, what kind of the structure element is responsible for the appearance of the super-paramagnetic state after deformations. Since no new phases are revealed by TEM studies after deformations, the strain-induced ferromagnetism is described using the term "magnetic cluster". It should be noted that the above results were obtained after cold deformation (e.g., cold rolling; Takahashi and Ikeda 1983).

Magnetic methods for analyzing of the deformed state of nickel superalloys. In this section, we discuss possible applications of the magnetic methods of non-destructive testing to study turbine blades made of nickel superalloys. Measurements of magnetic susceptibility are done using a new experimental device with improved magnetic sensitivity, developed at the Institute of Metal Physics, Ural Branch of the Russian Academy of Sciences (Ekaterinburg), patented and currently used in the industry (Rigmant et al. 2000, 2005).

The next generation of devices for measurements of the magnetic susceptibility is compact, portable, has an improved sensitivity ($\pm 10^{-4}$), and computer software for processing the measurement results for low-magnetic materials. Power of gas turbines is improved mainly by increasing their operating temperatures and rotation speeds. This is the global trend in enhancements of the thermal efficiency and output capability of the power generation (Scobie et al. 2015, Tsukagoshi et al. 2007). It is typical for a standard regime that the turbine blades are never used at the maximum of the working temperature range of the alloy. There must remain a margin of 50-100°C, ensuring a structural stability of the alloy in case of an uncontrolled overheating. For example, for blades made of the ChS-70V and EP-800 superalloys (with 40% of γ'-phase), whose upper temperature boundary is 900°C, the working temperature is (as a rule) about 800°C (Maslenkov 1983). Chemical composition of the Russian EP-800 superalloy is similar to that of Rene 80 (USA).

The energy industry actively attempts to increase the power of gas turbine plants. In this section, we discuss results of the study of the gas turbine blades after an accident. The blades were taken from an industrial turbine which produced the power four times as large as that of the usual turbines by increasing the working temperature and the rotation speed.

In such a regime, the blades made of the ChS-70 and EP-800 superalloys work under extreme thermal and stress conditions (Stepanova et al. 2015, 2016, Davydov et al. 2015a).

Strain-induced magnetism in the EP-800 superalloy after high-temperature deformations. The standard working regime of a gas turbine prescribes their exploitation at 800°C and their operation time of 27000 hours. An experimental regime involves an increased temperature of 880°C and a forced regime requires operation time of 9000 hours with 17 turbine starts. Chemical composition of the EP-800 superalloy is given in Table 1.42. The sample used as reference was subjected to the standard stepwise heat treatments: annealing at 1160°C for 5 hours, annealing at 1060°C for 2 hours, annealing at 1000°C for 2 hours, and annealing at 850°C for 16 hours. Air cooling was applied after each step of the annealing process. Structure of the EP-800 superalloy consists of the nickel solid solution, approximately 40% of the strengthening γ'-phase (Ni_3Al), and small amounts of carbides (about of 2%). All of these phases are paramagnetic at room temperature. Turbine blades made of the EP-800 superalloy retain paramagnetic state for the entire standard exploitation period. After operating of the turbine in the experimental regime during 9000 hours at 880°C, an increase in the magnetic susceptibility χ of blades was observed. Distribution of magnetic susceptibility on the surface of the turbine blade is shown in Fig. 1.21.

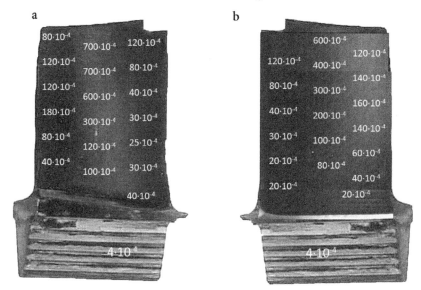

Fig. 1.21 Results of the magnetic susceptibility measurement superimposed on the photograph of the blade from the both sides: (a) convex part; (b) concave part (Stepanova et al. 2015).

Magnetic susceptibility of the reference sample is $\chi = 4 \cdot 10^{-4}$. The increase of χ was not uniform in different parts of the blade. The maximum magnitudes were obtained in the convex feather part of the blade, where the temperature and the dynamic stress level were at maximum (Sun et al. 2015). The back part of the feather is a narrow zone running along the axis of the feather on its convex side at the place of maximum curvature; another critical part of the blade is the feather's edge. In the locking part of the blade, which is mainly subjected to the thermal effects, magnetic susceptibility χ does not change in comparison with that of the reference sample $\chi = 4 \cdot 10^{-4}$. For the EP-800 superalloy, the changes of χ are considerable in the convex feather part (Fig. 1.21): from $4 \cdot 10^{-4}$ up to $7 \cdot 10^{-2}$. The stepwise annealing restores χ in the samples taken from the convex feather part to the initial level of $4 \cdot 10^{-4}$.

In this study, the magnetic susceptibility measurements were performed directly on the blades, without preliminary preparation of the surface according to the method of non-destructive magnetic testing. The blade surface was found to be oxidized, making it necessary to trace the oxidation role in the changes of magnetic properties. The high chromium concentration of 12.2% leads to the formation of a protective Cr_2O_3 oxide layer on the surface, which is antiferromagnetic, with low χ values. The diffusive redistribution of alloying elements in the feather cross section was also determined. The sample was cut such that one of its sides was the back part of the feather. Three scanning directions were used: (1) close to the concave side of the feather, (2) the middle part of the feather, and (3) near the convex part. The fourth direction was on the external side of the feather cross-section, running through the oxide layer on the back of the feather. From 12 to 16 measurements in the sample were taken along each direction for statistical processing.

Table 1.45 (Stepanova et al. 2015) shows results of the chemical redistribution of alloying elements in the superalloy after the application of high temperatures and stresses. The results demonstrate that the diffusive redistribution of chemical elements inside the blade was significantly suppressed by the alloying. Apparently, the magnetic effect is not associated with the redistribution of chemical elements under deformations.

Table 1.45   Chemical composition of different parts of the blade feather after exploitation, wt.%

Parts	Ni	Cr	Fe	Al	Mo	W	Co	Nb
1	7.6	12.2	1.1	4.5	6.0	7.1	9.5	2.2
2	6.7	1.9	1.0	4.2	6.8	7.7	9.1	2.3
3	6.5	12.3	1.0	4.4	6.4	7.4	9.6	2.2

X-ray microanalysis revealed the redistribution of chemical elements in the back part of the feather surface. Table 1.46 (Stepanova et al. 2015) presents the chemical composition of the oxide layer. A local increase of the iron concentration and decrease of the chromium concentration by diffusion under stress led to the ferromagnetic Fe_3O_4 oxide formation.

Table 1.46   Chemical composition of the feather oxide layer on the back surface of the feather, wt.%

Part	Ni	Cr	O	Fe	Al	Mo	W	Co	Nb
4	3.6	4.6	13.5	6.2	7.2	3.6	1.3	3.2	2.3

Iron is not an alloying element in the studying material and is present as an impurity. The increase of the iron concentration from 1 to 6% and the decrease of the chromium content from 12.2 to 4.6% were found in the oxide layer. The iron enrichment leads to the formation of the ferromagnetic iron oxide. As a result, those places of the blade that are most prone to stress suffered from surface corrosion. The formation of the ferromagnetic iron oxide led to increasing magnetic susceptibility (χ). The magnetic method did not allow us to separate all factors responsible for the χ increasing. However, these factors indicated degradation of the alloy structure under high temperature stress.

Electron microscopic studies were performed on the samples cut from the different parts of the blades after their operation in the experimental regime. It should be noted that the electron microscopic analysis did not reveal formation of any new phase. In the samples cut from the convex part of the feather, the studies reveal large numbers of planar defects in both the solid solution and in the intermetallic γ'-phase (Ni$_3$Al) particles (Fig. 1.22).

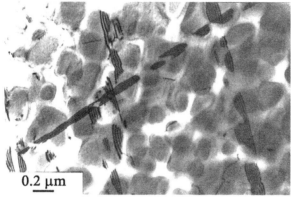

Fig. 1.22 High density of stacking faults in the back of the feather blade after the operation at 880°C for 9000 hours, the bright-field image (Stepanova et al. 2015).

The stacking faults were observed inside the deformed particles of the intermetallic γ'-phase found with the dark-field images. In the locking part of the blade, the planar defects inside the γ'-particles were not observed, just a coagulation of the intermetallic phase under heating was found. The stepwise annealing of the deformed samples led to the restoration of the defect-free state inside the γ'-phase particles. A carbide reaction was also found in the feather part of the blade, when the NbC carbide was replaced by the carbide (Cr,Mo,W)$_{23}$C$_6$. Additional Me$_{23}$C$_6$ carbides were located on the grain boundaries and planar defects inside the grains. Thus, in the turbine blade investigated, the same processes of structural degradation take place as in the blades operating in the standard regime. However, under high stress and temperatures, these processes are accelerated.

Deformation dislocation bands appear in the structure during the blade operation and defects are accumulated inside the particles of the intermetallic phase.

The coalescence of the γ'-phase particles occurs in the areas of maximum stresses (in the back of the feather). Carbide reactions form the chromium-based carbide $Me_{23}C_6$ instead of the MC carbides (NbC in EP-800). Separate fatigue cracks appear on the blade surface (Davydov et al. 2015a). The structural degradation is the real problem, because the increasing operation time in the experimental regime leads to a catastrophic destruction of the blade after only 9390 hours (13 months) (Davydov et al. 2016, 2017). The gas turbine blade studied was broken off during its operation by an impact of another blade fragment. In the areas of the high stress levels of the turbine blade, the magnetic susceptibility value increased from $2 \cdot 10^{-3}$ to $6 \cdot 10^{-2}$ (during operation). In the impact area, magnetic susceptibility increased up to $1.4 \cdot 10^{-1}$.

Independent magnetic measurements were performed with a vibrating magnetometer *Lake Shore* 7407. The measurements were done using certified methods with a verified device, but this method of study was destructive. Thin polished samples with a thickness of 0.3 mm were cut out from the inner part of the blade feather (far from the oxide layer) and were used for subsequent magnetic measurements. Magnetization curves for the lock and initial samples are the straight lines (Fig. 1.23), whereas the curves for the convex feather part and the impact area of the broken blade show saturation magnetization.

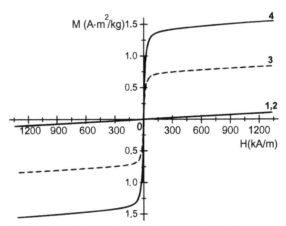

Fig. 1.23 Magnetic field dependence of magnetization M(H) of the blade: (1) initial; (2) locking part; (3) feather convex part; (4) impact area at the feather.

The saturation magnetization for the blade feather is increased in comparison with that of the lock part of the blade and the initial sample. Hysteresis loops are very narrow, which may suggest a superparamagnetic-like behavior in the samples. With increasing saturation, the hysteresis loops become wider, which may mean an increase of the number of magnetic clusters in the material.

It can be seen that after the operation at the forced mode, a second peak appears near 280 K in the curve, in addition to the main peak associated with the Curie temperature of the intermetallic phase. (Davydov et al. 2017). Thus, there is a correspondence between the number of the structure defects in various parts of the blade and magnetic susceptibility.

The temperature dependence of the ac-magnetic susceptibility in an alternating magnetic field is shown in Fig. 1.24.

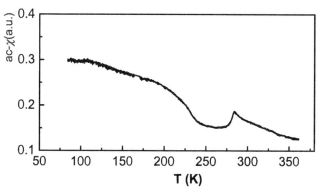

Fig. 1.24 The temperature dependence of the AC – magnetic susceptibility for the sample cut from the back of the feather of the turbine blade (Davydov et al. 2017).

It should be noted that the long-time operation of turbine blades made from superalloys in the standard regime does not lead to the formation of stable defect complexes inside the particles of the strengthening intermetallic phase. The defects are located preferably in the solid solution. Such structures do not lead to the development of ferromagnetic properties in the paramagnetic alloy. Thus, magnetic clusters can be formed by defect complexes inside the intermetallic phase.

The formation of the stable defect complexes inside the γ'-phase testifies for the γ'-phase softening. In this case, mechanical properties of the intermetallic phase approach those of a solid solution and γ'-phase loses its strengthening effect. Magnetic susceptibility measurements using devices with improved sensitivity allow detecting such defects inside γ'-phase and using the magnetic nondestructive testing methods to evaluate the power output capacity of turbine blades.

Structure and magnetic properties of the nickel superalloys after a high-temperature deformation. The ChS-70V superalloy was initially developed for the production of details for the marine turbines. The long-lasting strength of this Russian superalloy is similar to that of the In738 superalloy (USA). Stepanova et al. (2011c) and Kazantseva et al. (2014a) carried out studies of the samples cut out of different parts of the as-cast polycrystalline blade after its operation in a stationary power plant. The experimental regime of the operation for ChS-70V was the same as in the case of EP-800 (880°C). The blade made of Ch70V was destroyed in the result of an emergency accident, which occurred after 13 months of the turbine operation (with 17 starts). For the blades made of the ChS-70V superalloy that contained 40% of γ'-phase, the maximum operation temperature is 900°C and the normal working temperature is, as a rule, 800°C (Maslenkov 1983).

The EP-800 superalloy is more resistant to elevated operating temperatures and stresses, in comparison with ChS-70V. For the ChS-70V alloy, a number of factors leading to irreversible structure changes and contributing to its destruction impedes attainment of an increase in the operating time.

The accumulation of defects inside the particles of the intermetallic phase, stress-induced diffusion of alloying elements, and porosity formed during high-temperature operation reduce the resistance to fatigue cracks and lead to further destruction (Davydov et al. 2015).

The chemical composition of the ChS-70V superalloy, corresponding to technical standards (Maslenkov 1983), is given in Table 1.47 (Stepanova et al. 2011c). The stepwise annealing for a reference sample is: annealing at 1170°C for 4 hours (followed by cooling in argon), annealing at 1050°C for 4 hours (followed by air cooling), and annealing at 850°C for 16 hours (Maslenkov 1983). Increases of magnetic susceptibility were found to be non-uniform in different parts of the blade; the maximum magnitudes were obtained in the convex part of the blade feather (the back part of the feather). The blade feather was subjected simultaneously to high temperatures and stresses (alternating tensile, twisting, and compression under vibration). The maximum stresses were observed in the back part of the blade feather (Sun et al. 2015). In the locking part of the blade, which is mainly thermally affected, the magnetic susceptibility χ does not change in comparison with that in the initial sample, obtained with the stepwise heat treatments. The results are shown in the diagram in Fig. 1.25. The magnetic susceptibility changed from $2 \cdot 10^{-4}$ in the initial sample to $4 \cdot 10^{-3}$ in the blade feather near the locking part and to $3.6 \cdot 10^{-2}$ in the back part of the feather. The change of χ by two orders of magnitude can be interpreted as "the formation of about one percent of dispersed ferromagnetic phase in the alloy upon the deformation." However, the deformation process of the heat-resistant alloys does not form new magnetic phases.

Table 1.47 Chemical composition of the ChS-70V superalloy, wt.%

Alloy	Cr	Ti	Mo	W	Al	Co	Nb	C	Ni
ChS-70V	15.4	5.0	3.5	3.6	3.5	10.6	0.25	0.10	balance

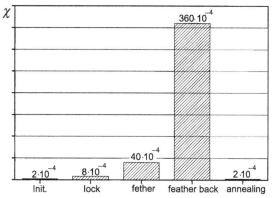

Fig. 1.25 Magnetic susceptibility χ in different parts of the turbine blade from alloy ChS-70 after operation and emergency destruction in the form of a diagram (Stepanova et al. 2011c).

The necessary condition for a successful long-term high-temperature operation of nickel superalloys is their stability to phase transformations. The reaction of substitution of $M_{23}C_6$ carbides by the initial MC carbides is the only transition that can take place in the heat-resistant nickel superalloys under exploitation. All of these carbide phases are paramagnetic. Chemical diffusion under stress was found to take place in the ChS-70V superalloy (Kazantseva et al. 2013). Detailed structural studies of the material in various parts of the blades did not reveal the formation of any new phases in the inner layers of the blade. At the same time, magnetic susceptibility within the blade increases relatively to that of the reference sample. The question arises, whether the increase of χ with the concentration inhomogeneity occurs in the alloy during its operation. The phase and chemical contents of the surface layer are also interesting.

The feather cross-section was used to study the diffusive redistribution of the chemical elements in the blade after operation. The sample was cut out such that one of its sides was the back part of the feather. Table 1.48 shows the results of chemical redistribution of alloying elements in the ChS-70V superalloy affected by high temperatures and stresses.

Five scanning directions were selected: (1) close to the concave side of the feather, (2)-(4) in the middle part of the feather, and (5) near its convex part. The 6th direction was selected through the oxide layer on the back part of the feather. Between 12 and 16 measurements along each direction were used for statistical processing. As can be seen from Table 1.48, in the central area of the sample (3), the concentration of the elements is almost identical to that for the reference sample (Kazantseva et al. 2013).

The greatest changes in the chemical composition were observed in the 6th direction, at the surface on the back part of the feather (Table 1.49) (Kazantseva et al. 2013). Fluctuations in the nickel content (balance) and the alloying elements, such as cobalt, titanium, and chromium, were observed during the withdrawal from the central part of the sample. The iron concentration also changed (initially, iron was present in the alloy as an impurity). There one can note a very high oxygen content and increased concentration of iron, cobalt, and titanium. Concentrations of the alloying elements, such as tungsten and molybdenum, were reduced; the nickel content (balance) was reduced by more than a factor of two.

Table 1.48 Average chemical composition in different areas of the blade feather made of the ChS70-V superalloy after operation, wt.%

Element/direction	1	2	3	4	5
B	0.6	23.6	2.2	2.1	8.1
C	0.2	2.9	0.3	0.2	1.1
O	0.9	1.2	0.9	0.9	2.7
As	0.25	0.6	0.4	-	-
Al	1.5	1.1	1.5	1.6	2.0
Mo	3.5	2.8	3.2	3.2	3.7
S	0.01	0.03	0.00	0.01	0.00
Ti	4.4	3.1	4.3	4.6	3.7
Cr	14.7	10.4	14.3	14.7	11.6
Fe	0.8	1.0	0.7	0.7	0.9
Co	10.6	7.6	10.3	10.4	8.8
Ni	57.1	40.8	56.7	56.3	52.0
W	5.6	4.9	5.3	5.4	5.4

Table 1.49 Average chemical composition of the feather surface for the ChS-70V superalloy after operation, 6th direction of scanning, wt.%

B	C	O	As	Al	Mo	Ni
4.7	1.6	17.0	3.9	1.8	2.7	Bal.
Ti	Cr	Fe	Co	Ni	W	Bal.
14.4	5.2	5.7	15.5	20.9	1.6	

The X-ray diffraction data are in a good agreement with the chemical composition measurements in different areas of the feather of the blade. Values of the lattice parameter of γ/γ'-phases in those parts of the feather calculated from the X-ray results are: $a = 0.36060$ nm in the inner part, $a = 0.36070$ nm in the central part, and $a = 0.36040$ nm in the surface zone (errors $\pm$ 0.00005 nm). This change in the lattice parameters may indicate different stress levels in these parts and changes in the γ'-phase chemical composition.

According to the X-ray analysis, no additional diffraction lines are found in the samples cut out of the inner part of the blade feather.

There are only the γ- and γ'-phase lines in the diffraction patterns. The surface of the back part of the feather (6th direction) shows some diffraction lines of the complex spinel-type oxides (AB_2O_4), including the $NiFe_2O_4$, $CoFe_2O_4$, and Fe_2TiO_4 oxides; the diffraction lines of the γ and γ' phases are also seen. In addition, the diffraction pattern includes lines of the oxides with the rutile-type structure, such as TiO_2 and CrO_2, as well as the complex oxides having a perovskite (ABO_3) structure, such as $FeTiO_3$. Among all possible oxides that can form on a surface of the nickel-based superalloy, only few of them can increase magnetic susceptibility. These complex oxides have a spinel structure (AB_2O_4) and they include simple or complex oxides of iron or cobalt, such as Fe_3O_4, Fe_2TiO_4, Co_3O_4, and $CoFe_2O_4$. The iron γ-Fe_2O_3 oxide and the chromium CrO_2 dioxide with rutile structure are also ferromagnetic.

When gas turbine blades have no protective coating, their surfaces are oxidized, because of the interaction with an external aggressive atmosphere. Thus, the diffusion of alloying elements, such as chromium, titanium, cobalt, nickel, etc., from the alloy volume to the surface occurs, enabling the formation of different complex oxides. The counter-diffusion of sulfur and oxygen to the blade material occurs from the combustion products atmosphere. As a result, the surface layer of the alloy loses its strength. Dependence of the chemical concentration on stresses is known as the Gorsky effect (Gorsky 1935). The effect describes the increasing atomic diffusion in a non-uniform stress field, when the tension locations are enriched by atoms with large ionic radii and compressed places are enriched by the atoms with small ionic radii.

According to the Gorsky theory, in elastic bending of the crystal, which is a substitution solid solution, the driving force is proportional to the difference of the atomic volumes of the impurity atom and the lattice atom. If this is the case, we should expect a significant increase in the concentrations of titanium, iron, and cobalt in the surface layer, which is indeed observed (see Table 1.50). In any case, the mechanical removal of the oxidized layer from the surface of the deformed sample (0.15 mm thickness) does not eliminate the ferromagnetic properties of the blade studied. Independent magnetic measurements were performed with the vibrating magnetometer *Lake Shore* 7407 at a frequency of 82 Hz, with a vibration amplitude of 1.5 mm and a relative measurement error of 1%. This method used standard equipment and was certified. Thin polished samples cut out of the blade (far from the oxide layer) were used for the measurements. The reference sample showed the paramagnetic state. The field dependence of magnetization $M(H)$ of the locking part of the blade had the same behavior as that in the reference sample, represented by a straight line (paramagnetic state). In the convex feather part and the feather impact area of the broken blade, the field dependence of magnetization is a curve with a saturation, typical for the ferromagnetic materials. This magnetic behavior looks like that in the above case of blades made of the EP-800 superalloy.

It should be noted that in the ChS-70V superalloy (like in EP-800), operated under extreme conditions, formation of new phases was not observed, neither by the electron-microscopic analysis, nor by the structural neutron diffraction method.

Evidently, the level of working stresses (also the quantity of structural defects) is related to the level of magnetic susceptibility. In the locking part of the blade, which has a minimal stress level, temperatures induce coalescence of the γ'-phase particles, and no defects are observed inside the particles. In the electron-microscopic images obtained from the blade feather after its operation under extreme conditions, an increased concentration of defects is observed. In particular, the images reveal slip dislocation bands with a high dislocation density inside them. In the back part of the feather (convex part of the blade), where the stresses are at maximum, both slip bands and stacking faults exist inside the intermetallic particles (Stepanova et al. 2011c).

When the gas turbine blades work in the standard operation regime, the deformations preferably occur in the solid solution of the blade material. A heat-resistant alloy may be considered as a composite material consisting of a matrix (γ-solid solution) and a strengthening γ'-phase (Ni_3Al). Both of them participate in the deformation, but the resistance to deformation of these phases is different. The solid solution can deform up to a higher degree of plastic deformation. In the ordered crystal lattice of the intermetallic phase, the dislocation motion is accompanied by the formation of high-energy defects, such as the antiphase boundaries. When the dislocations run through the particles of the intermetallic phase, stable defect complexes do not form inside the γ'-phase particles. This process does not accompany the appearance of the ferromagnetic properties.

Electron-microscopic studies of the turbine blade demonstrate that both solid solution and the strengthening phase are deformed on completion of the experimental operation regime. In this case, one can observe the appearance of magnetic properties of the blade, which persist for a long time (for example, on storage for a year at room temperature). It was found that the high-temperature annealing of the blade eliminates the defects inside the intermetallic phase and the ferromagnetic state disappears (Fig. 1.25). Apparently, the magnetic clusters may be considered as volume complexes of the defects formed inside the particle of intermetallic phase. Such objects must have small (nanoscopic) sizes.

It should be noted that, in the standard operation regime, the blade is always removed from the turbine before stable defects appear inside the particles of intermetallic phases, because such defects in the strengthening phase are the main factor of structural degradation. Increased stresses and temperature levels in the experimental operation regime bring the mechanical properties of the superalloy to the limits of its working capacity. In this case, the long-time high-temperature loading spreads deformations into the strengthening phase. The formation of stable defects inside the γ'-phase indicates that the intermetallic phase loses its strength (Jahangiri and Abedini 2014). The properties of the intermetallic compound approach those of a solid solution, and the intermetallic phase stops being the strengthening component. Changes in magnetic susceptibility allows detection of such defects in γ'-phase. This allows the use of magnetic nondestructive testing methods to evaluate the working capacity of turbine blades and their timely replacement to eliminate any possibility of catastrophic events.

Dependence of the magnetic properties on plastic deformation of the Ni_3Al intermetallic compound is not so obvious. For example, decreasing magnetic susceptibility is found after cold rolling that maintains the initial paramagnetic state, whereas the mechanical milling of the alloy leads to the super-paramagnet state formation (Zeng and Baker 2007). High-temperature deformation of Ni_3Al-based alloys also leads to their ferromagnetic properties. Kazantseva et al. (2016) investigated the structure and magnetic properties of a single crystal of $Ni_3(Al, Fe, Cr)$ after its high-temperature deformation at 850-900°C to failure by tensile tests. In the fracture zone of the alloy tested at 900°C, the crystal fragmentation is accompanied by bending of the atomic layers and by changes in the chemical composition in the sub-lattices of nickel and aluminum. Two Curie temperatures are detected by magnetic studies in the samples cut out of the fracture zone.

Magnetic properties of Ni_3Al after deformation by shear under high pressure were investigated by Kazantseva et al. (2014a); they studied samples of the Ni_3Al, $Ni_3(Al,Co)$, and $Ni_3(Al,Fe)$ alloys. The chemical compositions of the studied alloys are given in Table 1.50 (Kazantseva et al. (2014a).

Table 1.50 Chemical composition of the alloys, at.%

Alloy	Ni	Al	Fe	Co
Ni_3Al	75.5	24.5	-	-
Ni_3Al-Fe	74.0	18.5	7.5	-
Ni_3Al-Co	67.0	25.0	-	8.0

After annealing at 1100°C for 5 hours, the samples have a single-phase (Ni_3Al, γ') state. The dislocation density inside the grains is low; the SAED patterns have the point diffraction reflexes.

Formation of the nanocrystalline state in the Ni_3Al samples after severe plastic deformation by shear under high pressure was studied in detail by Korznikov et al. (2001). Initially, crystal fragments with sizes of about of 100 nm form, separated by small-angle boundaries. Next, the local disordering takes place, the dislocation density increases, and a large number of the deformation twins appear. The processes occur under non-uniform deformation, propagating from the surface of the specimen to its volume and finishing by complete disordering. As a result, nanostructures form, with an average grain size of ~20 nm and the shear deformation degree $e = 7$. A partial recovery of the long-range order begins when heating samples up to 350°C.

These above results are consistent with those obtained by Kazantseva et al. (2014a). High dislocation density and deformation micro twins are observed inside the grains. Structural changes in the Ni_3Al, Ni_3Al-Co, and Ni_3Al-Fe alloys are similar to each other. An average grain size after the shear under high pressure (10 turns of the Bridgman anvils) in both doped alloys is ~20 nm, whereas for Ni_3Al it is ~30 nm. Perhaps, this similarity is due to the fact that both iron and cobalt are the plasticizing elements. The SAED patterns display the ring diffraction reflexes. The ring, corresponding to the (001) superstructural diffraction reflection, is absent in Fig. 1.27, indicating the loss of the long-range order. The lattice parameters calculated from X-ray data increase from 0.3565 to 0.3591 nm in Ni_3Al, from 0.3571 to 0.3590 nm in Ni_3Al-Co, and from 0.3560 to 0.3585 nm in Ni_3Al-Fe.

Specific magnetization (M) was measured at the room temperature with *Faraday Balance*. The relative measurement error for the external magnetic field is within 0.5% and that for the magnetization is within 1.5%. The magnetic field inhomogeneity in the sample area does not exceed one percent. Figure 1.28 shows changes of the specific magnetization M for the Ni_3Al alloy with increasing deformation degree. It is evident that a single anvil rotation is sufficient to significantly reduce specific magnetization. A shear under high pressure can completely destroy the ferromagnetic state of a Ni_3Al-based alloy.

The behavior of the field dependence of the specific magnetization $M(H)$ of the Ni_3Al-Fe alloy is very interesting. This alloy was in the ferromagnetic state before deformation with $T_C = 320$ K. Dependence of the Curie temperature T_C on the iron concentration in the $Ni_3(Al,Fe)$ alloys was given by Masahashi et al. (1987).

Deformation processes lead to a substantial decrease of the degree of long-range order, but in our case, complete disordering does not occur. The superstructural lines of the ordered γ'-phase are observed in the X-ray diffraction patterns of the Ni_3Al-Fe alloy after the shear under high pressure. The degree of the long-range order S^2 is equal to 0.94 in the initial state of this alloy. After the deformation (with 10 anvil rotations) S^2 decreased to 0.55. Superstructural diffraction reflexes are also visible in the SAED patterns of the deformed sample.

The Ni_3Al-Co alloy is in the ferromagnetic state initially (like Ni_3Al-Fe), although magnetization of this alloy is much lower. Deformation of the cobalt-doped alloy does not lead to complete disordering, despite the decreasing long-range order. The deformed Ni_3Al-Co sample shows the ferromagnetic state with decreased magnetization.

Magnetic properties of a nickel-based superalloy after deformation by shear under high pressure were investigated by Davydov et al. (2015b, 2016). The study of the structure and magnetic properties was done using samples of the nickel-based ChS-70V superalloy. Chemical composition of the ChS-70V superalloy was given in Table 1.40. Deformation by shear under pressure with Bridgman anvils under the pressure $P = 8$ GPa at room temperature was done using the samples with diameter of 5 mm and thickness of 0.3 mm and the rate of anvil rotation of 0.3 rotation/min.

According to the X-ray and TEM data, the diffraction patterns of the samples before and after deformation look the same. Intensities of the superstructure diffraction lines of γ'-phase are significantly reduced with an increasing degree of shear deformation.

The nickel solid solution and γ' intermetallic phase are coherent with each other. The lattice parameters of these phases are almost the same and the structural X-ray diffraction lines of these phases overlap each other. The superstructure diffraction lines belong to the ordered intermetallic phase (Ni_3Al). Thus, before shear deformation, the (100) superstructure diffraction line can be seen, along with the structural lines in the X-ray diffraction pattern. After the deformation by shear under high pressure with 10 anvil rotations, the structure becomes highly fragmented. The alloy is characterized by large numbers of planar defects, such as the dislocations and twins inside the crystallites, with an average grain size of 20 nm. The γ'-phase lattice parameter, calculated from the X-ray data, is increased from 0.3565 ± 0.0002 nm before deformation to 0.3591 nm after the deformation. The SAED pattern shows diffraction rings; the X-ray diffraction pattern does not show any superstructure reflections, indicating the disappearance of the long-range order in the intermetallic phase.

After the shear deformation of the ChS-70V superalloy, the field dependence of magnetization $M(H)$ remained linear. Magnetic susceptibility, estimated from the inclination angle of the $M(H)$ curves, are practically same as that before the shear deformation. The vibrating magnetometer *Lake Shore* 7407 was also used for the magnetic measurements. Measurements were done at a frequency of 82 Hz, with a vibration amplitude of 1.5 mm and a relative error of 1%.

Thus, deformations can lead to both increased and decreased magnetic susceptibility of the alloys based on Ni_3Al. Different behavior of the magnetic susceptibility of the Ni_3Al samples depends on the deformation type, apparently, due to the resulting structures.

Formation of the nano-scale structures leads to the formation of numerous defects inside the crystal fragments, large numbers of grain boundaries, and distorted areas along the grain boundaries. As a result, severe plastic deformations by shear under high pressure lead to a significant reduction of the long-range order in the alloy as a whole, which in turn, leads to decreasing magnetic susceptibility of the alloy. Deformations occurring inside intermetallic particles, produce complex local volume defects that destroy the long-range order of the Ni_3Al intermetallic compound. Such volume defects act as a magnetic cluster providing an increase of magnetic susceptibility of the alloy.

Structural elements are responsible for the appearance of ferromagnetic properties in nickel superalloys. Structural studies show large numbers of Superstructure Stacking Faults (SSF) in nickel superalloys, when their ferromagnetic properties appear (Stepanova et al. 2011c, Davydov et al. 2015).

Tests of the turbine blades made of the EP-800 and ChS-70V superalloys deformed at high temperatures showed a correlation between the deformation degree and amounts of the crossing superstructure stacking faults (SSF) inside the intermetallic particles. The higher the SSF density in the intermetallic particles, the higher the ferromagnetic effect . The crossing SSFs form the volume structural defects, within which the nearest atomic neighbors are exchanged. Such volume defects are the ferromagnetic clusters responsible for the ferromagnetic properties of the alloys. A complex of planar defects, which may also be responsible for the ferromagnetic properties in the nickel superalloys, is the antiphase boundaries (APB). Appearance of the long-period modulated structures in the intermetallic compounds was previously discussed by Wang et al. (1991), Bendersky et al. (1994), Makoto and Yasumasa (2006), and Christian et al. (2003). We can suggest that a self-organization of the stacking faults is responsible for their periodic arrangement to reduce the energy of the crystal, which increases substantially in a distorted area of the ordered alloy. Defective parts of a crystal may be formally described by an inclusion of long-period phases into the initial intermetallic crystal lattice. For Ni_3Al ($L1_2$), these may be long period modulated phases, such as $D0_{22}$ or $D0_{23}$. The long-period tetragonal superlattice $D0_{22}$ ($M = 1$) could be obtained, if the parallel plane of the stacking fault defect is introduced into each cubic plane M along the cubic axes $L1_2$.

The long-period structures consisting of periodic antiphase boundaries (APB) were discovered in some systems of ordered alloys and intermetallic compounds (Bendersky et al. 1994, Sato and Toth 1962, Ramesh et al. 1992, Bhattacharya and Ray 2000, Greenberg et al. 2006, Takahashi and Ikeda 1983) – for example, Cu-Au, Cu-Pt, Cu-Pd, Au-Cd, Au-Zn, Au-Mn, Au-Cr, Au-V, Pt-V, Ag-Mn, Ni-V, Ni-Ti, Ni-Cr,

Ni-Al, Ti-Al, Mo-Pt, and Ni-Ti. As a rule, periodic APBs are found in the alloys with a low energy of the long-period structure formation, whereas the Ni_3Al intermetallic compound has a high energy. Liu et al. (2015) carried out a theoretical calculation of the energy change of the crystal Ni_3Al upon the introduction of an APB. The energy value of the APB in the (001) cube plane of 80 mJ/m^2 was adopted. However, according to the experimental data, the minimum of the APB energy in the cube plane, calculated from the APB width, was 31-43 mJ/m^2, i.e., half of the adopted value. A significant decrease in the energy of a crystal can be obtained if one allows for the possibility of periodic APBs. According to the calculations, the formation energy of the structures with ordered arrangements of APBs is slightly different from that of the $L1_2$ structure (4.486 eV). For example, the formation energy of a long-period $D0_{22}$ structure ($M=1$) is 4.318 eV, whereas that of a long-period $D0_{23}$ ($M = 2$) is 4.401 eV.

Thus, the winning free energy of the Ni_3Al crystal under certain conditions (e.g., at deviations from stoichiometry or at strain) is profitable enough to transform its crystal lattice into a new ordered state. The instability of the intermetallic compound lattice Ni_3Al ($L1_2$) to the formation of Long-Period Structures (LPS) at the departure from stoichiometric composition was found in several experimental works (Glas et al. 1996, Duval et al. 1994). In some cases, the appearance of the hardening peak on the temperature dependence of yield stress $\sigma_{0.2}$ in Ni_3Al was explained by the $D0_{22}$ phase formation (Ramesh et al. 1996). The $L1_2$ crystal structure appears to transform to $D0_{22}$ crystal structure during heating. Such a transformation starts at a temperature around 700°C and seems to complete around 1100°C. In the temperature range 700-1100°C both phases coexist. Tetragonal distortion of the $L1_2$ lattice appears as the tweed morphology in TEM observations. The main idea of (Ramesh et al. 1996) is that the anomalous strengthening behavior appears not only due to the change of the dislocation type as it is generally accepted in the literature, but also due to the structural changes.

TEM images of the crossing stacking faults inside the intermetallic particles are shown in Fig. 1.26.

Glas et al. (1996) revealed the formation of the nano-scale (15 nm) areas with the $D0_{22}$ structure of nickel superalloys and with a high content of cobalt and chromium during annealing at 700°C.

With the annealing temperatures increasing to 760°C, the sizes of the LPS areas increase to 100 nm. Diffuse scattering in the [100] and [420] directions was observed in the neutron diffraction patterns of the samples. The size of the LPS areas obtained with the neutron diffraction data was 1.5 nm. An analysis of the structures by Duval et al. (1994) also showed the presence of the concentration fluctuations with a wavelength of 0.72 nm in the samples; this value is a doubled lattice period $2d$ of Ni_3Al.

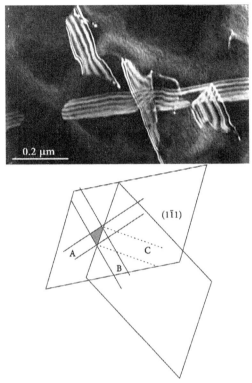

Fig. 1.26 Regions of interaction of stacking faults (SSFs) in the γ' phase, the dark-field image; the scheme of the formation of three-dimensional V-faults by intersection of SSFs; A and B are the SSFs in the $(1\bar{1}1)$ plane, and C is the SSF in the (111) plane.

It is known that the APBs and APB-tubes are easily formed by dislocation interactions during the deformation of Ni_3Al. As it was shown by Makoto and Yasumasa (2006), APBs are the places of seeds of a long-period modulated phase formed in the alloy after annealing. Rentenberger et al. (2003) discovered that numerous APB tubes appear under the deformation of single-crystal Ni_3Al (4% of strain). These APBs are clearly visible in both superstructural and structural reflexes of γ'-phase. The authors explained this surprising fact by additional atomic displacements inside the APB tubes.

The $D0_{22}$ superlattice formation in nickel superalloys by ordered APBs after severe plastic deformation by shock-wave loading was suggested by Kazantseva et al. (2006, 2014b) and Vinogradova et al. (2008). Tetragonal $D0_{22}$ superlattice is oriented on the basis of the $L1_2$ cubic crystal lattice, and because of that it is very difficult to reveal it using diffraction methods. Presence of the $D0_{22}$ phase can be determined, however, by the magnetic methods. Kazantseva et al. (2006, 2014b) and Greenberg et al. (2006) studied structure and magnetic properties of the single-crystal VKNA-4U superalloy after shock-wave loading. With a maximum pressure of 20 GPa on the sample surface of the VKNA-4U superalloy, magnetic susceptibility increased. Changes of the magnetic susceptibility with deformation are manifested more clearly in the sample of VKNA-4U superalloy after shock-wave loading with a maximum pressure of 100 GPa on its surface. In the latter case, the magnetic susceptibility increased by almost two orders of magnitude. In addition to this, the pronounced anisotropy of the magnetic properties was noted, related to the orientation of the single crystal, which was absent in the initial sample and appeared only after deformation. The <001> crystallographic axes are equivalent in the $L1_2$ structure, therefore the appearance of such anisotropy after deformation may also be related to the changes of the crystal lattice after deformation. After annealing of the samples at 900°C for 1 hour and cooling with a furnace, magnetic properties return to their initial values, indicating thermal instability of the magnetic state formed in the superalloy after deformation.

A neutron diffraction study was done using the multi-detection consoles for shooting a single crystal, allowing one to get all the lines as if it was a polycrystalline material. The appearance of a diffraction line doublet may indicate that the additional diffraction line does not belong to the cubic lattice of γ'-phase. After the shock loading with maximum pressures of 20 GPa and 100 GPa, the sample retained a single crystalline state. The maximum dislocation density after the loading, calculated according to PCA, reached 10^{-12} cm^{-2}. The neutron diffraction patterns of the VKNA-4U sample after shock loading with maximum pressures of 100 GPa showed the same additional lines coinciding with the positions of those for the long-period $D0_{22}$ phase. After the annealing, the intensity of the additional lines became reduced.

High values of magnetic susceptibility were also obtained after cold rolling (up to 38%) of the single-crystal EI-437B nickel superalloy (Kazantseva 2006), with the chemical composition shown in Table 1.51. This nickel superalloy contains about 9% by volume of γ'-phase and it has heat-resistant mechanical properties, analogous to such nickel superalloys, as NCF80A (Japan) and NiCr$_{20}$TiAl (Germany). The ferromagnetic effect is found to be less pronounced in the EI-437B superalloy, because of small quantities of the intermetallic phase in the alloy.

Table 1.51 Chemical composition of the EI-437B nickel superalloy, wt.%

Al	Cr	Ti	Fe	Mn	Si	C	Ni
1.0	22	2.8	4	0.4	0.6	0.07	balance

The presence of two values of the Curie temperature T_C determined by magnetic studies of nickel superalloys (Davydov et al. 2016, Rhee Joo Yull et al. 2003) indicates the existence of two magnetic phases in the alloy. *In situ* neutron diffraction patterns show additional lines near the temperatures of magnetic transitions. It was found that the modulation period varies with temperature (Idzikowski et al. 2003).

Ultrasonic testing of single- and polycrystalline articles made of the nickel-based heat-resistant alloys. Turbine blades and disks are the most heavily loaded parts of aircraft engines and stationary gas-turbine energy facilities, which largely determines their characteristics, service life, and reliability. Such parts operate at high temperatures and they have variable cross-sections with numerous concentrators of stresses (cooling channels, mounting holes, blade latches, etc.). These parts operate under long-duration static tensile stresses, accompanied by the combined action of the thermal stresses and vibrational loads. This leads to the appearance of fatigue cracks. A short-time increase of temperatures, overshooting the range of operating temperatures (a temperature jump) also has a negative influence on the stability of an alloy structure.

Nickel superalloys must have a high strength throughout the entire operating temperature range, high plasticity under both long- and short-term loading, high corrosion resistance, and they must be easy to process and resistant to low-cycle fatigue. The presence of γ'-phase increases the heat resistance, but with increasing of its content alloys become prone to the formation of hot cracks during processing and operation.

Results of acoustic nondestructive testing (NDT) applied for crack detection in model samples and real details made of nickel superalloys in poly- and single- crystalline state are presented below (Rinkevich et al. 2009, 2011).

Cracks can appear during production of details, in particular, owing to the presence of metallurgical flaws (Reed et al. 2000, Habicht et al. 2004, Wen et al. 2017). The problem of detecting cracks appearing during operation is extremely important.

The nondestructive testing (NDT) becomes especially urgent because of a developing trend toward extending the operational lifetime of a detail, when its service life declared by a manufacturer has expired. Under such conditions, undetected cracks may cause failures.

Ultrasonic testing is capable of detecting hazardous flaws; examples of its application to testing the quality of heat-resistant materials are well known. In particular, Stupino Metallurgical Company (Russia) and

SONOTEC (Germany) are using this method to detect invisible cracks, inclusions, voids, and other discontinuities in metals, plastics, ceramics and composites, and stamped billets made of heat-resistant alloys.

However, developments of the technology of ultrasonic testing of heat-resistant alloys require preliminary investigations. There are factors limiting the testability of these materials; high elastic anisotropy leading to the high damping of ultrasonic waves in polycrystalline alloys is one of these factors. In addition, the several natural ultrasonic waves are simultaneously excited, propagating in different directions. Therefore, the ultrasonic methods for testing single-crystal heat-resistant alloys have specific features. The problems of detectability of artificial reflectors in single-crystal heat-resistant alloys, the resolution, and the signal-to-noise ratio were considered by Rinkevich (2003, 2006, 2008) as the necessary initial stages for designing the testing technique. They presented an example of detecting a real flaw in a component manufactured of a polycrystalline alloy. The study by Rinkevich (2008) was based on previous detailed investigation of the acoustic properties of single-crystal and polycrystalline heat-resistant alloys, model ternary and industry alloys.

Single-crystal samples of TsNK-8MP and VKNA-4U superalloys were used for the research. The chemical composition for ChS-70V is given in Table 1.48 and the one of the alloy TsNK-8MP is presented in Table 1.52. Some samples were cut from a turbine blade made of the ChS-70V superalloy after its long-time operation at 880°C and emergency destruction. An optical study reveals a structure of a casting alloy with large grains of 100-800 µm. The grains have a dendritic internal structure with eutectics in inter-dendrite areas. The γ'-phase volume content in TsNK-8MP is 47%.

Table 1.52 Chemical composition of the TsNK-8MP superalloy, wt.%

C	Cr	Ti	Al	W	Co	Mo	Ni
0.02	12	4	4.2	7	9	0.5	balance

The samples were obtained by the Bridgman method with a crystallization rate of 1 mm/min and a temperature gradient of 80°C/cm. The distance d between the secondary branches of dendrites was 20 µm in the TsNK-8MP superalloy. The average size of the γ'-phase particles in the samples of TsNK-8MP and ChS-70V superalloys was 400 nm; in the sample of VKNA-4U, the particle sizes were up to 1000 nm. Such objects are too small to detect in acoustic testing in the frequency range used.

The sample prepared for acoustic measurements had dimensions of $10 \times 14 \times 8$ mm, with two parallel planes, from which waves were excited and received. The acoustic path length for a sequence of ultrasonic echo pulses varied from 0.8 to 8 cm. Measurements were performed at transverse waves (TWs) at the frequencies of 5 and 15 MHz. An L15M normal transducer (produced by Krautkrämer) at an operating frequency f = 15 MHz and a P112-5.0 dual transducer at 5 MHz with an 8-mm-diameter plate were used in experiments with Longitudinal Waves (LWs). The wave velocities and attenuation coefficients were measured using a PCUS10 microprocessor board installed as a device on a personal computer (PC). This board was developed at the Institute of Nondestructive Testing of the Fraunhofer Society (Germany). Some of the experiments were performed on the RAM-5000 facility (Ritec company). The velocities of waves were determined from precise measurements of the pulse propagation times. The measurement error of the wave velocities was ~0.5%. The attenuation coefficient was measured from the relative decrease in the amplitude of echo pulses that traversed different distances in a material. The attenuation coefficient measurement error depends on the accuracy of the amplitude measurements, on the number of echo pulses, for which measurements were performed, on the quality of the specimen (parallel faces of the sample), and other factors.

Four model samples with artificial reflectors were used, two of which were cut out of the TsNK-8MP superalloy and two were made of the VKNA-4U superalloy. A flat-bottom reflector with a diameter of 2 mm was made in one specimen of the VKNA-4U superalloy shaped as a disk with a diameter of 20 mm. The normal direction to the sample flat surfaces corresponded to the [100] crystallographic axis. Another sample of the VKNA-4U superalloy had two plane-parallel faces. A hole with a diameter of 2 mm served as the reflector. Two samples of the TsNK-8MP superalloy had the same shapes and dimensions. One of the samples was cut out of a single-crystal ingot and the other of a polycrystalline ingot. A pronounced pattern of the echo pulses with a low noise level and false signals was obtained in the single-crystal sample.

Rinkevich et al. (2011) studied the possibility of detecting artificial reflectors and flaws in the single-crystal and polycrystalline states of the heat-resistant nickel-based alloys. Ultrasonic flaw detection technique for single-crystal materials has substantial peculiarities. In operation with single crystals, one should take into account that three elastic waves usually arise simultaneously in an anisotropic material at an oblique incidence of an ultrasonic wave on the sample boundary. The ratio of the amplitudes of these waves, the refraction angles, and the polarizations of these waves depend on the angle of incidence and the polarization direction of the incident wave with respect to the axes of the single crystal. In principle, two approaches are possible in developing the ultrasonic testing technique. The first approach involves detailed consideration of the crystal anisotropy and elaboration of the conditions under which anisotropy helps select the desired characteristic of a wave in the material. This is an efficient but very laborious procedure requiring knowledge of the elastic constants of single crystals and fulfillment of preliminary experimental work. The second approach involves selection of only the normal incidence of waves on the boundary of a single crystal with the further propagation of ultrasound along a crystallographic axis of a rather high order. For example, in a crystal with cubic symmetry, these axes must be of at least the second order. This simpler approach is possible only if the sample is oriented along the (100) crystal axis. In this study, the second approach was adopted. With a normal incidence, the waves are not transformed and no visible refraction of rays is observed. As a rule, TWs are polarization-degenerate, i.e., their velocities do not depend on the polarization direction in the plane perpendicular to the propagation direction.

In this case, between the probing signal and the first bottom echo signal, as well as between the first and second bottom echo signals, there are pulses reflected from the artificial reflector. Figure 1.30 shows an A-scan obtained for a dual transducer operating at f = 5 MHz and set in the position 1 above the reflector. At this lower frequency, a signal reflected from the FBH is also observed, but the resolution is insufficient for reliable determination of its characteristics. At the testing parameters chosen, both reliable detection of the artificial reflector and good resolution along the propagation direction are ensured.

Figure 1.27 shows an A-scan obtained for the same transducer set in position directly above the reflector.

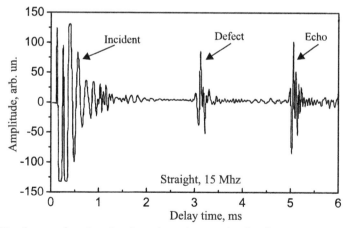

Fig. 1.27 A-scan showing the detection of an artificial reflector in a disk-shaped specimen of the VKNA alloy above the reflector, position of the dual transducer, the gain Kamp = 44 dB (Rinkevich et al. 2011).

To determine the minimum distance, at which a flaw of the FBH type can fit the diameter is <2 mm, the corresponding DGS diagrams were calculated. These diagrams relate the amplitude of a signal reflected from the flaw to the distance at which this flaw is located and to its diameter. To calculate the DGS diagrams for a flaw of the FBH type, when a normal dual piezoelectric transducer with a circular aperture is used, the technique expounded by Danilov and Ermolov, 2000; Danilov, 2009 was used.

As an example of detecting a flaw, consider the data from a study of a turbine blade from an electrical power station. The blade was operated at $T = 880°C$ for 9000 hours (17 thermal shifts). Typical locations of the fatigue cracks are the leading edge of the feather and the end face of the locking part. The end face of the lock was stiffly attached to the turbine disk and the lock was deformed under restricted conditions, which initiated fatigue cracks. In this specimen, the cracks were 3-15 mm in length and 0.2-0.4 mm in width along almost their entire length. One crack had a larger size, with a length of 10 mm and a width of 1-2 mm.

Such surface flaws are additional stress concentrators, accelerating the damage of the turbine blade. Discontinuity flaws can also lead to a non-uniform temperature field inside a detail. To detect the flaws, a P121-5.0-65° angle transducer was used with an operating frequency of 5 MHz and an injection angle of 65° (for steel). Due to the complex shape of the blade, the possibility of mounting a transducer and scanning was limited. We managed to mount the transducer only in a position, where the acoustic-emission axis in the blade was approximately parallel to the crack plane. Although this position is unfavorable for efficient testing, it was possible to detect a signal from the flaw. The signal from a crack is on the segment of the scan with a delay of 26-27 μs from the beginning. The amplitude of the reflected signal is higher by only a factor of 2 than the amplitude of the noise-like signals present on the scan and having no relation to the crack. By changing the incidence angle of the acoustic beam on the crack within a narrow range, it was found that the signal belongs to the crack. The signal amplitude abruptly decreased when the beam axis was diverted from the crack. In addition, the measured delay time of the signal corresponds to the calculated time for a signal reflected from the crack.

Figure 1.28 shows one of the A-scans.

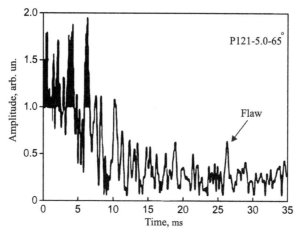

Fig. 1.28 A-scan showing the detection of a crack on a blade; a P121-5.0-65°
transducer, f = 5 MHz (Rinkevich et al. 2011).

1.2 Analysis of the Strength and Vibrations of Cooled Single-Crystal Blades

Modern trends in the improvement of gas turbine engines are associated
with an increase of the working gas flow parameters. In this case, the
blading experiences the influence of ever higher intensities of dynamic
loads and temperature fields. This necessitates the use of working
blades of gas turbines from heat-resistant single-crystal alloys with
internal cooling (Nozhnitsky and Golubovsky 2006). However, such
materials are heterogeneous in their strength and thermo-mechanical
properties (Shalin et al. 1997). The stress-strain state of such blades
under static loads and especially under vibrations, depends substantially
on the crystallographic orientation of the materials. The choice of the
crystallographic orientation is associated with a number of factors, such
as the blade geometry, the temperature field, the influence of centrifugal
forces, and the peculiarities of the blade vibration modes (Kopelev 1983,
Melnikova et al. 2001, Pridorozhny et al. 2006, Pridorozhny et al. 2008,
Pridorozhny et al. 2013, Vorobiov et al. 2014a, Vorobiov et al. 2014b,
Zhondkovski et al. 2013).

Cooled working blades of gas turbines are widely used, because
the capabilities of modern heat-resistant materials to work at high
temperatures are noticeably lagging behind the emerging needs of
modern gas turbine constructions (Nozhnitsky and Golubovsky 2006).
There exist various cooling systems for the working blades of gas
turbine engines. The most common are the cooling systems with internal
channels, allowing circulation of the cooling air. Additional external

cooling is used for the air coming through the internal channels and exiting through the holes, which creates a thin film of air on the blade surface. The external film-cooling is often used only for a part of the blade surface, for example in the region of the outlet edge. The resulting temperature gradients in the blade body cause high temperature stresses. The greatest static and vibrational strength is possessed by modern high-temperature single-crystal alloys, whose mechanical properties depend on the orientation of the crystallographic axes. Some works (Hino et al. 2000, Caron et al. 2011, Maclachlan and Knowles 2002) are devoted to perfection of such alloys. The complex geometric shape of the cooled blades contributes to the concentration and localization of stresses (Nozhnitsky and Golubovsky 2006, Maclachlan and Knowles 2002, Maggerramova and Vasiliev 2011). An analysis of the vibrational and thermoelastic stress-strain states of such blades is a problem associated with the properties of single-crystal materials.

For the cooled blades of gas turbines, chromium-nickel-based heat-resistant alloys are the most commonly used ones. Blades of complex shapes are made by means of precise casting. A complex process of the casting cooling provides a directional crystallization of the blade material. The use of directional crystallization in the cast enables forming a structure consisting of columnar grains, whose boundaries are parallel to the direction of the principal tensile stresses. The structure of a single crystal with the crystallographic orientation [001] along the axis of the blade consists of dendritic branches that are parallel to that direction and oriented along the heat flux. The structure of such alloys is a face-centered cubic lattice (Shalin et al. 1997). There are a number of crystallographic directions in such single crystals. Three mutually orthogonal crystallographic axes are [001], [010], and [100].

Preliminary studies have shown that the axial-orientation of the single-crystal cooled blades of the first stage, operating in high-temperature fields with large temperature gradients, is preferred, with the longitudinal axis of the blade z aligned with the crystallographic direction [001]. The modulus of elasticity in this direction is minimal for the material, but the thermal stresses are lower than for the other crystallographic orientations. Stresses from the centrifugal forces are much lower than thermal stresses. The strength characteristics of such blades remain sufficiently high, especially since the blades are comparatively short and massive (Pridorozhny et al. 2013).

Blades with the axial orientation [001] of the material have the greatest rigidity for torsion, since the shear resistance in the transverse direction has the maximum value. The geometric shape of the cooled blades is such that spinning from the action of the centrifugal forces is small. As a result, the tension at torsion is not great for such blades. Significantly more important is the possibility of localization of stresses on the internal surface of the cooling channels. Therefore, the first stages of gas turbines have cooled blades with the axial orientation [001] of the single-crystal material. This is because of the provision of their static strength under high-temperature conditions and their high gradients (Nozhnitsky and Golubovsky 2006).

The main problem of the strength reliability of single-crystal working blades at high gas flow temperatures was considered by Nozhnitsky and Golubovsky (2006) on the basis of experimental and theoretical studies. General properties of single-crystal nickel superalloys and their dependence on the orientation of the crystallographic axes were analyzed by Shalin et al. (1997) in their monograph. Pridorozhny et al. (2013) considered changes in the stress-strain state of a single-crystal uncooled blade as a function of the location of the crystallographic axes. Detailed studies of the dependence of the stress-strain state in the volume of a single-crystal working blade of gas turbine engines, containing a system of direct cooled channels, on the orientation (including the azimuthal one) of the crystallographic axes were described by Maggerramova and Vasiliev (2011) and Pridorozhny et al. (2013). The influence of the azimuthal orientation on stress distribution in the system of perforations in a single-crystal cooled blade was considered by Pridorozhny et al. (2013). The features of the temperature state of cooled blades were studied by Kopelev (1983) and Zhondkovski et al. (2013). These works showed a significant effect of the structural features of the blade and of the orientation of the crystallographic axes on the thermoelastic stress-strain state of a single-crystal blade. Zhondkovski et al. (2013) and Vorobiov et al. (2016a) studied the thermoelastic state of a single-crystal blade with a new vortex cooling system in a temperature field with a constant orientation of the crystallographic axes. Effects of the orientation of the latter on the vibrational characteristics and distribution of the vibrational stresses for the blades with direct cooling channels were considered by Pridorozhny et al. (2006, 2008) and Melnikova et al. (2001). For the blades with a vortex cooling system, the effects were considered by Vorobiov et al. (2014b). It was shown that changes in the position of the crystallographic axes strongly affect the distribution and magnitude of the thermoelastic and vibrational stresses.

Therefore, it is important to investigate the effects of changes in the azimuthal orientation of the crystallographic axes on the thermoelastic and vibrational stress-strain states of a blade with a vortex cooling system.

1.2.1 Object of the study and calculation model

Cooled blades have different cooling systems. To study the main properties of the thermoelastic and vibrational states of the blades, it is advisable to choose such a blade structure, in which the effects of all features of the cooling system on the stress-strain state of the blade manifests itself most clearly.

The selected blade structure is used in gas turbine engines, which should have a long operational lifetime. For that, an effective cooling system is required, which would limit the maximum temperatures and maximum stresses in the blade. We consider a cooled single-crystal blade with an effective, but complex vortex cooling system. Figure 1.29 gives an idea of the blade structure, cooling channel system, location of the crystallographic axes, and direction of their rotation in the study. Near the outlet edge, there are channels for the exit of cooling air, which provides external cooling of the thin output edge. The resulting temperature field with high temperature gradients causes a thermal expansion of the blades and high temperature stresses. Therefore, it is necessary to solve the interrelated problems of determining the temperature and thermoelastic state of a single-crystal cooled blade.

The complexity of the blade geometry and the heterogeneity of the material (both the initial and the developing in the temperature field) require a choice of the type and size of the finite element. Therefore, several models of the blade, differing in the density and type of the finite elements, and consequently by the number of degrees of freedom, were investigated, the characteristics of which are summarized in Table 1.53 (Vorobiov et al. 2016a).

Table 1.53 Characteristics of the finite element models of a cooled blade

Finite element type	Number of nodes	Number of degrees of freedom
4-node tetrahedron	83 910	242 367
	132 908	382 806
	244 713	702 561
	436 228	1 264 614
	580 031	1 704 114
	935 793	2 747 526
10-node tetrahedron	1 072 656	3 217 968
	1 739 242	> 7 500 000
	3 117 290	> 9 000 000

Tetrahedral finite elements provide a more accurate description of the blade geometry. It has been found that, starting with 3 217 968 degrees of freedom, the values of the characteristics of the stress-strain state of the blade do not change. Therefore, the finite-element model of the blade of 10 nodal tetrahedral elements with that number of degrees of freedom was chosen to solve the problem. At the same time, the minimum size of the finite element was $3.7713 \cdot 10^{-5}$ mm, sufficient to describe all the structural features of the blade and stress fields.

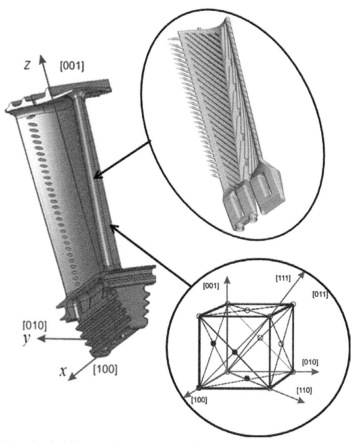

Fig. 1.29 Cooled blade with the designation of single-crystal directions and its cooling system in the form of a vortex matrix.

Convergence results for a series of finite-element models with different sampling levels were studied by Vorobiov et al. (2014a). In view of the complexity of the geometric shape of the blade, the most suitable finite elements were 10-node tetrahedral elements, taking into account the heterogeneity of the material. Since the frequency is an integral characteristic, the results convergence in calculating it with a changing number of freedom degrees of the model is faster than in the analysis of the stress state (Table 1.54). The chosen calculation scheme provides a relative error in the natural frequencies of the order of 1% at 1.7 million degrees of freedom, as well as a fairly clear picture of the fields of relative stresses, with their strong gradients (Vorobiov et al. 2014b, 2016a). The calculation scheme ensures a good agreement of the numerical frequencies with the known ones (Vorobiov et al. 2014b) from experimental data (Table 1.55).

Table 1.54 Influence of the number of degrees of freedom on the relative error in determining the natural frequency

Frequency no.	Number of degrees of freedom				
	242367	382806	702561	1264617	1704114
1	2.1	2.0	1.8	1.4	0.2
2	2.5	2.3	1.9	1.5	0.3
3	3.5	3.0	2.1	1.6	0.31
4	4.1	3.1	2.3	1.6	0.33
5	4.8	3.4	2.8	1.7	0.4
6	6.5	8.5	3.8	3.8	0.5
7	11.0	8.5	6.1	4.6	0.6
8	11.2	8.6	6.3	4.7	0.7

Table 1.55 Numerical and experimental values of the natural vibration frequencies of the blade of an isotropic material

Frequency No.	Mode	Frequency value, Hz	
		Numerical	Experimental
1	1st bending	737.89	740-780
2	2nd bending	1712.5	-
3	1st torsional	2551.6	2545-2740
4	3rd bending	3789.5	-

1.2.2 The temperature condition of the blade

In general, determination of the temperature state of a cooled blade is a boundary-value problem of mathematical physics in three-dimensional space. The thermal conductivity process is described by a nonlinear partial differential equation

$$\frac{\partial}{\partial t}(\rho c T) = div(\lambda\, grad T) + Q, \tag{1}$$

where T is the temperature, λ is the coefficient of thermal conductivity, c is the specific heat, Q is the internal source or runoff characteristic, ρ is the density of the blade material, and t is time.

This blade has temperature differences both in the longitudinal direction and in its cross sections. Therefore, the problem should be solved in three dimensions. For such problems, it is possible to use boundary conditions of the first and third kinds (Kopelev 1983). If the temperature distribution on the surface of the blade and the cooling channels is specified, it is possible to use boundary conditions of the first kind. However, such data are not always known, and it is usually difficult to obtain them with a sufficient degree of accuracy. The boundary conditions of the third kind reflect convective heat transfer on the surfaces of the blade and the cooling channels (Kopelev 1983); their use is also associated with inaccuracies. Therefore, the problem of the temperature state of the blade was solved in two ways. Comparison of the results opens a possibility of obtaining more adequate results.

Temperatures of various sections of the blade are often set as the initial data. The temperature field of the blade is determined by averaging. In addition, the heat conduction problem is often solved for the case, when the temperature of the gas on the blade surface and the temperature of the cooling air at the inlet to the cooling system are set, corresponding to the operating mode of the blade. In this case, we consider the stationary heat conduction problem, which reduces to the variational equation $\delta J = 0$ for the functional (Vorobiov 1988)

$$J = \frac{1}{2} \iiint_V \lambda \left[\left(\frac{\partial T}{\partial x} \right)^2 + \left(\frac{\partial T}{\partial y} \right)^2 + \left(\frac{\partial T}{\partial z} \right)^2 \right] dx\,dy\,dz + \frac{1}{2} \sum_k \iint_{(S_k)} h_k (T - T_k)^2 dS_k, \quad (2)$$

where V is the body volume, S_k is the surface areas of the blade and cooling channels, h_k is the coefficient of heat transfers in the areas of the blade, and k is the number of the surface areas of the blade and its cooling channels.

The variational equation (2) reflects the heat conduction equation and the boundary conditions of the third kind at different sections of the blade and the cooling channel. These conditions have the form:

$$h_k (T_{gk} - T) = \lambda \frac{\partial T}{\partial n},$$

$$h_k (T - T_{aк}) = \lambda \frac{\partial T}{\partial n},$$

where T_{gk} is the gas temperature, $T_{aк}$ is the air temperature in different sections of the cooling channels, and n is the normal to the surface of the blade or the cooling channels. The cooling air temperature in different sections of the channels is selected according to the initial data.

The average value of the coefficient of heat transfer from the working fluid to the blade is most commonly defined at the entrance edge (h_1), in the middle part of the profile (against the cooling channels) (h_2), and at the output edge (h_3), as well as on the surface of the cooling channels $(h_{aк})$. In the absence of sufficient information on the nature of heat transfer, the average integral values of heat transfer coefficients are used, found from empirical relationships (Kopelev 1983).

The solution of the temperature problem on the basis of equation (2) with boundary conditions of the third kind enabled calculation of the temperature distribution over the volume of the blade. The predetermined temperature field and that obtained as a result of the solution of the heat conduction problem were compared and averaged. The final temperature field for one of the operating modes is shown in Fig. 1.30 in the longitudinal and cross sections of the blade (Vorobiov et al. 2015).

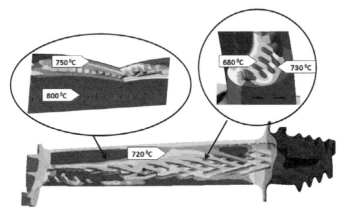

Fig. 1.30 Temperature distribution in different sections of the blade.

It can be clearly seen that the largest temperature gradients are observed in the cross sections, which significantly exceed those over the length of the blade. The knowledge of the temperature field makes it possible to determine the temperature changes in the blade. The longitudinal change in the size of the blade corresponds to a given value of the gaps in the labyrinth seal and is practically independent of the orientation of the crystallographic axes.

1.2.3   Analysis of the thermal stress of the blade

General equations of thermoelasticity for a homogeneous body in the $x\,y\,z$ coordinate system have the form:

$$\frac{\partial \sigma_x}{\partial x} + \frac{\partial \tau_{xy}}{\partial y} + \frac{\partial \tau_{xz}}{\partial z} + X = \frac{\alpha E}{1-2v}\frac{\partial T}{\partial x},$$

$$\frac{\partial \tau_{yx}}{\partial x} + \frac{\partial \sigma_y}{\partial y} + \frac{\partial \tau_{yz}}{\partial z} + Y = \frac{\alpha E}{1-2v}\frac{\partial T}{\partial y}, \tag{3}$$

$$\frac{\partial \tau_{zx}}{\partial x} + \frac{\partial \tau_{zy}}{\partial y} + \frac{\partial \sigma_z}{\partial z} + Z = \frac{\alpha E}{1-2v}\frac{\partial T}{\partial z}.$$

where σ_x, σ_y, σ_z are the components of normal stresses, τ_{xy}, τ_{xz}, τ_{yz} are the tangential stress components, α is the coefficient of thermal expansion, X, Y, Z are the components of mass forces, and E, v are the elastic constants, which depend on the temperature and orientation of the crystallographic axes.

For an anisotropic material, one can write the components of the temperature deformation tensor in vector form:

$$\{\varepsilon^{th}\} = \Delta T \{a_x, a_y, a_z, 0, 0, 0\}^T, \tag{4}$$

where a_x, a_y, a_z are the coefficients of thermal expansion in the corresponding directions and ΔT is the difference between the current and initial temperatures.

For the production of modern cooled blades, high-temperature single-crystal alloys that have anisotropic properties are used. The structure of a single crystal of such alloys is a face-centered cubic lattice, with three independent constants (Shalin et al. 1997). In general, the compliance matrix for a single-crystal blade is completely filled (Shalin et al. 1997). In this case, the crystallographic axes [100], [010], and [001] coincide with the axes of the blade x, y, z.

For anisotropic materials, characterized by nine independent constants (more general in comparison with a single-crystal alloy with a face-centered cubic lattice), compliance matrix S has the form

$$S = K^{-1} = \begin{bmatrix} s_{11} & s_{12} & s_{13} & 0 & 0 & 0 \\ s_{12} & s_{22} & s_{12} & 0 & 0 & 0 \\ s_{13} & s_{12} & s_{33} & 0 & 0 & 0 \\ 0 & 0 & 0 & s_{44} & 0 & 0 \\ 0 & 0 & 0 & 0 & s_{55} & 0 \\ 0 & 0 & 0 & 0 & 0 & s_{66} \end{bmatrix}, \tag{5}$$

which can be rewritten using technical elastic constants as

$$S = K^{-1} = \begin{bmatrix} \dfrac{1}{E_x} & \dfrac{-v_{xy}}{E_x} & \dfrac{-v_{xz}}{E_x} & 0 & 0 & 0 \\[2mm] \dfrac{-v_{yx}}{E_y} & \dfrac{1}{E_y} & \dfrac{-v_{yz}}{E_y} & 0 & 0 & 0 \\[2mm] \dfrac{-v_{zx}}{E_z} & \dfrac{-v_{zy}}{E_z} & \dfrac{1}{E_z} & 0 & 0 & 0 \\[2mm] 0 & 0 & 0 & \dfrac{1}{G_{xy}} & 0 & 0 \\[2mm] 0 & 0 & 0 & 0 & \dfrac{1}{G_{yz}} & 0 \\[2mm] 0 & 0 & 0 & 0 & 0 & \dfrac{1}{G_{xz}} \end{bmatrix}, \tag{6}$$

for which the equalities hold:

$$\frac{v_{xy}}{E_x} = \frac{v_{yx}}{E_y}; \frac{v_{zx}}{E_z} = \frac{v_{xz}}{E_x}; \frac{v_{zy}}{E_z} = \frac{v_{yz}}{E_y},$$

where E_x, E_y, E_z are the modules of elasticity under tension or compression in the direction of the axis indicated in the index, G_{xy}, G_{yz}, G_{xz} are the shear modules (double indices correspond to the axes between which a change of right angle occurs during the shift), and v_{xy}, v_{yz}, v_{xz} are the the coefficients of transverse deformation in the direction of the second of the axes indicated in the index, under the action of normal stresses in the direction of the first axis. These constants can be used, when taking into account the change in the mechanical properties of the material along the volume of the blade due to deformation of the material. This leads to additional iterations in the calculation.

Let us now consider the structure of a single-crystal cooled blade having a face-centered cubic lattice. This, as already noted, is a particular case of an anisotropic material, for which any three mutually orthogonal directions are equivalent. Then the compliance matrix (5) takes the form, where S is the compliance matrix, K is the stiffness matrix, s_{ij} are the coefficients of the compliance matrix, E is the tensile module, G is the shear module, v is the coefficient of transverse strain, and E, G, v are three independent constants.

$$S = K^{-1} = \begin{bmatrix} s_{11} & s_{12} & s_{12} & 0 & 0 & 0 \\ s_{12} & s_{11} & s_{12} & 0 & 0 & 0 \\ s_{12} & s_{12} & s_{11} & 0 & 0 & 0 \\ 0 & 0 & 0 & s_{44} & 0 & 0 \\ 0 & 0 & 0 & 0 & s_{44} & 0 \\ 0 & 0 & 0 & 0 & 0 & s_{44} \end{bmatrix}$$

$$S_{11} = \frac{1}{E}, \ S_{12} = -\frac{v}{E}, \ S_{44} = G$$

The nature of compliance matrices for a given problem, accounting for the properties of a single-crystal material, is described by Shalin et al. (1997) and Vorobiov et al. (2016a).

Dependences of the coefficients of the compliance matrix (8) on temperature T for a single-crystal blade may be represented in the form (9), where h_{ij} are the coefficients defined for a particular material.

$$s_{11} = h_{11} + h_{12}T + h_{13}T^2,$$
$$s_{12} = h_{21} + h_{22}T + h_{23}T^2, \tag{9}$$
$$s_{44} = h_{31} + h_{32}T + h_{33}T^2.$$

The solution of the problem of thermo-elasticity makes it possible to analyze the stress-strain state of the blade and compare the results with the data obtained only under the influence of centrifugal forces. The equivalent stresses are used for the analysis of stress-strain state.

Zhondkovski et al. (2013) and Vorobiov et al. (2016a) showed that centrifugal forces cause the maximum stresses of 10 MPa, while the greatest thermoelastic stresses are of the order of 500 MPa. There is a great heterogeneity in the distribution of stresses and manifestation of their localization, especially on the surface of the cooled channels and the holes for the exit of the cooling air. Figure 1.31 shows the characteristic distribution of the stresses on the blade surface and in the characteristic longitudinal and cross sections. The figure shows a general decrease in the equivalent thermoelastic stresses from the root to the periphery and significant changes in the stress fields in each section. In addition, the stresses localized near the exit holes for the cooling air. This is because of the presence of the cooling air outlets, although it significantly reduces temperatures at the exit edge of the blade, causing significant temperature gradients and a local stress increase.

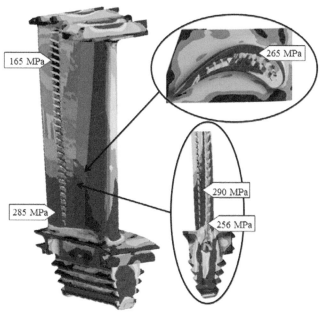

Fig. 1.31 Distribution of equivalent stresses on the blade surface and in the section of the working blade under the combined action of centrifugal forces and the temperature field.

The order of stresses from centrifugal forces does not change. In the initial position, the crystallographic axis [001] is aligned with the longitudinal axis z of the blade and the orthogonal axes x and y are aligned with the crystallographic axes [100] and [010], respectively. The change in the azimuthal orientation means rotation of the [100] and [010] crystallographic axes around the z axis, and the change in the axial orientation means deviation of the [001] axis from the z axis (in any direction).

It is known that the azimuthal orientation of the crystallographic axes significantly affects the distribution of stress fields (Vorobiov et al. 2015, 2016a, 2016b). For example, the effect of rotation of the crystallographic axes in the xy plane about the z axis corresponding to the crystallographic axis [001], which is determined by the angle φ, is shown in Fig. 1.31.

In the cross sections of the blade, there are large gradients of equivalent stresses and their localizations on the surface of the cooling channels. The distribution of thermoelastic stresses in the most stressed cross-section of the blade was investigated, at a distance of one-third of the length from the root at different positions of the crystallographic axes.

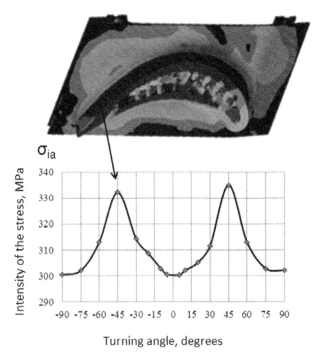

Fig. 1.32 Variation of the equivalent stress σ_{ia} at one point of the blade section, when the crystallographic axes [100] and [010] rotate around the axis z.

It is interesting to analyze the changes of equivalent stresses at certain characteristic points of the cross section of the blade, when the crystallographic axes around the z axis are rotated. Such points are chosen near the input and output edges of the blade section (Fig. 1.32); the point (a) lies near the cooling air outlet, where the stress concentration is observed. Figure 1.32 shows the dependence of stress at the point (a) on the azimuthal orientation of the crystallographic axes. The maximum stresses are seen at $\varphi = 45°$ and the dependence of the stress change on rotation of the crystallographic axes around the z axis is periodically repeated through 90° (Vorobiov et al. 2016a).

From the data presented, it can be seen that the greatest stress gradients in the cross sections exceed the stress gradients in the longitudinal direction. Multivariate calculations of the changes in the thermoelastic stress-strain state of the blade in different sections and on the blade surface were carried out, with changes in the orientation of the crystallographic axes. The results give an idea of the general distribution (quite complex) of stresses over the volume of the blade and their variations, depending on the orientation of the crystallographic axes.

With the vortex cooling system, the channels for cooling air are located at an angle to the longitudinal axis of the blade and the nature of stress distribution varies between its sections. When the orientation of the crystallographic axes is changed, all stresses redistribute and the entire picture of the stress-strain state of the blade changes. Since the maximum equivalent stresses can be in a new region of the blade, they also depend on the geometric features of that region and its temperature. Therefore, it is difficult to display them with simple graphical representations used the case of the influence of the orientation of the crystallographic axes on the natural frequencies of blades vibrations (Vorobiov et al. 2014b). However, general patterns may be identified. It would be useful to reveal some dependences reflecting the change in the greatest equivalent stresses, independent of their location, when the orientation of the crystallographic axes changes. Table 1.56 summarizes the changes in the maximum stresses when the azimuthal orientation varies between 0 and 45°. Note that the largest stresses in the volume of the blade depend on a large number of factors and therefore vary in a certain range. When the crystallographic axes [100] and [010] rotate around the [001] axis aligned with the z axis, the greatest stress intensities change with a period $\varphi = 90°$. When the crystallographic axes [100], [010], and [001] are aligned with the x, y, and z axes, the maximum intensities of thermoelastic stresses on the surfaces of the cooling channels reach 260-300 MPa. When the crystallographic axes rotate clockwise around the z axis, the maximum stresses tend to significantly increase to 450-500 MPa, whereas the minimum thermoelastic stresses decrease. When the crystallographic axes are rotated counterclockwise, the opposite changes occur, with the maximum thermoelastic stresses decreasing to 240 MPa.

Table 1.56 Variations of the maximum stresses with changes in the azimuthal orientation

Turning angle, φ	σ_{max}, MPa
45°	350-380
30°	425-435
20°	450-500
15°	350-360
10°	300-320
0°	260-280

These results may have deviations, given the complexity of the geometric shapes of the blade, the temperature fields, and the heterogeneity of the material. The general picture of the stress-strain state is cyclically repeated when the crystallographic axes rotate by $\varphi = 90°$, as shown by Maggerramova and Vasiliev (2011) and Pridorozhny et al. (2013) for the blades with a system of direct cooling channels.

To assess the orientation effects, the stressed-deformed states of the blade were studied, when the [001] crystallographic axis was rotated around the x and y axes. Table 1.57 shows the changes in the maximum thermoelastic stresses, as a function of the rotation angle. When the axial orientation of the crystallographic axes changes, the temperature stresses can increase by 80%, whereas when the azimuthal orientation changes, the stresses do not increase by more than 42% (Table 1.57).

Table 1.57 Changes of the maximum thermoelastic stresses σ^x_{max} and σ^y_{max}, when the [001] crystallographic axis rotates around the x and y axis, respectively

Turning angle	σ^x_{max}, MPa	σ^y_{max}, MPa
40°	356	254
30°	351	265
20°	341	275
10°	272	253
0°	250	250
–10°	271	245
–20°	342	234
–30°	351	246
–40°	357	255

Taking into account the effects of temperature fields and centrifugal forces, the largest stress intensities do not exceed 500 MPa. This is well below the static strength limit for single-crystal materials at 800°C (Shalin et al. 1997). In this case, the maximum stress intensities are in the places of the maximum temperature gradients. As a rule, this corresponds to the locations of cooling channels and cooling air outlets on the surface. Localization of thermoelastic stresses often coincide with the localization of vibrational stresses (Vorobiov et al. 2014b), which creates additional dangers.

Therefore, in order to evaluate the strength of the cooled single-crystal blades, it is necessary to perform a complex analysis of their dynamic stressed state under vibrations and a static thermoelastic state. It is advisable to limit the increase in thermoelastic stresses by 320-350 MPa (30%), which corresponds to the rotation of the crystallo-graphic axes by 12° ± 15° (Nozhnitsky and Golubovsky 2006). In this case, the equivalent stresses at the characteristic points near the entrance edge have the smallest values, and near the output edge the greatest. The same restrictions on the rotation of the crystallographic axes are recommended for blades of other designs (Maggerramova and Vasiliev 2011), and also correspond to the conditions for the spread of natural frequencies of the blades by no more than 8-10% (Pridorozhny et al. 2013, Pridorozhny et al. 2006, 2008, Melnikova et al. 2001).

1.2.4 Analysis of vibrations of a cooled single-crystal blade

In the cooled blades, the localization of vibrational stresses often occurs on the inner surface of the cooling channels, which presents a great danger. For such blades, it is important to analyze the spectra of their natural frequencies, vibration modes, and distribution of vibrational stresses, especially stress localization. Without a detailed analysis of the vibrational characteristics of single-crystal blades and the distribution of relative stresses, taking into account the geometric shape, temperature fields, centrifugal forces, and orientation of a single-crystal material, it is impossible to draw substantiated conclusions about the choice of rational design parameters.

The use of single-crystal materials with anisotropic properties leads to changes in the natural frequencies of the blades. To assess this effect, the frequency spectra of the blades of single-crystal and polycrystalline material were studied. The modulus of elasticity of a polycrystalline material was assumed to be equal to that of a single-crystal material, when the material [001] axis was oriented along the z direction. The spectra of the natural vibration frequencies of a blade made of polycrystalline and single crystal materials are compared in Table 1.58 that shows the relative differences of the natural frequencies for these materials (Vorobiov et al. 2014b). One can see an increase in the natural frequencies of the blade made of a single-crystal material, which is to be expected, based on the character of the anisotropy of the material with a face-centered cubic lattice (Fig. 1.31).

Anisotropy of the material has the greatest influence predominantly on the torsional frequencies. Peculiarities of the mechanical properties of a single-crystal material make it clear that the deviation of the crystallographic axes from the initial state must significantly affect the frequency characteristics of the blades.

Table 1.58 Comparison of the natural frequency spectra of blades made of polycrystalline and single-crystal materials

No frequency	Values of natural frequencies of blades made of alloys, Hz		The relative frequency difference, %
	Polycrystalline	Single-crystal	
1	576	609	5.8
2	1326	1478	11.3
3	1943	2916	50
4	2940	3261	10.9
5	4881	6477	32.7
6	6067	7510	23.8
7	6733	7817	16.1
8	7560	8646	14.4
9	8524	11152	30.8
10	8934	11319	26.7

To determine the effects of deviations of crystallographic axes on the vibrational parameters, a series of calculations of the natural vibrations was carried out. All calculations and results can be ordered by considering rotations around the three axes x, y, z. Variations of the crystallographic orientation in numerical calculations were done from $-45°$ to $+45°$ with increments of $5°$. Similar investigations were presented by Vorobiov et al. (2014b), however, with cooling channels located along the longitudinal axis of the blade. A vortex cooling system has not yet been considered in solving this problem.

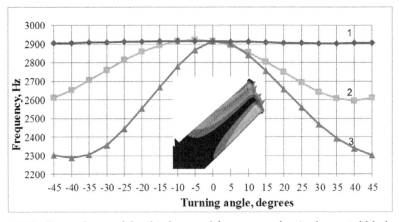

Fig. 1.33 Dependence of the third natural frequency of a single-crystal blade on the rotation angle of the crystallographic axes relative to the axes z (1) x (2) y (3).

The effects of deflection of orientation of the crystallographic axes of a single-crystal material from the original one, coinciding with the global coordinate system of the blade for the first 10 natural frequencies, was studied. For example, Fig. 1.33 shows the influence of the angles of deflection of the crystallographic axes on the third natural frequency. It can be seen that the deviations of the axial orientation of the crystallographic axes have the greatest effect on the third natural frequency. The deviations of the azimuthal orientation of the crystallographic axes exert considerably less influence on the third frequency. The same trend is observed for a number of natural frequencies (Table 1.59).

Table 1.59 Variations of the natural frequencies (ω_k^x and ω_k^y) with changes in the axial orientation of the crystallographic axes

Turning angle	ω_1^x, Hz	ω_1^y, Hz	ω_5^x, Hz	ω_5^y, Hz	ω_7^x, Hz	ω_7^y, Hz
45°	764	750	6651	5801	8951	8297
35°	742	741	6780	5710	8802	8151
25°	671	668	6790	6051	8553	8052
15°	632	624	6701	6250	8151	7950
5°	620	628	6502	6412	7815	7821
0°	615	615	6450	6450	7810	7810
–5°	625	626	6410	6470	7901	7911
–15°	645	623	6481	6280	8220	8212
–25°	685	671	6501	6112	8651	8553
–35°	746	741	6602	5821	8902	8490
–45°	763	752	6651	5802	8951	8300

The frequencies in Table 1.59 are denoted as ω_k^x, ω_k^y, where the subscript indicates the frequency number, and the superscript denotes the axis around which the crystallographic axes are rotated.

Changes in the azimuthal orientation have a noticeable effect on the natural frequencies for only some forms of oscillations. For the blade, such frequencies are the sixth, ninth, and tenth. The influence of the axial and azimuthal orientations on the natural frequencies for the sixth and ninth waveforms is given in Table 1.60.

The first two forms of vibration of the cooled blade are predominantly the bending ones. From the data in Table 1.60, it is evident that for the first form of oscillations, the deviation of the crystallographic axes of the material from the axes of the blade, regardless of direction, leads to an increase in the natural oscillation frequency of the blade. A similar change in frequency also occurs for the fourth form, which is predominantly the bending one with one nodal line.

Table 1.60 Influence of the axial and azimuthal orientation of the crystallographic axes on the sixth and ninth natural frequencies.

Turning angle	ω_6^x, Hz	ω_6^y, Hz	ω_6^z, Hz	ω_9^x, Hz	ω_9^y, Hz	ω_9^z, Hz
45°	7825	7825	7501	11215	10650	10770
35°	7730	7850	7490	11170	10840	10710
25°	7605	7750	7470	11160	11010	10770
15°	7540	7610	7450	11090	11060	10870
5°	7470	7510	7471	11060	11150	11010
0°	7480	7480	7480	11055	11055	11055
−5°	7490	7410	7490	11055	10910	11105
−15°	7580	7420	7505	11080	10640	11150
−25°	7700	7550	7515	11120	10420	11070
−35°	7800	7720	7506	11180	10440	10900
−45°	7825	7825	7500	11205	10648	10775

The third form of oscillation is predominantly the first torsional one. In this case, deviations of the crystallographic axes from the blade axes lead to a decrease in the natural frequencies (see Fig. 1.33). For the fifth (predominantly the second torsional) form of oscillations, an increase in the natural frequency is observed, with the rotation of the crystallographic axes around the x axis and its decrease with rotation around the y axis. An increase in frequencies is also observed with other forms of oscillations. However, the dependences for different axes may have a different character and the maxima and minima may not coincide.

The data obtained for the ninth waveform of the blade show a significant effect on its rotation frequency around the z direction, while the effect of rotation relative to the x axis is much weaker. It must also be noted that, beginning with the seventh form of oscillations, these dependences acquire a complex character. The largest deviations of the natural frequencies of the blade when rotating the crystallographic axes around different directions are given in Table 1.61.

The greatest influence on the natural frequencies is caused by the deviations of the orientation of the crystallographic axes. Moreover, the effect is maximized, when the crystallographic axes around the y-axis turn (reaching 27%), whereas with a rotation around the x axis, the maximum frequency change reaches 26%.

The deviations of the azimuthal orientation of the crystallographic axes have a much smaller effect on the natural frequencies of the blade. These results are compatible with the data obtained by Pridorozhny et al. (2006, 2008) and Melnikova et al. (2001) for single-crystal blades with other cooling systems. From the results of such studies, one can make conclusions for manufacturing single-crystal blades.

Table 1.61 The greatest relative deviations of natural frequencies, depending on variations in the orientation of the crystallographic axes relative to the axes x, y, z

Natural frequency no.	The magnitude of the largest relative deviation of the frequencies when rotating around directions x, y, z, %		
	x	y	z
1	26	23	0.2
2	21	12	0.01
3	13	27	0.4
4	17	19	0.2
5	5.6	11	1.0
6	4.7	5.8	1.1
7	15	9	0.2
8	17	16	1.6
9	1.4	7.1	3.5
10	1.6	5.6	0.8

To limit the spread of their natural frequencies to less than 10% (and taking into account other technological deviations), it is necessary to ensure the axial orientation of the crystallographic axes to within 10°-15° (Nozhnitsky and Golubovsky 2006, Vorobiov et al. 2014b). The crystallographic axis [001] should be in a cone with an elliptical section parallel to the xy plane, the angles of deviation of this axis varying from 10° to 15°. The limits on the deviations of the azimuthal orientation are less severe (Shalin et al. 1997, Hino et al. 2000).

For rigid cooled blades, an increase in temperature causes a decrease in natural frequencies by more than 10%. The increase in the natural frequencies of such blades under the action of centrifugal forces is noticeably lower and the cooling frequencies of the cooled blades are observed at operating conditions. As a result of numerical studies, it was found that the natural frequencies of the blade (ω_T), heated to operating temperatures (hot), are much lower than the natural frequencies (ω_0) of the blade at a constant temperature of 20°C (cold). The changes of the natural frequencies of the blade as a function of temperature, with and without the account of the rotor rotation, amounted to an average of 10% for different frequencies (Table 1.62).

The temperature field reduces oscillation frequencies for a non-rotating blade by 10% or more. Practically the same effect of the temperature field was also noted for the rotating blade, when the frequencies are reduced by 8-10%.

The lowest influence of temperature is observed on the first, ninth, and tenth forms of oscillations. At the same time, the difference in frequencies for a cold blade without rotation and for a hot blade with rotation (ω_T) is approximately 4-11%, which should be taken into account when analyzing the resonance modes of the blades without the bench and in the conditions of operation of the gas turbine engine.

Table 1.62 The natural frequencies of the blade at 20°C and in the operating mode (the rotor rotation with Ω = 7790 rpm taken into account)

Frequency no.	Frequency at 20°C, Hz	Frequency at operating temperatures, Hz	The magnitude of the relative difference between the natural frequencies of oscillations, %
1	688.66	633.15	8.06
2	1512.4	1355.1	10.40
3	2932.2	2625.6	10.46
4	3341.3	3000.3	10.21
5	6526.2	5843.5	10.46
6	7598.4	6799.5	10.51
7	7824	6975.5	10.84
8	8677	7779.7	10.34
9	11232	10135	9.77

Of great interest are the results of the analysis of vibration modes and stress distribution. The complexity of the geometric shape of the cooled blades also causes the complexity of the forms of their oscillations (Vorobiov et al. 2014b). However, the most interesting is an analysis of the distribution of the relative equivalent stresses. Examples of the location of the localization zone of relative stress intensities are shown in Figs. 1.34 and 1.35 (Vorobiov et al. 2014b). The results of numerical studies show that the zones of localization of relative stresses are not only on the surface of the blades, but also on the inner surface of the cooling channels and on the holes for the cooling air exit (Figs. 1.34 and 1.35). These circumstances should be taken into account, because of possible damage in the areas inaccessible to visual inspection and strain gauging. Figures 1.34 and 1.35 also display examples of the appearance of cracks and damage in the blades during their testing on laboratory stands (Vorobiov et al. 2014b). It can be seen that the zones

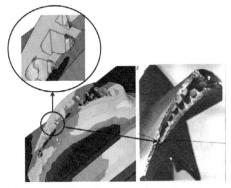

Fig. 1.34 Calculated zones of localization of relative stresses of oscillations of a cooled blade and destruction of a cooled blade in laboratory tests.

of localization of vibrational stresses, obtained in numerical analysis, are the places of crack initiation. The numerical studies of the distribution and localization of vibrational stresses explain the nature and location of damage and destruction of the blades, found during laboratory tests of blades in production (Figs. 1.34 and 1.35).

Nozhnitsky and Golubovsky (2006) and Melnikova et al. (2001) noted that in the single-crystal blades, the propagation of the main crack corresponds to the arrangement of the crystallographic axes. However, it is also known that with the main trends observed, the specific spread of a number of cracks is more complex, as is seen in Fig. 1.35.

In general, the thermal state of the cooled blade depends on the gas temperature, the cooling air, and the design of the blade. It affects the length of the blade, which should be taken into account when determining the gaps in the labyrinth seals.

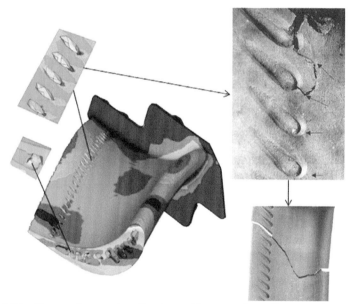

Fig. 1.35   Zones of stress localization with natural oscillations of the blade and the appearance of cracks in these zones and destruction of the blade.

The main contribution to the stress-strain state of the cooled blade is made by thermoelastic stresses. Voltages from centrifugal forces reach significantly lower values when turning crystallographic axes. However, the total maximum stress intensities are significantly lower than the static strength limit for single-crystal materials.

The control of stresses on the surface of the cooled channels is difficult. But systematic calculations of the stressed-deformed state of the blade with allowance for various factors make it possible to relate the stresses on the surface of the blade with the stresses on the surface of the cooled channels, which enables estimation of the magnitude of the latter. This makes it possible to monitor and diagnose the maximum stresses.

The azimuthal orientation of the crystallographic axes exerts a significant influence on the entire picture of the stress-deformed state of the blade. The overall picture of the stress-strain state is cyclically repeated when the crystallographic axes are rotated by 90°. At the same time, the maximum stresses significantly change up to 80%, therefore some limitations on the deviation of the crystallographic axes are required. When manufacturing single-crystal blades, one should strive that the rotation of the crystallographic axes in the xy plane does not exceed 12-15°. The change in the axial orientation of the crystallographic axes causes an increase in the thermal stresses at 42°. It is necessary to strive for smoothing the surface of the angular shapes in the cooling channels and the holes for the exit of the cooling air.

The regularities of the influence of the crystallographic axes of blade material on their natural frequencies with respect to the x, y, z axes over a wide range of deflection angles are revealed. The change in the axial orientation (deviation from the z-axis direction) has the greatest effect on the natural frequencies. The change in the azimuthal orientation has some effect on the higher frequencies (especially starting with the ninth) of the natural vibrations. However, the change in the azimuthal orientation greatly affects the stress-strain state of the blade, caused by the temperature field. General restrictions on the change in crystallographic axes with allowance for the thermoelastic and vibrational stresses, and the spectrum of natural frequencies are schematically shown in Fig. 1.36 (Vorobiov et al. 2017).

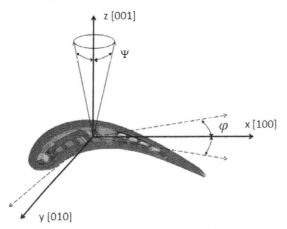

Fig. 1.36 Scheme of restrictions on changing the orientation of the crystallographic axes of the cooled blade. The crystallographic axis [001] should not extend beyond the cone. The angle ψ is 14°-15°. The angle φ varies from 12° to 15°.

The influence of the crystallographic orientation on the vibrational characteristics depends on the shape of the oscillations. For predominantly bending shapes, the deviation of the crystallographic axes causes the greatest increase in the natural frequencies, and for predominantly torsional axes, their decrease is possible.

The zones of stress localization that correspond to different forms of oscillations and specific configuration of the structure are identified:

– Surface of the cooling channels inside the structure. They are dangerous in that the nucleation and growth of a crack is problematic to determine by visual inspection or strain gauging;
– Holes for the cooling air outlets;
– Shelves, legs, and shanks.

The distribution of the intensities of dynamic stresses depends on the oscillation modes. Predominantly the bending vibration patterns are (as a rule) accompanied by the localization of stresses at the output and input edges, in the shank, and around the outlets of the cooling channels. In the predominantly torsional oscillation modes of the blade, the maximum stresses may appear on the back.

Temperature fields reduce oscillation frequencies for non-rotating blades by 10% or more. For the rotating blades, the frequencies are reduced by 8-10% due to the temperature effects. The least influence of temperatures is observed on the first, ninth, and tenth forms of oscillations.

The frequency difference for a cold blade without rotation and a hot blade with rotation is approximately 4 to 11%, which should be taken into account, when analyzing the resonant operating conditions of the blades at the stand and in the operational conditions of the gas turbine engine.

The temperature fields have a little effect on the shape of the oscillations and a somewhat higher effect on the stress intensity distributions. An absolute increase in the stress level under the influence of temperature fields at the same intensity of external loads is to be expected. Localization zones of the vibrational stresses in the presence of explicit concentrators remain in the same places as in the cold blades, but they can be displaced on the internal channels. Even in the analysis of free vibrations, the zones of localization of vibrational stresses are clearly visible, which (as experience shows) are practically preserved under forced oscillations. Therefore, the analysis of free oscillations is very effective in the design of blades to identify hazardous areas and the selection of rational options for the design of the cooling system and of the blade as a whole.

When analyzing the distribution of stresses on the natural blades, special attention should be paid to the localization of stresses on the internal surface of the cooling channels, since the occurrence of possible cracks or other damages cannot be detected by a visual inspection of the blades.

To ensure the requirements for the permitted spread of the natural frequencies of single-crystal blades during their manufacture, it is necessary to control the angles of deviation of the axial orientation of the crystallographic axes.

Temperatures affect the natural frequencies of the rigid cooled blades more than the centrifugal forces, and in the operating modes, the natural frequencies of such blades are reduced.

It is necessary to take into account the danger of occurrence of zones of localization and increase of relative vibrational stresses on the internal surface of the cooling channels and cooling air outlets, which leads to damage in places inaccessible for visual inspection.

Trunk cracks in monocrystalline vanes extend mainly along the crystallographic axes, although the propagation of specific cracks may be more complex.

Analysis of the singularities of vibrations and thermoelastic stresses of single-crystal blades is quite complex and requires detailed studies.

The final requirements for the manufacture of specific cooled blades can be identified with the joint work of representatives of scientific organizations and industrial enterprises (Nozhnitsky and Golubovsky 2006, Caron et al. 2011, Maclachlan and Knowles 2002, Vorobiov et al. 2014b).

CONCLUSION AND FUTURE DEVELOPMENTS OF THE NICKEL HEAT-RESISTANT INTERMETALLIC ALLOYS

Nickel superalloys are a unique class of materials resistant to high temperatures. For such alloys, many key points of structure formation are understood and a high-strength condition is attained on the basis of a perfectly balanced complex of hardening mechanisms. Possibilities for pushing up the operating temperature limits for the nickel superalloys are almost exhausted. The immediate goal of the world nickel superalloy metallurgy is the design of alloys based on nickel intermetallic compounds with a single-crystal structure, operating at temperatures of up to 1300°C, and having an increased resistance to thermal fatigue.

Different low-alloying nickel superalloys are showing the promising mechanical and physical properties. The directionally solidified Ni_3Al-based IC6 superalloy (USA) has been developed for advanced jet-engine turbine blades and vanes operating within the temperature range of

1050-1100°C. This superalloy contains only aluminum, molybdenum, and boron as alloying elements. Stress-rupture strength at 1100°C/100 hours of IC6 is 100 MPa, i.e., approximately by 20 MPa higher than the similar Russian VKNA-1Y superalloy and American EX-7 superalloy (Jozwik et al. 2015). The IC-50 superalloys (USA) contain only aluminum (nearly 11%), zirconium, and boron as alloying elements. For most intermetallic-based alloys, such as IC-221M, IC-396, and IC-438, only zirconium, chromium, and molybdenum are included in their composition, along with aluminum. All of these alloys are also microalloyed with hafnium and boron (Jozwik et al. 2015, Li et al. 2012).

Among the directions of further work, it is important to further improve strength at a given temperature (Erofeev et al., 2016). In Russia, intermetallic alloys of the VKNA type are alloyed more economically, compared with high-temperature nickel superalloys. Carbon-free alloys of the VKNA type are suitable for single-crystal castings, alloys with 0.12-0.18% of carbon in composition are designed for accurate casting of details with the equiaxed structure. In the creation of the Russian intermetallic alloys, there is a process of elimination of boron from the composition of the alloying elements. It is believed that the introduction of boron leads to a decrease of the ductility of the alloy in the temperature range of 300-850°C, because of the formation of the boron eutectics with a low melting temperature. Russian alloys are more complex in chemical composition and they contain some amounts of heavy refractory elements (W, Re, Ta), that brings their composition closer to the modern nickel-based superalloys (Bazyleva et al. 2014). The introduction of such elements in the alloy composition contributes to the increasing short-term strength and creep resistance at temperatures above 1000°C.

Alloying with rhenium and ruthenium allows one to achieve a significant hardening of the superalloy (Petrushin et al. 2010, Kablov et al. 2015). Simultaneous presence of these metals in the alloy composition allows one not only to improve the mechanical properties, but also to avoid the development of undesirable TCP phases. Alloying with iridium or hafnium also has interesting results (Yokokawa et al. 2016). Nickel superalloys of the fifth generation, such as the Russian VZhM-6 and Japanese TMS-162 alloys, have a very high limit of the long-term strength: 130 MPa at 1100°C on the basis of 1000 hours. The new Russian superalloy VZhM-8 shows high values of long-lasting strength, such as 140 MPa at 1100°C for 1000 hours and 150 MPa at 1150°C for 100 hours.

On the other hand, the increase in the number of heavy refractory metals in nickel superalloys quickly leads to an increase in its specific gravity. The density of VZhM-6, TMS-162, and VZhM-8 superalloys are 8.9, 9.04, and 9.06 g/cm^3, respectively. Currently, the most important parameter for the development of new air-space alloys is the reduction in the density of blade material without loss of its strength. The new nickel superalloy VZhL-20 has a density of 8.4 g/cm^3, while its mechanical properties correspond to the nickel superalloys with a density of 8.7 g/cm^3 (Erofeev et al. 2016). Also important is the technological problem of decreasing the impurity content in the batch (Kablov et al. 2017a, Min et al. 2017).

One of the ways of increasing the heat resistance is to create the so-called natural composites, based on eutectic compositions with the γ/γ'-NbC structure (Petrushin et al. 2016). For example, the nickel superalloys VCLS-20 and VCLS-20R were developed on the basis of such a eutectic. These alloys, obtained by directional crystallization contain the nickel-based solid solution, 60% of the γ'-phase and 3-4% of the ideal whiskers of niobium carbide or tantalum carbide. Directed nickel eutectic has approximately the same melting point (eutectic temperature) as that of the nickel superalloys, but it has unsurpassed record temperature strength due to the combined hardening. Unfortunately, the wide use of such superalloys is hampered by a too long process of structure formation. One single crystal blade made of nickel-base superalloy generally crystallizes at speeds of 1 mm/min, but the preparation of such eutectics with microscopic flat front of growth, ensuring the formation of oriented whiskers along the growth axis, requires an extremely low growth rates of the order of 1 mm/hour. Therefore, for blade industrial production from the nickel eutectic composites, should be installed in foundries several lines with multiple (and very expensive) units, each of which grows only a single blade. For this reason, and also because of the relatively low melting point (~1400°C), nickel eutectic alloys with a directed structure are not considered as alternative construction materials of the future. Replacement of single crystal nickel superalloys with dispersion hardening should be sought to among eutectic alloys based on refractory metals and composite hardening intermetallic compounds (Li and Peng 2007, Zhang et al. 2015, Petrushin et al. 2016, Semenov et al. 2016, Maji et al. 2017). The refractory transition elements IVA, VA, and VIA of the Periodic Table, such as Ti, Zr, Hf, V, Nb, Ta, Cr, Mo, and W, can serve as a matrix of such composites, and silicide of these elements can be considered as intermetallic hardeners. Silicides of the transition elements M_5Si_3 (5:3) have high melting points in the temperature range 2500-2800 K and low density. Among the binary systems, such as the refractory metal-silicon, only three systems exhibit stability between silicide 5:3 and a metal, namely Nb_5Si_3, Re_5Si_3, and W_5Si_3. Nb_5Si_3 is liberated by its combination of the highest melting point and the lowest density. According to theoretical calculations, the weight of the rotor of a promising high-pressure turbine decreases by 20% after replacement of the nickel superalloy blades with lighter Nb-Si composite blades.

A number of European countries (France, Great Britain, Austria, Czech Republic, Italy, Germany) joined their efforts to develop new high-temperature composite materials based on niobium and molybdenum for a promising gas turbine engine with a lower specific fuel consumption (by 20%) than the existing one and smaller emission levels of nitrogen oxides (by 80%) and carbon dioxide (by 20%). Extensive research in the field of high-temperature Nb-Si composites is also carried out now in China, India, and Japan. Precision investment casting method is widely used in industry to produce cooled blades of gas turbine engines. However, for melting of the high-temperature composites, in particular for Nb-Si, the practical implementation of this method is associated with the problem of increasing the inertia and deformation stability of ceramic rods and molds at casting temperatures above 1700°C. A new step in this direction is the directional solidification conditions with variable controlled gradient (Kablov et al. 2017b). This method allows one to obtain the homogeneous fine-dendritic structure (<200 microns) with low porosity and reduced dendritic segregation and leads to high strength properties of superalloys, including the directional solidification of the eutectic superalloys of the "niobium-silica" system. When the Nb-Si binary eutectic is doped with titanium, aluminum, hafnium, and chromium, the viscosity not only increases the strength further, but also increases the fracture toughness of the composite.

The production of blades, made of Nb-Si composites, will allow raising gas turbine operating temperatures up to 1350°C. This is 200°C higher than the temperature capability of modern blades made of the single-crystal nickel superalloys, and this is certainly a revolutionary leap, opening the application of the new class of superalloys.

Chapter Two

Cobalt-based Superalloys

INTRODUCTION

Developments in the area of gas turbines are aimed to increase their efficiency by raising gas stream temperature. For nickel superalloys, this has been achieved by increasing the γ' fraction, the solvus temperature, or the content of refractory elements. However, the potential for improvement of nickel superalloys has been almost exhausted. Many researchers are attempting to develop new superalloys with higher temperature capabilities to replace the conventional nickel-based superalloys.

Cobalt is known to have the highest Curie temperature among all metals. Cobalt (refractory) metal alloys have potential for increasing both Curie temperature and high-temperature strength (Ashbrook et al. 1968). Unlike the nickel superalloys, the cobalt-based superalloys have a good oxygen resistance and they can be cast in air or under argon. The high-temperature cobalt superalloys containing chromium and tungsten have been developed as the magnetic construction materials for electric motors and generators, capable of operating at high temperatures. High magnetic permeability is required to efficiently generate electric power, along with high stress materials, because of the high rotational stresses involved. Cobalt alloys based on Co-W system (e.g., VM-103, VM-108) were suggested as potential materials for stator vane application in advanced turbine engines. Cobalt-based superalloys with high nickel content (e.g., X 40, X 45, and FSX-414) are primarily used in manufacturing of all first-stage nozzles. In some turbines, they are used in the last stage, because of their good weld ability and hot corrosion resistance (Luna Ramírez et al. 2016).

The industrial cobalt-based alloys, unlike the nickel superalloys, are not strengthened by a coherent, ordered precipitate. Because of that, the cobalt-based superalloys are not as widely used at high temperatures as the nickel superalloys. These superalloys consist of the cobalt solid solution (FCC) matrix and strengthening carbides, such as MC (where $M = \{Ti, Zr, W, Cr, Co\}$), M_3C_2, M_7C_3, or $M_{23}C_6$ (Luna Ramírez et al. 2016). Alloying with different elements (Ni, Al, W, Cr, Ti, Fe, Zr, Re) is used to increase the workability, solid-solution strengthening, oxidation, and hot corrosion resistance, also providing the precipitation of the intermetallic phases such as Ni_3Al, Co_3W, Co_2W, Co_7W_6, CoAl, Co_3Ti, and $Co_2(Ta, Nb, Ti)$.

The $Co_3(Al,W)$ intermetallic compound with an $L1_2$ structure was discovered experimentally by Sato et al. (2006) in a Co-Al-W ternary phase diagram. The existence of this strengthening intermetallic phase demonstrates the possibility of obtaining high-temperature cobalt-based superalloys, similar to the nickel-based superalloys. The $Co_3(Al,W)$ intermetallic compound has many attractive mechanical properties, such as high-temperature resistance, high strength, and high Young modulus, as well as good plasticity and casting properties (Suzuki et al. 2007, Tanaka et al. 2007, Miura et al. 2007). A great deal of research efforts has focused on developing this new family of alloys. Alloys based on the intermetallic compound $Co_3(Al,W)$ are considered as the promising high-temperature materials that may be viable candidates in high-temperature applications, particularly in the land-based turbines. The strength properties of the alloys based on the $Co_3(Al,W)$ intermetallic compound are not inferior to the nickel-based superalloys. On the other hand, unlike the nickel-based superalloys, the $Co_3(Al,W)$-based intermetallic alloys show ferromagnetic properties in the range of the expected operating temperatures (Kazantseva et. al. 2017a). This fact opens perspectives for wide application of these new materials as high-temperature magnetic materials and/or materials for recording media.

This chapter describes the main features of the $Co_3(Al,W)$ intermetallic compound and alloys based on this compound, together with their advantages and drawbacks, and suggests their best industrial usage.

2.1 Structure and Mechanical Properties of the $Co_3(Al,W)$-Based Alloys

2.1.1 Effects of alloying with Al or W in the cobalt-based alloys

First of all, we consider the alloying of cobalt with aluminum and tungsten, because these elements have a strong impact on the structure and properties of Co-based superalloys and also take part in formation of the $Co_3(Al,W)$ intermetallic compound.

Features of martensitic transformation in Co-Al alloys. Alloying with Al increases the oxidation resistance and decreases the density of cobalt superalloys. The Co-Al-system alloys attract significant scientific and practical interest as the materials with a Magnetic Shape-Memory Effect (MSME), in which stresses can be controlled by an external magnetic field. Among these alloys are Ni-Mn-Ga (Heusler alloys), Co-Ni-Al, Ni-Fe-Ga, Fe-Pd (Pt). In contrast to the known materials with the Shape-Memory Effect (SME), such as NiTi or CuAlNi, the MSME materials show no effect of thermal inertia (Vasil'ev et. al. 2003). It is assumed that the shape memory effect in Co-Al alloys is due to the presence of a high volume fraction of the martensitic phase. In these materials, the high Curie temperature of the α phase makes it possible to consider them as high-temperature materials with the magnetic shape-memory effect (Omori et al. 2003).

The martensitic $\alpha \to \varepsilon$ transition was found in the alloys with an aluminum content of up to 5.5 wt. % (11.2 at.%) after water quenching from 1100-1300°C. The temperature at the start of the martensitic transformation is reduced with increasing content of aluminum in the alloy. In the alloys with the higher Al content (5.5-9 wt. % or 11-18 at.%), the complete stabilization of the high-temperature α-Co FCC phase is observed after quenching and no martensitic transformation occurs even on strong overcooling (to –196°C) (Nikolin and Shevchenko 1981). It was shown by Greeshenko and Koval (2000) that, on a cyclic loading of a Co-5 wt. % Al, the shape-memory effect was fully preserved (100%). It should be taken into account that the $\alpha(\text{FCC}) \to \varepsilon(\text{HCP})$ phase transition in cobalt occurs only via a martensitic mechanism, and the presence of a martensitic transformation in the cobalt alloys can have a significant effect on the equilibrium transformations.

The phase transition $\alpha \to \varepsilon$ in Co-Al system was studied by Kazantseva et. al. (2016b) in the Co-9 at.% Al alloy. The Co-9 at.% Al alloy was obtained by the Bridgman method, the rate of the directional crystallization was 0.83 mm/min, and the temperature gradient at the crystallization front was 80 K/cm. The ingots were subjected to a homogenizing annealing at 1250°C for 24 hours with a subsequent slow furnace cooling. The thermal X-Ray Diffraction (XRD) analysis was conducted with heating to 950°C, both on heating and cooling (with steps of 50 K). The thermal analysis was carried out in a setup for the synchronous thermal analysis (*STA 449 C Jupiter*) by the method of Differential Scanning Calorimetry (DSC). The heating was performed to a temperature of 1200°C. The rate of heating was 0.3 K/s, while the rate of cooling was 0.8 K/s. The measurements were conducted in the argon atmosphere, with a linear heating and cooling (dynamic regime).

In the X-ray thermogram of the alloy, in the initial state, there are lines of the ε phase and of an unknown phase (supposedly, the

4H phase); no lines of the *B*2 (BCC, CoAl) phase were revealed. On heating to 500°C, in the thermogram of the alloy, there are lines of α-Co (FCC) that do not coincide with the positions characteristic of the HCP phase, and lines of the *B*2 phase, whereas the additional line of the unknown phase (near the (102)ε line) disappears. At 550°C, the lines of the ε phase (HCP) disappear completely. In the thermogram, there are only lines of α-Co and of the *B*2 phase; the latter disappear completely on reaching 850°C. In the temperature range of 850-950°C, only the α-Co lines are present. On cooling of the sample, the single-phase state of α-Co is retained down to 100°C; below 100°C, the ε phase lines appear in the thermogram. At room temperature, the two-phase $\alpha + \varepsilon$ state is revealed in the alloy. The results of the thermo-XRD analysis agree with the data of the usual X-ray diffraction analysis obtained at room temperature before and after heating the sample. Table 2.63 gives the parameters of the crystal lattices of the phases obtained from the data of XRD analysis. The crystal lattice parameter of the α phase are determined only after heating and cooling of the alloy, since in the initial state in the diffraction pattern, only reflections of the ε phase are observed, including those that coincide with the position of the α phase reflection; no separate lines of the α phase are revealed.

Table 2.63 Results of X-ray diffraction analysis

Parameters of the crystal lattices of the phases	α-phase, nm ±0.0005	ε-phase, nm ±0.0003
initial state	-	$a = 0.2521$
after heating to 950°C and cooling	$a = 0.3561$	$a = 0.2562$ $c = 0.4111$

A comparison of the results with the Co-Al phase diagram (McAllster 1989) indicates that the slow cooling of the investigated alloy after the homogenization did not make the equilibrium state correspond to the phase diagram for the alloy of this chemical composition.

It is known that the initial part of the peaks in the DSC thermograms characterizes the beginning of the process, whereas the maximum (minimum) of the peak corresponds to its end. The reproducibility of the peaks on cooling and subsequent heating of a sample indicates the reversibility of the phase transition. In the thermogram of heating of the initial sample, four phase temperature ranges may be distinguished: 280-290, 550-600, 800-900, and 950-1000°C. In the thermogram of the sample cooling, only three temperature ranges are observed: 180-200, 750-800, 880°C-... (Fig.2.37). In the latter range, the process does not complete, which is probably related to the limited temperature range of the device used for measurements (up to 1200°C) (Fig. 2.37).

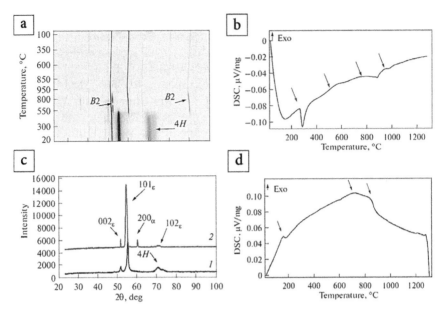

Fig. 2.37 Results of the thermo-XRD analysis (a-b) and differential scanning calorimetry (b, d): (a) total thermo-XRD pattern of heating and cooling; (c) XRD patterns at temperatures (1) 20°C before the heating and (2) after heating to 950°C and cooling; (b) heating and (d) cooling.

Thus, taking into account shifts of the thermogram, caused by a change in the rate of cooling toward the lower temperatures, three phase transitions are reproduced well. This behavior may indicate the irreversibility of the phase transition at temperatures of 550-600°C, which is related, e.g., to the formation of the $B2$ phase from the field of the supersaturated solid solution upon heating. According to the data of thermo-XRD, this phase is absent, when the sample is cooled.

According to the data of the elemental analysis, the chemical composition of the alloy is uniform; no regions enriched or depleted in aluminum are revealed (Table 2.64(a)).

Table 2.64(a) Average chemical composition (SEM)

Element	Wt. %	At.%
Al K	4.4	9.2
Co K	95.6	90.8

The SEM analysis shows that the structure of the alloy consists of coarse grains with thin martensitic plates. In the TEM micrographs, the regions with twins of the ε martensite (2H, HCP) and the regions of the residual α-Co phase with stacking faults are revealed (Fig. 2.38).

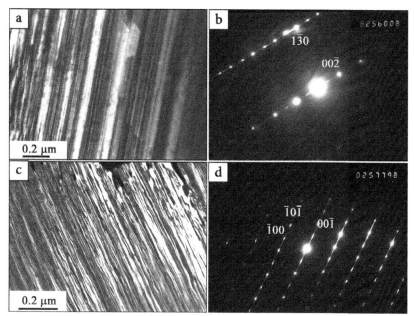

Fig. 2.38 Microstructure of the initial alloy (TEM), the defect region of the HCP (4*H*) phase: (a) the bright-field image, (b) the SAED pattern taken from region (*a*), zone axis [310]4*H*; (c) the dark-field image taken in the ($\bar{1}0\bar{1}$) reflection; (d) the SAED pattern to (*c*), zone axis [010]ε.

In the selected electron diffraction (SAED) patterns taken from defect regions, reflections of twins are present; calculations indicate that the twinning plane can be the ($1\bar{1}2$) plane of the HCP cobalt. Regions of ε martensite with an enhanced content of stacking faults are also revealed. In the SAED patterns obtained from such regions, diffuse streaks in the direction (001) of the ε phase and the same extra reflections are observed. The reflection positions in the diffraction pattern are interpreted as those corresponding to the reflections of the zone axis of the long-period (4*H*) HCP phase. In the dark-field images obtained from such defect regions, a characteristic alternation of regions with different densities of stacking faults can be seen. It is known that the reconstruction of the crystal lattices upon the FCC-HCP phase transition in cobalt occurs via the shifts of atomic layers in each second plane of the FCC lattice with the formation of the hexagonal 2*H* ε phase. In the alloys with a low energy of stacking faults or limited solubility of the alloying elements, the so-called *long-period structures* can form. In the Co-Al system, the formation of long-period phases is observed during quenching of single crystals with an Al content of 4-5.5 wt. % from 1100-1300°C. At smaller contents of aluminum, the usual ε phase is observed in the quenched alloys (Nikolin and Shevchenko 1981). Based on the results of X-ray diffraction analysis, Nikolin and Shevchenko (1981) proposed considering the resultant long-period structures as phases that have a trigonal crystal structure (structure type $R\,m$), only with different numbers of close-packed layers in the unit cells. Using electron-microscopic methods of the direct resolution of the crystal lattice, Dutkiewicz and Kostorz (1991) showed that a long-period HCP structure ($4H$) with the high density of stacking faults is formed in the Co-W alloys, in the interlayers between the imperfect plates of the $2H$ martensite. Thus, the results obtained by Kazantseva et al. (2016b) may indicate that the appearance of a long-period $4H$ martensitic phase is a specific feature of the binary cobalt alloys.

Features of martensitic transformation in Co-W alloys. The same martensitic transformation may be observed in the Co-W system. The polymorphic transformation of the disordered α-phase (FCC (Co,W)) into the disordered ε-phase (HCP (Co,W)) starts at 422°C and occurs by the martensitic way for tungsten contents of up to 8 at.%W (Dutkiewicz and Kostorz 1991). An increase in the tungsten content above 8 at.% lowers the temperature of the martensitic transformation M_s to room temperature. Thus, a slow cooling of the alloys with the tungsten content exceeding 8 at.% can provide the equilibrium structure with grains of the α-phase and plates of the $D0_{19}$ phase. A rapid cooling shifts the phase transitions toward the non-equilibrium transformations with the formation of the martensitic α'-FCC and ε-HCP phases. According to Dutkiewicz and Kostorz (1991), a sample with 14 at.% W, cooled in liquid nitrogen, has a martensitic microstructure. The non-equilibrium phases produce a significant effect on the properties of the cobalt-based alloys. For example, it has been found that the structure with the martensitic HCP phase and fine precipitates of a cobalt-rich ferromagnetic phase dispersed in a cobalt-rich paramagnetic matrix is one of the reasons for the excellent magnetic recording properties of the Co-W-Cr alloy thin films (Sato et al. 2005).

Three kinds of martensite, HCP ε(2H), double HCP ε'(4H), and ε''(NR) had been observed in bulk Co-W alloys, depending on the cooling rates and aging time (Dutkiewicz and Kostorz 1991, Guillermet 1989). The transition between the different kinds of martensite in the Co-W alloys may be very smooth. It has been shown that the martensitic area of the water-quenched (from 1250°C) cobalt alloys, containing up to 9.6 at.% W, consisted of arrays of $2H$ structure separated by narrow layers (up to 1.5 nm) of $3R$ martensite. After aging for 510 s at 600°C, followed by water quenching, narrow lamellae of $4H$ structure were observed within the $2H$ martensite. It was suggested (Dutkiewicz and Kostorz 1991) that such transitions resulted from a reordering of stacking periodicity, leading to a multilayer structure.

It was found that the microstructures, martensitic transformation behavior, and physical and mechanical properties of rapidly solidified alloys are quite different from those of the bulk material. The fast cooling process of the bulk materials always depresses the phase boundaries away from equilibrium and toward lower temperatures (Zhao 1999). For example, in bulk pure Co and Co-W alloys with the increasing of cooling rate, the M_s temperature could be depressed (down to 300 K, Zhao 1999). However, the martensitic transformation temperatures of the Ti50Ni40Pt10 as-spun ribbon were by 100 K higher than those of the bulk material with the same chemical composition. The same situation was found in the rapidly solidified Ti-50 at.% Ni-18 at.% Zr and Ti-50 at.% Ni-18 at.% Hf, which also had M_s by 150-200°C higher than that of the bulk materials with the same chemical composition (Inamura et al. 2006).

According to the theory of phase transformations, the FCC-HCP transition can be also described as the slip of the Shockley partial dislocations, which drag stacking faults through every other FCC close-packed planes (Christian 1951). Fast quenching creates a lot of the defects in the crystal, and one of the ways to decrease the crystal energy is the defect ordering.

Martensitic transformation was studied in Co-10 at.% W alloys by Kazantseva et al. (unpublished work[1]). The Co-10 at.% W alloys were prepared by arc melting from pure Co (mass fraction of 99.98%) and pure W (mass fraction of 99.9%), under Ar atmosphere, followed by single roller melt spinning onto copper wheels at a tangential speed of 15 m/s. The ribbons of 0.03 m in width and 6.0×10^{-5} m in thickness were subsequently subjected to furnace annealing at 900°C and vacuum of 10^{-3} Pa, followed by quenching into a cold water. According to the phase diagram, our samples were annealing above the Curie temperature in the paramagnetic area and they crossed it during quenching. An annealing followed by water quenching is the usual way for studying the phase composition of different areas in the equilibrium phase diagram. According to the diagram, the temperature of 900°C is within the two-phase $D0_{19} + \alpha$ area (Okamoto 2008). Under slow cooling, the α phase transforms into ε-phase, and the equilibrium state of the alloy should be $\varepsilon + D0_{19}$ at room temperature (Okamoto 2008, Guillermet 1989). The X-ray analysis of the initial melt spun sample shows the diffraction lines of the disordered face-centered-cubic (FCC) α-phase and disordered hexagonal (HCP) ε-phase. Some of the diffraction lines of the disordered hexagonal ε-phase can coincide with the lines of the cubic α-phase; usually, the

---

[1] The TEM study was conducted by N.V. Kazantseva during a trainee course at the NIST (USA), thanks to the Thermodynamic and Diffusion Group (Dr. C. Cambell, Dr. E. Lass, Dr. K-W. Moon), who provided the melting and heat treatments of the samples.

presence of the HCP phase in the two-phase FCC-HCP structure can be established by the intensity of the diffraction lines. Microstructure of the samples consists mainly of equiaxed grains, with thin plates of DO_{19}-phase dispersed in the matrix and along the grain boundaries. Grain sizes in the sample after 1 hour of annealing vary between 10 and 50 nm. The plates increase in size with increasing time of annealing, covering all grains. The process of the plate growth is very fast: after just 4 hours, the plate size and thickness increase more than by a factor of 3; the plates form both on the grain boundary and inside the grains. Twins of the α phase passing through the α grains can be seen in the TEM micrographs of the melt spun ribbons in the initial state (Fig. 2.39a). A typical FCC twinning with the {111} twining planes is found. The α grains with intrinsic stacking faults are also visible in the micrographs (Fig. 2.39b).

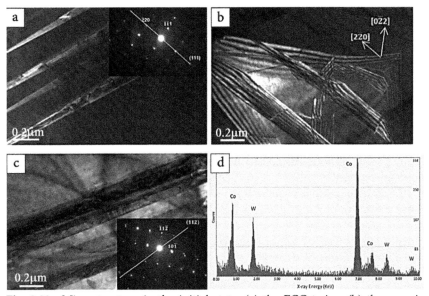

Fig. 2.39 Microstructure in the initial state: (a) the FCC twins; (b) the α-grain with the intrinsic stacking faults; (c) the martensitic lamellae of the ε phase; (d) TEM EDS spectrum from the martensitic area (c).

TEM structure in Figure 2.39c shows the thin martensitic lamellae of the ε phase. The (112) twinning plane of the ε martensite is found as the usual twinning plane for the HCP metals. The orientation relationship between the α and ε phases in the Co-10 at.% W melt-spun ribbon is similar to the crystal orientation of the Shoji – Nishiyama relationship for the martensitic transformations between the α and ε phases in pure cobalt, i.e., $(0001)\varepsilon \parallel (1\bar{1}1)\alpha$ and $[11\bar{2}0]\varepsilon \parallel [1\bar{1}0]\alpha$ (Wu et al. 2005).

According to these orientation relationships, we have calculated matrixes of the zone axis and planes for the α/ε two-phase structures:

$$\begin{pmatrix} h \\ k \\ l \end{pmatrix}\alpha = \begin{pmatrix} \bar{1} & 0 & 1 \\ 0 & 1 & \bar{1} \\ 1 & 1 & 1 \end{pmatrix}\begin{pmatrix} h \\ k \\ l \end{pmatrix}\varepsilon, \quad \begin{pmatrix} u \\ v \\ w \end{pmatrix}\varepsilon = \begin{pmatrix} \bar{1} & 0 & 1 \\ 0 & 1 & 1 \\ 1 & \bar{1} & 1 \end{pmatrix}\begin{pmatrix} u \\ v \\ w \end{pmatrix}\alpha.$$

The presence of the ε martensite in the melt-spun structure shows that the sample crosses the $\alpha \rightarrow \varepsilon$ M_s temperature under rapid cooling conditions. The high quantity of the random stacking faults in our melt-spun ribbons allows us to suggest that the Co-10 at.% W alloy has a low energy of the faults formation, which is close to that for pure cobalt. Figure 2.39 shows the ε-martensite area and the TEM EDS and EELS spectra obtained from the ε-martensitic area. According to the EELS analysis, the martensitic area in the melt-spun ribbon is enriched in W and has 90.48 at.% Co and 9.54 at.% W.

Figure 2.40 shows the TEM results of the melt-spun ribbon after long-time annealing at 900°C, followed by water quenching.

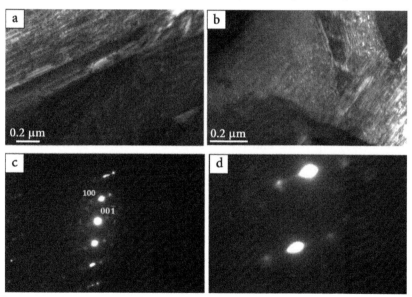

Fig. 2.40 Microstructure of the ε martensite: (a) annealing of 168 hours; (b)-(*d*) annealing of 1000 hours; (c) SAED pattern to (*b*), zone axis [010]ε; (d) diffuse streaks, magnified area.

These samples have no α-twins in the structure. Defective ε-martensite structures are formed in the samples. The latter fact indicates that water-quenching the sample at 900°C does not prevent the low-temperature

$\alpha \rightarrow \varepsilon$ polymorphic phase transformation. The phase composition from the high temperature area (900°C) of the Co-W equilibrium phase diagram is not obtained.

The multiple stacking faults form larger areas in the structure of the samples after the annealing; the reflexes on the diffraction patterns from such areas are calculated as those of the zone axis of the HCP ε-phase. The stacking faults cause the weak diffusion streaks in the (0001) 2*H* direction (Fig. 2.40d).

The large round particles, as well as the thin plates of the ordered $D0_{19}$ phase, were also observed in the sample structure. The diffusion streaks in the (0001) 2*H* direction, which usually assist the stacking faults defects, were observed near the ε-martensite reflexes in the samples after a prolonged annealing time at 900°C, followed by water quenching (Fig. 2.40d). Such a type of the defective martensite structure was described as multilayer 2*H*/3*R* martensite (Dutkiewicz and Kostorz 1991). However, no diffuse streaks were observed that could indicate multilayer structures on the martensite reflections in our melt-spun sample. Thus, one can say that melt spinning suppresses the defect ordering in the HCP structure. Finding the kind of martensite with the highest M_s temperature (according to Zhao 1999) suggests that the rate of cooling, used in the melt spinning method, did not decrease the M_s temperature in our Co-W sample.

It is known that the twinning process does not depend on temperature, it depends only on the strain levels in the crystal lattice. According to our knowledge, the FCC twins have never been found before in the Co-W alloys. The deformation FCC twins are usually observed in pure cobalt (Markstom et al. 2005). In the FCC crystal lattice, the twin boundary is also the (111) slip plane and the FCC twins are formed by overlapping stacking faults. In the FCC structure, the stacking faults on the (111) plate are formed by two Shockley partial dislocations with the dislocation reaction (Christian 1951):

$$\frac{1}{2}\left[1\bar{1}0\right] \rightarrow \frac{1}{6}\left[2\bar{1}\bar{1}\right] + \left[1\bar{2}1\right] + \mathrm{SF}.$$

The presence of such intrinsic stacking faults is visible inside the α grains of the melt-spun sample (Fig. 2.39b). Thus, the experimental results presented above testify that the appearance of long-period martensitic phases consisting of planar defects is a specific feature of the binary cobalt alloys.

2.1.2 *Phase transitions in Co-Al-W system (Co-rich corner)*

High temperature isothermal sections of Co-Al-W phase diagram have been experimentally constructed by Dmitrieva and Sato (Dmitrieva et al. 2005-2006, Sato 2006). Dmitrieva et al. studied the liquidus temperatures of this system and calculated the temperature boundaries of the different phases from liquidus to 900°C. The isothermal section of Co-Al-W phase diagram at 900°C suggested by Sato et al. shows the region of the intermetallic compound $Co_3(Al,W)$. Sato was a pioneer of new $Co_3(Al,W)$ intermetallic compound in Co-Al-W system (Sato et al. 2006). He found that the γ'-phase ($Co_3(Al,W)$) is stable at 900°C for 72 hours and metastable at 950, 1000, and 1050°C (Sato et al. 2006, Ishida 2009). The stability of this phase at 900°C was studied in different works (see Table 2.64 (b)). It was found that this phase is coarsened under annealing, however it still may be found as small precipitations even after annealing at 900°C for 4000 hours (Lass et al. 2014). This fact is consistent with the solvus temperature for this phase. Because of the diffusion mechanism of $\gamma' \rightarrow \gamma$ transformation, the γ'-phase dissolves in the temperature range. A low rate of the tungsten diffusion causes the prolonged period of time for this transition. According to DSC studies, the starting temperature of the $\gamma' \rightarrow \gamma$ phase transition is close to 850°C and the final temperature is about of 1000°C (Kazantseva et al. 2017a). According to many researchers, γ'-phase may be observed in the alloys after annealing at 900°C; the phase is in balance with others phases (γ, Co_3W, (β)$B2$, Table 2.64(b)). As can be seen from Table 2.64 (b), the results obtained by different researchers correlate with each other. In binary Co-based alloys, the Co-based solid solution is called α-phase, in contrast to the ternary Co-Al-W alloys, where the Co-based solid solution is called γ-phase. Small precipitations of the high-temperature Co_7W_6 (μ-phase) may be conserved in the alloys under homogenization at 1250°C. Two-phase $\gamma + \gamma'$ region is found within very narrow concentration limits near 9 at.% Al and up to 11 at.% W.

Table 2.64(b) Phase composition of the alloys at 900°C (at.%)

Alloy	Phase composition	Reference
Co-9Al-7.5W	$\gamma + \gamma'$	Sato et al. 2006
Co-9.2Al-9	$\gamma + \gamma'$	/-/
Co-9Al-3W	γ	/-/
Co-9.2Al-9W	$\gamma + \gamma'$	/-/
Co-9Al-3W	γ	/-/
Co-5.5Al-10W	$\gamma + Co_3W$	/-/
Co-5W	αCo (ferro) + Co_3W	Sato et al. 2006
Co-10W	αCo (ferro) + Co_3W	/-/
Co-23.6W	Co_3W	/-/
Co-47.1W	Co_7W_6	/-/
Co-10Al-12W	$\gamma + \gamma' + Co_3W$	Miura et al. 2007

Table 2.64(b) Cont...

Table 2.64(b) (Contd.) Phase composition of the alloys at 900°C (at.%)

Alloy	Phase composition	Reference
Co-4Al-4W	γ	Dmitrieva et al. 2008
Co-7Al-7W	γ	/-/
Co-15Al-15W	$\gamma + Co_7W_6 + \beta(B2)$	/-/
Co-21Al-21.7W	$Co_7W_6 + \beta(B2)$	/-/
Co-25Al	$\gamma + \beta(B2)$	/-/
Co-12.5Al-12.5W	$\gamma + \beta(B2) + Co_7W_6$	/-/
Co-15Al-10W	$\gamma + \beta(B2) + Co_7W_6$	/-/
Co-18.5Al-6.7W	$\gamma + \beta(B2)$	Dmitrieva et al. 2008
Co-22Al-3.1W	$\gamma + \beta(B2)$	/-/
Co-4Al-37W	$B2 + \gamma + Co_7W_6$	Kobayashi 2009
Co-9.4Al-5W	γ	/-/
Co-37.4Al-2.7W	$B2$	/-/
Co-1.7Al-22.9W	Co_3W	/-/
Co-10W	$\gamma + Co_3W$	Lass et al. 2014
Co-10Al-7.5W	$\gamma + \gamma' + Co_3W$	/-/
Co-8.25Al-10W	$\gamma + \gamma' + B2 + Co_3W$	/-/
Co-12.5Al-7.5W	$\gamma + \gamma' + B2 + Co_3W$	/-/
Co-12Al-15W	$\gamma + \gamma' + B2 + Co_3W$	/-/
Co-9.5Al-10W	$\gamma + \gamma' + B2 + Co_3W$	/-/
Co-10W	$\gamma + Co_3W$	/-/
Co-10Al-7.5W	$\gamma + \gamma' + Co_3W$	/-/
Co-8.25Al-10W	$\gamma + \gamma' + B2 + Co_3W$	/-/
Co-12.5Al-7.5W	$\gamma + \gamma' + B2 + Co_3W$	/-/
Co-12Al-15W	$\gamma + \gamma' + B2 + Co_3W$	/-/
Co-9.5Al-10W	$\gamma+\gamma'+B2+Co_3W$	Lass et al. 2014
Co-9Al-4.6W	$\gamma+\gamma'$	/-/
Co-7.9Al-6.8W	$\gamma + \gamma'+Co_3W$	Kazantseva et al. 2017a
Co-7.9Al-8.5W	$\gamma + \gamma'+Co_3W$	/-/
Co-8.2Al-8.5	$\gamma + \gamma'+Co_3W$	/-/
Co-8.7Al-10W	$\gamma + \gamma' +Co_3W$	/-/
Co-8.2Al-12.6W	$\gamma +\gamma' + Co_3W + Co_7W_6$	/-/
Co-9.2Al-9W	$\gamma + \gamma'$	Pollock et al. 2010
Co-9.4Al-10.7W	$\gamma + \gamma'$	Suzuki et al. 2008
Co-10Al-7.5W	$\gamma + \gamma' + B2 + Co_3W$	Tsukamoto et al. 2010
Co-9.5Al-5.1W	γ	/-/
Co-37.6Al-2.6W	$B2$	/-/
Co-1.9Al-22.6W	Co_3W	/-/
Co-5.5Al-10W	$\gamma + Co_3W$	Xue et al. 2011
Co-9Al-7.5W	$\gamma + \gamma'$	/-/
Co-9Al-9W	$\gamma + \gamma'$	Bauer et al. 2010
Co-10Al-35W	$B2 +Co_3W + Co_7W_6$	Cui et al. 2011
Co-15Al-30W	$B2 + Co_3W + Co_7W_6$	Cui et al. 2011
Co-16.3Al-7.3W	$\gamma + \gamma'+B2$	/-/

The crystal structure and parameters of the crystal lattice of the different phases are given in Table 2.65.

Table 2.65 The crystal structure of the phases in Co-Al-W alloys (Co-rich corner)

Phase	Structure	Lattice parameters, nm
γ-Co(W,Al)	A1, FCC, Fm-$3m$	$a = 0.358$
γ'-Co$_3$(Al,W)	L1$_2$, Cu$_3$Au-type, Fm-$3m$	$a = 0.357$
Co$_3$W	DO$_{19}$ Ni$_3$Sn-type, $P63/mmc$	$a = 0.511$, $c = 0.410$ $c/a = 0.80$
μ-Co$_7$W$_6$	D8$_5$, W$_6$Fe$_7$-type, R-$3m$ (topologically closed packet (TCP) phase)	$a = 0.473$, $c = 2.55$ $c/a = 5.39$
β-CoAl	B2, CsCl- type, Pm-$3m$	$a = 0.288$

Figure 2.41 shows combined experimental results for the phase boundaries in Co-rich corner of the Co-Al-W phase diagram at 900°C obtained by different researches (the alloys compositions and references are in Table 2.64(b)).

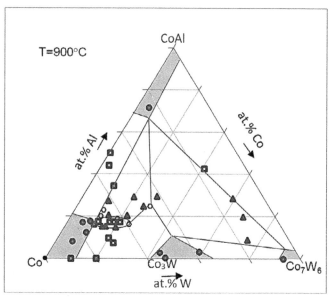

Fig. 2.41 The Co-rich corner of the Co-Al-W system. Schematic isothermal section at 900°C with experimental data, ●– single phase, □– 2 phases, ▲– 3 phases, ○– 4 phases.

Alloys with the $\gamma + \gamma'$ structure are interesting for practical applications. Such alloys are considered as promising heat resistant materials.

Two-phase $\gamma + \gamma'$ alloys have a cuboidal structure like Ni$_3$Al-based alloys (Fig. 2.42a). The $B2$ phase may be observed in the structure of the alloys as small particles or regions (Fig. 2.42b). The Co$_7$W$_6$ intermetallic compound (μ-phase) usually show a platelet structure and is in equilibrium with $B2$- or γ-phase (Fig. 2.42c-d). The Co$_3$W intermetallic phase may be observed as a thin plate or small areas (Fig. 2.42e-f). Depending on the concentration of the aluminum and tungsten, the Co$_3$(Al,W)-based alloys may be multi-phases ones (Fig. 2.42).

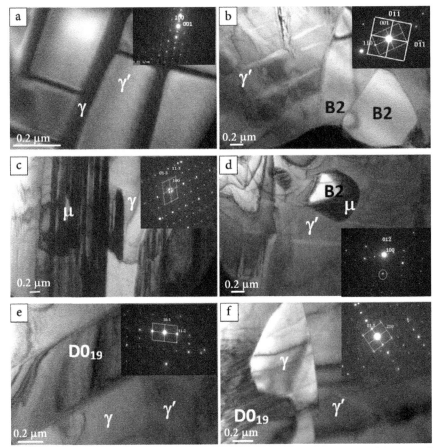

Fig. 2.42 Morphology of the phases in Co$_3$(Al,W)-based alloys: (a) $\gamma + \gamma'$ cuboidal structure; (b) grains of $B2$ phase; (c) platelet Co$_7$W$_6$ (μ)-phase; (d) $B2/\mu$ particle; (e) plate of Co$_3$W; (f) γ- and Co$_3$W -areas.

The orientation relationships between the phases are presented in Table 2.66.

Table 2.66 Orientation relationships between the phases in Co-Al-W alloys

Phases	Orientation relationships	Reference
γ/γ'	<100>‖<100>, {100}‖{100}	Sato et al. 2006
$\gamma'/\gamma/B2$	(002)FCC‖B2 (002); [011]FCC‖B2 [001]	Baker et al. 2011
γ/μ	(111)FCC‖(0001)μ [$\bar{1}$10]FCC‖[11$\bar{2}$0]μ	Zhao et al. 2005
$\gamma/D0_{19}$	(111)FCC‖(0001)$D0_{19}$ [$\bar{1}$10]FCC‖[11$\bar{2}$0]$D0_{19}$	Carvalho et al. 2002

Four phases were found: γ (Co-based solid solution), γ' (Co₃(Al,W)), $D0_{19}$(Co₃W), and μ (Co₇W₆) in the Co–Al–W system near the composition range of existence of the Co₃(Al,W) intermetallic compound (Kazantseva et al. 2016a). Table 2.67 shows results of the thermal X-ray diffraction analysis of the initial alloys: at room temperature, after heating to 950°C, and after subsequent cooling.

Table 2.67 Phase composition of studied alloys at 30°C, after heating to 950°C, and after cooling to 39°C

Alloy, at.%	T = 30°C	T = 950°C	T = 39°C
Co-9Al-4.6W	γ	$\gamma + \gamma'$	$\gamma + \gamma'$
Co-7.9Al-6.8W	$\gamma + \gamma' + \mu$(traces)	$\gamma + \gamma' + \mu + D0_{19}$	$\gamma + \gamma' + \mu + D0_{19}$
	$\gamma + \gamma' + \mu$(traces)		
Co-7.9Al-8.5W	$\gamma + \gamma'$	$\gamma + \gamma' + \mu + D0_{19}$	$\gamma + \gamma' + \mu + D0_{19}$
	$\gamma + \gamma' + \mu$(traces)		
Co-8.2Al-8.5W	$\gamma + \gamma' + \mu$(traces)	$\gamma + \gamma' + D0_{19}$	$\gamma + \gamma' + D0_{19}$
Co-8.7Al-10W		$\gamma + \gamma' + \mu + D0_{19}$	$\gamma + \gamma' + \mu + D0_{19}$
Co-8.2Al-12.6W		$\gamma + \gamma' + \mu + D0_{19}$	$\gamma + \gamma' + \mu + D0_{19}$

X-ray diffraction patterns were taken without sample rotation in vacuum during heating to 950°C and subsequent cooling to room temperature in steps of 50°C. X-ray diffraction patterns at selected temperatures were taken in steps of 0.05°C; the exposition time was 5 s/step. Prepared polycrystalline ingots were subjected to homogenizing annealing at 1250°C for 24 hours and subsequent slow cooling in the furnace (50 deg/hours).

The homogenized alloys have cellular dendritic structures, consisting of morphologically homogeneous areas that differ in chemical composition. The areas in alloy with 4.6 at.% W differ substantially in their composition; as the tungsten content increases (alloy 6.8 at.% W), the chemical compositions of the areas become more uniform.

γ- Phase, Co-based solid solution. Reflections of the γ phase were found for all alloys over the entire studied temperature range. However, using the X-ray diffraction data obtained under the applied conditions, we failed to separate the common reflections of the γ (FCC) and γ' (L12) phases. According to the literature data (Sato et al. 2006, Bocchini et al. 2013), this fact may indicate the closeness of the lattice parameters of the phases that differ in the tungsten content. The X-ray diffraction pattern for alloy with 4.6 at.% W in the homogenized state (X-ray diffraction analysis at 30°C) only indicates the common reflections for the γ'/γ phases. Table 2.68(a) shows variations in the cubic lattice parameters for the alloy in the initial state (before heating) depending on the tungsten content, calculated using the common fundamental (200), (111), and (220) reflections for γ/γ'. As can be seen, the cubic lattice parameter for γ/γ' increases with the tungsten content in the alloy. According to transmission electron microscopy (TEM) micrographs, no areas of disordered γ phase, which were free from the γ' phase, were found in the structure of studied alloys in the homogenized (initial) state.

Table 2.68(a)   Dependence of γ/γ' crystal parameters on tungsten content

at.% W	4.6	6.8	8.5	8.5	10	12.6
Crystal parameter of γ/γ'-phase, $\pm$ 0.00005 nm	0.3572	0.3571	0.3575	0.3577	0.3578	0.3580

γ'-*Phase, $Co_3(Al,W)$ intermetallic compound.*   The $Co_3(Al,W)$ intermetallic compound was discovered in the Co-Al-W system by Sato et al. (2006). The compound has the $L1_2$ supercell, where Al and W atoms alternately occupy the cube-corner layers along the <001> direction. The "ideal" chemical composition of $Co_3(Al,W)$ intermetallic compound is assumed to be equal amounts of Al and W mixing in the B sublattice, i.e. $Co_3(Al_{0.5}W_{0.5})$. The experimentally determined compositions tend to be slightly Co-rich and Al-poor relative to this "ideal" composition.

Kazantseva et al. (2016a) found that the size and shape of nanoparticles of the γ' ($Co_3(Al,W)$) intermetallic phase and its degree of long-range order depend on the tungsten content in the alloy. At a tungsten content of 10 at.%, the γ'-phase particles were cuboids; as the tungsten content increases, the shape of particles becomes close to rounded. Two superlattice reflections, such as (110)γ', corresponding to an angle $2\theta = 41.04°$, and (210) γ', corresponding to an angle $2\theta = 67.6°$, were observed in all X-ray diffraction patterns for the initial alloys, excluding the alloy with 4.6 at.% W. The lattice parameter of γ' was not estimated using the (110) reflection, since it corresponds to the low angle and, in this case, the estimates suffer from large errors. After heating and cooling, the intensity of superlattice reflections corresponding to the ordered γ' phase ($Co_3(Al,W)$) increased, in particular for the alloy 12.6 at.% W. After heating, the X-ray diffraction pattern of the alloy with 4.6 at.% W exhibited the presence of a weak superlattice (110) γ' reflection. The areas, containing nano-scale γ'-phase particles (or clusters, to be more precise), differ in shape and closely cover γ-phase solid solution matrix. The electron diffraction patterns for these areas indicate superlattice reflections, corresponding to the ordered crystal lattice of the $Co_3(Al,W)$ intermetallic. The appearance of such nano-scale clusters confirms the assumption about their homogeneous nucleation. Figure 2.43 shows the dark-field images of the γ'-phase nanoparticles, taken in superlattice reflection.

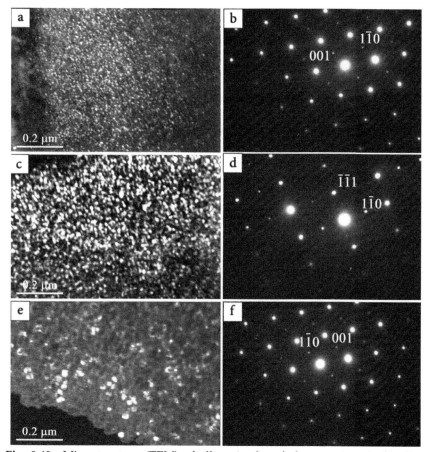

Fig. 2.43 Microstructure (TEM) of alloys in the γ'-phase region: (a, b) alloy with 4.6 at.% W, dark-field image taken in the (001) γ' reflection and SAED pattern, zone axis [110] γ'; (c, d) alloy with 6.8 at.% W, dark-field image taken in the ($\bar{1}$10) γ' reflection and SAED pattern, zone axis [112] γ'; (e, f) alloy with 12.6 at.% W, dark-field image taken in the (001) γ' reflection and SAED pattern, zone axis [110] γ'.

An analysis of the TEM pictures showed that the morphology and size of the γ'-phase particles depend on the tungsten percentage in the alloy. In the case of a low tungsten content, particles were fine and rounded in shape; as the tungsten content increased, the particle sizes also increased. Particles became cuboidal in shape, and stacking faults were observed within the particles. The shapes of the γ'-phase particles in the alloy with 12.6 at.% W again became close to rounded (Fig. 2.43e).

The intensity of superlattice reflections in SAED patterns also increased with the increasing of the tungsten content in the alloys. This fact may indicate an increasing degree of the long-range order of the ordered γ'-phase with increasing tungsten content in the alloys. The latter may explain the low intensity of the superlattice reflections in the X-ray diffraction pattern of the alloy with 4.6 at.% W, which is the lowest tungsten content among the studied alloys.

We analyzed phases using a polythermal section that was constructed experimentally for ternary alloys (Dmitrieva et al. 2008) with varying cobalt contents and equiatomic tungsten and aluminum contents that match (approximately) the alloys with 8.5 at.% W. Dmitrieva et al. (2008) studied the phase diagram only at high temperatures (above 900°C) without the TEM and she did not find the ordered $\gamma'(Co_3(Al, W))$ phase. The existence of the γ' phase was shown in the isothermal section of the system at 900°C by Sato et al. (2006). In our study with the TEM, the γ' phase was found in the Co-Al alloy containing 4.6 at.% W. Thus, the composition range of the γ'-phase is wider than that determined by Sato et al. (2006). This shows that the TEM allows finding the ordered γ' phase with a low degree of the long-range order and small particle sizes. The literature data present studies of some compositions to determine the boundaries of stability of the γ'-phase only. For example, the solvus temperature of the γ' phase in the Co-8.1 at.% Al-8.4 at.% W alloy, aged at 900°C for 200 hours, is 963°C (Bauer et al. 2010). The chemical composition of the alloy is consistent with that of the alloy with 8.5 at.% W in our study. The thermal stability of the γ'-phase was also studied by Sato et al. (2006); the thermal analysis curve for the Co-9 at.% Al-9.8 at.% W alloy displays the endothermal effect at $T = 990°C$ (1263 K), corresponding to the solvus of the γ' phase. In our case, this alloy is close to our alloy with 10 at.% W.

$D0_{19}$ phase, (Co_3W) intermetallic compound. Alloying with W in Co-based superalloys promotes the hardening precipitations. Except the strengthening of the solid solution, tungsten enters the M_6C carbide composition and intermetallic phases. Two ordered phases, Co_3W ($D0_{19}$) and μ-phase Co_7W_6, are present in the Co-W binary equilibrium phase diagram (Okamoto 2008). Both Co_3W and Co_7W_6 are the embrittling phases in this alloy system.

Usually, the Co_3W phase forms in a Widmanst'atten morphology. The morphology of the Co_3W intermetallic phase precipitations was studied in the Co-10 at.% W melt-spun ribbons[2]. The lines of the $D0_{19}$ phase (Co_3W) appear in the diffraction patterns of the melt-spun samples after heat treatment. In this study, no evidence is obtained for the presence of μ-phase (Co_7W_6) in the samples after the heat treatment. According to the SEM EDS results, the chemical composition of the $D0_{19}$ particles changes with increasing annealing time. After 1000 hours, the chemical composition of Co_3W ($D0_{19}$) is similar to the stoichiometric one. These results are consistent with the results obtained by Carvalho et al. (2002). The APBs and lots of planar defects inside the precipitations of the $D0_{19}$ phase were not found in the samples after a prolonged annealing at 900°C. This fact may also point to a state, similar to the equilibrium $D0_{19}$ phase, formed as the plates inside the cobalt matrix.

It was shown that the $D0_{19}$ (Co_3Al) phase is formed in a temperature range of 740-780°C and remains in the alloys after a rapid cooling. According to the X-ray diffraction data, the $D0_{19}$ intermetallic phase is formed at temperatures above 740°C. This phase is absent in the alloy with 4.6 at.% W; in other alloys, it appears during heating and remains after cooling from 950°C down to room temperature. The intense (011) reflection of $D0_{19}$ is observed at the angle $2\theta = 346°$. This behavior of the reflection may indicate that the phase is formed at temperatures of 740-950°C and conserved by rapid cooling. The presence of the $D0_{19}$ phase along with $\gamma' + \gamma$ was found in the Co-10 at.% Al-12 at.% W alloy annealed at 900°C (Miura et al. 2007). This is in a good agreement with our results for the alloy with 12.6 at.% W (Table 2.67).

μ-Phase, Co_7W_6 intermetallic compound. Topological close-packet (TCP) μ-phase (Co_7W_6) has a hexagonal structure, with the {0001} face in this structure having the lowest surface energy when acting as the basal plane. Two-dimensional nucleation and growth of the μ-phase precipitations occur only at the edges, not on the basal planes, which shows twinning plate-like morphology of the phase. In the nickel superalloys, the μ-phase precipitations occur during long term heat treatments or during service at high temperatures. The TCP μ-phase Co_7W_6 usually forms at grain boundaries, which promotes crack initiation and propagation that lead to low-temperature brittle failures of the alloys, as well as the loss of solid-solution strengthening, as the refractory elements are drawn out of the matrix. Therefore, this phase is undesirable and the composition of the alloys must be carefully controlled to avoid its formation. In the cobalt superalloys, μ-phase is formed by alloying with large amounts of refractory elements, such as Mo, W, Re, Ru, and Ta. The presence of μ-phase decreases the ductility, impact resistance, and thermal fatigue of the cobalt superalloys (Luna Ramírez et al. 2016).

---

[2] The TEM images were obtained by N.V. Kazantseva.

In the Co-Al-W system μ (Co_7W_6) phase exists above 950°C (Kazantseva et al. 2016a). The weak reflections of the μ phase, which are observed for Co-Al-W alloys with 6.8 at.% W, 8.5 at.% W and 10 at.% W, 12.6 at.% W in the initial state, become clearer after heating the alloys to 950°C and cooling; no μ phase is found in Co-9Al-4.6W and Co-8.2Al-8.5 alloys after heating and cooling (Table 2.67). This fact might indicate separation in the initial alloy and a small amount of the phase, as the chemical compositions of alloys with 8.5 at.% W are very similar (Table 2.67). Moreover, this fact also indicates that the µ phase cannot be dissolved during heating to 950°C, i.e., that it is the high-temperature phase formed by diffusion nucleation and growth at temperatures substantially above 950°C. Large μ-phase precipitates within grains and at their boundaries were only found in homogenized alloy with 12.6 at.% W. No $D0_{19}$-phase precipitates were found. Chemical composition of μ-phase was determined in the alloy with 12.6 at.% W with the SEM analysis (Table 2.68(b)). The phase composition of the Co-10 at.% Al-13 at.% W alloy annealed at 1200°C is $\gamma + \mu$ [8]; this also agrees well with the results obtained in our work. We assume that the high-temperature µ phase in the studied alloys remains after homogenizing annealing at 1250°C.

Table 2.68(b) Chemical composition of μ-phase precipitations in the alloy with 12.6 at.% W

Element	Co	Al	W
at.%	52.56	4.76	42.68

CoAl intermetallic compound. The cobalt alloys are widely applied and, depending on the chemical composition, are considered to be the promising functional materials with a magnetic shape-memory effect (MSME), as structural high-temperature, high-strength materials, and they can also be used as the basis for the high-temperature coatings. An important role in the cobalt alloys belongs to the intermetallic phases – in particular, to the intermetallic compound CoAl with an ordered cubic (*B*2) structure. Silicon enters the composition of protective coatings made of the heat-resistant intermetallic alloys NiAl and CoAl. Alloying with silicon increases the corrosion resistance and describes the porosity of the intermetallic CoAl alloys (Pugacheva 2015, Novak et al. 2011). In the intermetallic compound CoAl, silicon (a substitutional atom) enters the sublattice of aluminum (Bozzolo et al. 1998). In the Co-Al-Si system, the homogeneity range of the intermetallic compound CoAl (*B*2) is significantly wider, i.e., 48-78.5 at.% Co. The solubility of silicon in CoAl of stoichiometric composition at room temperature reaches 15 at.% (Huber et al. 2011). To significantly reduce the porosity in the intermetallic compound CoAl, the content of silicon in the alloy should exceed 5 at.% (Novak et al. 2011).

The low-temperature part of the Co-Al phase diagram is based on calculations; the boundaries of the existence of equilibrium phases are drawn by dashed lines (McAllster 1989). According to the diagram, the temperature of the α(FCC) $\rightarrow$ ε(HCP) phase transition, with an increase in the content of aluminum in the alloy, is reduced relative to the transition temperature in pure cobalt (422°C).

The high-cobalt alloys (96.8 at.% Co) undergo a eutectoid transformation at 300°C. At low temperatures, the equilibrium alloys with a cobalt content of 46-49 at.% should have two phases (ε + B2). The intermetallic phase CoAl (B2) forms from the liquid and at room temperature it has a homogeneity range in the limits of 49.3-53.8 at.% Al (extrapolated, dashed lines in the experimental diagrams; McAllster 1989, Lyakishev 1996), but no such region is present in the calculated diagrams. The maximum solubility of aluminum in the equilibrium FCC α phase is 16 at.% at 1400°C (McAllster 1989). The solubility of aluminum in the equilibrium ε phase is, in fact, zero (Lyakishev 1996). With a content of cobalt of less than 50 at.% in the alloy, the intermetallic compound CoAl (B2 phase) is paramagnetic; with a content of more than 50 at.%, it is ferromagnetic. The Curie temperature of the intermetallic phase CoAl in the alloy with a content of cobalt equal to 57.9 at.% is ~153°C (Wachtel et al. 1973).

In the binary Co-Al system, the intermetallic compound CoAl (B2, CsCl type) forms from the liquid phase and, at a temperature of 200°C, it has a homogeneity range of 47-57 at.% Co (McAllster 1989, Lyakishev 1996). The crystal structure of the β' phase, which is ordered, according to the B2 (CsCl) type, represents two primitive cubic lattices occupied by the atoms of cobalt and aluminum each and shifted by half of the lattice parameter. A specific feature of the B2 ordered crystal structures is the possibility of producing vacancies in different sublattices and anti-structural (anti-site) replacements of atoms. According to theoretical calculations, the number of thermal vacancies in the intermetallic compound CoAl of stoichiometric composition is 1.55×10^{-3} at $T = 1273$ K and increases with increasing temperature to $\sim5 \times 10^{-3}$ at $T = 1500$ K (Breuer and Mittemeijer 2003).

Experimental investigations of the intermetallic compounds have shown that the aluminum vacancies appear at the aluminum content in the alloy of 47-50.5 at.%. With higher contents of aluminum (51-54 at.%), cobalt vacancies form in the cobalt su blattice. In alloys of stoichiometric composition, the numbers of vacancies in both sublattices are almost equal (Araki et al. 2002). According to Bester et al. (1999), in nonstoichiometric alloys Co_xAl_{1-x}, cobalt anti-site atoms form at $x > 0.5$, cobalt vacancies form at $x < 0.5$, the latter being somewhat more favorable energetically than the aluminum vacancies. Tamminga (1973) also indicates that, upon the deviation from the stoichiometry in

CoAl, the character of the occupation of the crystal lattice by the atoms changes. With a deficit of aluminum, there occurs an anti-structural occupation of aluminum positions by cobalt atoms; a deficit of cobalt, leads to structural vacancies.

Structure of the $B2$ (CoAl) phase was studied by Kazantseva et al. (2016b), who calculated the degree of the long-range order in the $\beta'(B2)$ phase in the Co-Al-Si alloys. Allowance was made for the fact that, in the intermetallic CoAl compound, the intensity of the superlattice (100) lines is higher than that of the (200) lines. This relationship between the line intensities, which is described by the appearance of atomic scattering factors of different signs, is observed in many compounds of Ti, Ni (e.g., TiNi), and Mn with the CsCl structure. The ambiguity of the choice of the sign of the atomic scattering factor f is related to the fact that this factor is calculated using the amplitude of the scattering atoms via the following formula (Cantor and Schimmel 1980):

$$f(S) = \pm I(S),$$

where $f(S)$ is the atomic scattering factor or the form factor, $I(S)$ is the amplitude of the scattering atoms (in the experiment, this is the intensity of scattering by a separate atom), and S is the scattering angle of the incident beam. With the allowance for the change in the intensity ratio of the fundamental and superlattice lines to the opposite, the degree of the long-range order is calculated via the following formula:

$$S^2 = \frac{I_{sl}}{I_f} \cdot \frac{Lp_f(f_{Cof}\exp(-M_{Cof}) + f_{Alf}^*\exp(-M_{Alf}^*))^2}{Lp_{sl}(f_{Cosl}\exp(-M_{Cosl}) + f_{Alsl}^*\exp(-M_{Alsl}^*))^2},$$

where f_f and f_{sl} are the atomic factors of the elements in the compound, corresponding to the fundamental and superlattice lines, respectively, I_f and I_{sl} are the integral intensities of the fundamental and superlattice lines, respectively, Lp_f and Lp_{sl} are the parameters including the Lorentz and the polarization factors, which take into account the recording geometry, and M is the temperature factor. In the case of the ternary compound, the atomic factor of silicon in the sublattice of aluminum was also taken into account:

$$f_{base}^* = \frac{c_{base}}{c_{base} + c_x} f_{base} + \frac{c_x}{c_{base} + c_x} f_x,$$

where c_{base} and c_x are the concentrations of aluminum and silicon, respectively. Results of the XRD analysis are given in Table 2.69. For comparison, the data for the stoichiometric alloy of the composition CoAl are also given (Richter and Gutierrez 2005). As can be seen from Table 2.69, the simultaneous reduction in the content of cobalt and silicon in the alloy leads to changes in the lattice parameter and in the degree of the long-range order (LRO) of the β' (B2) phase. These results agree with the literature data on the effects caused by silicon.

Table 2.69 Results of X-ray diffraction analysis

Alloy	Lattice parameter, ±0.00005 nm	Degree of long-range order, S^2
Co-50 at.% Al (as reference)	0.28611	1
Co-39.5 at.% Al-2.9 at.% Si	0.28591	0.62
Co-40.7 at.% Al-2.7 at.% Si	0.28582	0.56
Co-43.4 at.% Al-1.9 at.% Si	0.28564	0.32

Alloying with silicon exerts a strong influence on the lattice parameter of CoAl (Richter and Gutierrez 2005); in the Co-40 at.% Al-10 at.% Si alloy, the lattice parameter of the β'(B2) phase decreases to 0.28451 nm. The deviation from the stoichiometric composition of CoAl toward the excess of aluminum or cobalt also affects the lattice parameter and the long-range order (LRO, Cooper 1963). However, the substantial change in the degree of the LRO, observed in this work, may hardly be explained by alloying alone. This may also be related to the highly defective structure of the intermetallic CoAl compound in the alloys. Prismatic dislocation loops were found in the Co-39.5 at.% Al-2.9 at.% Si alloy. They have a characteristic cubic shape with sides oriented along the (010) and (001) faces of the cubic lattice (Fig. 2.44). An analysis of the extinctions of the dislocations in different reflections has shown that the prismatic dislocation loops are formed by edge dislocations with the Burgers vectors $a\langle 100\rangle(100)$. Figure 2.44 demonstrates a typical extinction of dislocations ($gb = 0$) in the reflection ($0\bar{1}1$).

The prismatic loops are observed upon the irradiation of metals, when the arising excessive interstitial atoms form planar pileups or on the growth of crystals from the melt. Near the melting temperature, the supersaturation of the crystal lattice by thermal vacancies occurs and the prismatic loops are formed as a result of the collapse of vacancy disks.

The prismatic vacancy-type dislocation loops, with a Burgers vector <100> lying in the {100} planes, were studied in the NiAl single crystals with nickel contents of 47 and 49.7 at.% during *in situ* experiments at various temperatures. The study revealed that the slow cooling of the alloy after a preliminary high-temperature annealing at 1175°C for 3 hours favors the formation of prismatic loops, whereas the quenching of the alloy from the high temperatures prevents their formation. In the slowly-cooled samples, the loops have a specific square shape with edges lying along $\langle 100\rangle$ (Marshall and Brittain, 1976). Prismatic vacancy dislocation loops are also found after annealing at 800 and 1200 K in the thin CoAl and NiAl films prepared by the method of spinning from the melt, with sizes of the loops of 20-50 nm (in CoAl). The formation of prismatic loops occurs only during annealing; no similar defects were revealed in the quenched samples (Marshall and Brittain 1976). In contrast to the intermetallic NiAl compound, the prismatic loops in CoAl are fewer and smaller; the authors explain this by the low diffusion mobility of atoms in CoAl, compared to NiAl (Novak et al. 2011). However, no detailed analysis of the loops was performed by Nakamura et al. (2007).

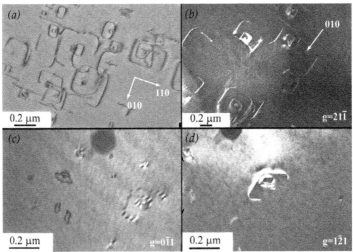

Fig. 2.44 Microstructure of the Co-39.5 at.% Al-2.9 at.% Si alloy (TEM): (a) bright-field image of prismatic dislocation loops, (b) dark-field image of prismatic dislocation loops, taken in the reflection (21$\bar{1}$), (c) dark-field image of prismatic loops, taken in the reflection (0$\bar{1}$1), (d) dark-field image of prismatic loops, taken in the reflection (1$\bar{2}$1).

An investigation of the shape of prismatic loops was done by Fitzgerald and Yao (2009). The authors performed a simulation and an *in-situ* electron-microscopic investigation of the prismatic loops in α-Fe (BCC) and showed that the change in their shape from hexagonal to cubic was related to the character of the dislocations that formed those loops. The loop sides are extended along the directions having the lowest energy. For the [100](100) loops, which have a cubic shape, these are the (001) and (010) directions; for the 1/2[111](111) loops, which have a hexagonal shape, these are the (11$\bar{2}$), (2$\bar{1}\bar{1}$), and (1$\bar{2}$1) directions (Fitzgerald and Yao 2009). In our case, we also observed thermal, rather than quenching-induced, prismatic dislocation loops, which confirms the conclusions made by Nakamura et al. (2007). In the Co-39.5 at.% Al-3 at.% Si alloy, these loops are formed during slow cooling with a furnace after a high *a* temperature-homogenizing annealing. The revealed [100] (100) loops have cubic shapes with sides oriented along the (010) and

(001) planes of the crystal lattice of the $\beta'(B2)$ phase. Comparing our results with the literature data on NiAl (Marshall and Brittain 1976, Nakamura et al. 2007), one notes the similarity of the morphology and the character of the formation and annihilation of thermal vacancies in the structures of the intermetallic CoAl and NiAl compounds.

2.1.3 Deformation behavior of $Co_3(Al,W)$-based and Ni_3Al-based alloys

Deformation behavior of the crystal with the $L1_2$-type structure is well known in the case of Ni_3Al intermetallic compound. Plastic deformation of the Ni_3Al intermetallic compound with the $L1_2$-type superstructure occurs with the formation of the different types of planar defects, such as antiphase boundaries (APBs) with high formation energy and Superlattice Intrinsic Stacking Faults (SISFs) with low formation energy. As is known, several variants of various dislocation reactions, involving the presence of planar defects, can be realized during deformation of Ni_3Al (Greenberg and Ivanov 2002). This is the well-known M.J. Marcinkowski reaction:

$$a[011] \rightarrow \frac{a}{2}[011] + \frac{a}{2}[011] + APB, \tag{1}$$

where *APB* is the antiphase boundary. Another type of reactions involves the dissociation of a perfect dislocation into partial dislocation dipoles with the formation of the intrinsic or extrinsic super structural defects (Kear et al. 1970):

$$a[011] \rightarrow \frac{a}{3}[121] + \frac{a}{3}[\bar{1}12] + SSF. \tag{2}$$

As it was shown by Greenberg and Ivanov (2002), the reaction (2) in alloys with high energies of *APB* formation does not require any additional splitting of the partial dislocations. A numerical calculation of the stacking fault energy is correlated with the experimentally observed results and shows that intrinsic super structural defects are more likely than extrinsic ones for the Ni_3Al intermetallic compound (Pettinary et al. 2002). Several modern mechanisms for the SISF formation in $L1_2$-type superstructure have been proposed, one of which involves the interaction of dislocations sliding along different planes, the other one is associated with the formation of a SISFs in the dislocation loops (Baker et al. 1987). The effect of the second mechanism is justified by the fact that SISFs are observed for both large deformations (before failure) and very small deformations ($\varepsilon = 0.06\%$). When the dislocation density is still low, the possibility of SISFs forming by the dislocation interaction is excluded. It is likely that, depending on the conditions and nature of loading, both mechanisms can be realized in the alloy structure.

Deformation behavior of Ni₃Al single crystals after deformation to failure in the temperature range 1200-1250°C was studied by Kazantseva et al. (2010). Single crystals of Ni₃Al (75.5 at.% Ni and 24.5 at.% Al) with the [001] orientation were grown by the Bridgman method. The cast samples were homogenized at 1100°C for 100 hours. Mechanical tests were carried out with a Heckert FP-100/1 equipment. The loading rate was 1.32 mm/min (2 · 10⁻⁵ m/s). After deformation to failure at 1200°C, a low-angle subgrain structure was found in the alloy. A subgrain rotation of 2-3° was determined from the SAED patterns. Inside the subgrain, a high dislocation density and numerous packing defects were observed. The same structure was found in the alloy after a deformation

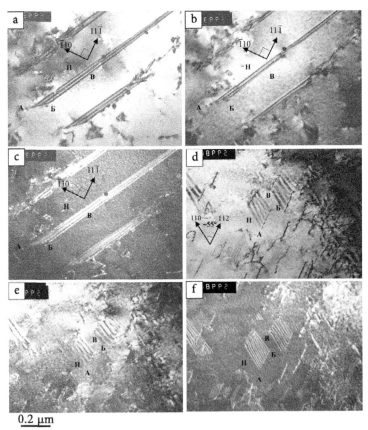

0.2 μm

Fig. 2.45 Superstructural stacking faults, deformation at 1200°C (T, B – upper and bottom defect edges, AC – partial dislocation): (a)-(c) (1̄11) plane of defect; (a) bright-field image in the reflex +g = 11̄1̄; (b) dark-field image in the reflex +g = 11̄1̄; (c) dark-field image in the reflex −g =11̄1̄; (d)-(e) (11̄1) plane of defect, (d) bright-field image in the reflex +g =1̄1̄1̄; (e)-(f) dark-field images in the reflexes 1̄1̄1̄ and 11̄1̄, correspondingly.

at 1250°C; the subgrain rotation was also 2-3°. Inside the subgrains, a high dislocation density intersecting in two directions at an angle of 110° was detected. Dislocation directions were determined by the trace analysis of the bright-field images and SAED patterns as: [112] and [11$\bar{2}$]. In addition, large twins were found in the fracture zone, with the twinning plane corresponding to the ($\bar{1}$11) Ni$_3$Al twinning plane.

The dislocation mechanism of the twin formation in the FCC metals and $L1_2$-structures includes a superposition of the intrinsic stacking fault, which appears as a result of the motion of helical partial twinning dislocations. In the $L1_2$ superstructure, passages of $\frac{a}{3} < 112 >$ dislocations create twins of the order (the so-called super-twinning), which do not require an additional "shuffle" of atoms (Vinogradova 2008). In Figure 2.45, the stacking faults and dislocations are well distinguishable in reflections of ($\bar{1}\bar{1}$1) and (11$\bar{1}$) type.

Standard calculation procedure (Bushneev et al. 1990) determined that these defects were super-structural intrinsic stacking faults with $R = -\frac{1}{3}[111]$ displacement vector (Table 2.70).

Table 2.70 Phase shift α for SSF on the ($\bar{1}$11) plane, (Fig. 2.45a-c)

Displacement vector, R	Reflex, g	
	$(-g)\bar{1}\bar{1}1$	$11\bar{1}(+g)$
$-\frac{1}{3}[111]$	$\dfrac{2\pi}{3}$	$-\dfrac{2\pi}{3}$

The defect plane was calculated to be ($\bar{1}$11). The direction of the traces of partial dislocations (Figure 2.45a-c) bordering the stacking faults is within 30° from the direction. Taking into account the cubic crystallography, the Burgers vectors of these dislocations may be $\frac{a}{3} < 112 >$. Further calculation was performed taking into account the peculiarity of the formation of the deformation contrast in the image when the dislocations were parallel to the surface of the crystal $gb = n$, where n was the order of the image (Kosevich 1976).

Since the dislocations bordering the SSF are visible as weak single lines in the Fig. 2.45a-c, according to the Frank criterion (b^2) and calculated the ($\bar{1}$11) slip plane, the following set of partial dislocations is possible: $\frac{a}{3}[\bar{1}1\bar{2}]$ and $\frac{a}{3}[12\bar{1}]$, constituting the $a[01\bar{1}]$ dislocation. At inclination of the sample another series of SSFs having a much smaller width was found (Fig. 2.45d-f). This may indicate the allocated direction of the partial dislocations forming these defects, or the allocated $a <110>$ dislocation direction composed by these partial dislocations. Taking into account the crystallography of the cubic superlattice $L1_2$ and the possibility of the appearance of another {111} SSF plane, this plane was also calculated by the traces method. We found that, in this case, the intrinsic superstructural stacking faults were formed too. The direction of exit of the partial dislocations bordering these defects in the (110) plane is within 55° with (see Fig. 2.45d).

At temperature of 1250°C, the stacking faults inside the dislocation loop were found. The appearance of a contrast from the loops in the reflexes $+g = 200$ and its change to the opposite in the reflex $-g = \overline{2}00$ may indicate that the loops are the vacancy loops (Bushneev et al. 1990, Pettinari et al. 2002). The changing contrast inside the loop can also indicate the presence of a stacking fault. It means that these loops are formed by partial or super-partial dislocations. Analyzing the loop contrast in the reflections (200), (002), and ($1\overline{1}1$), we can assume that the loops are formed by partial dislocations of the $\dfrac{a}{3} < 112 >$ type.

The results of the image analysis are presented in Table. 2.71.

Table 2.71 Phase shift α and gb for dislocations and stacking faults in ($1\overline{1}1$) plane, (Fig. 2.45d-f)

Reflex g	Displacement vector $R = -\dfrac{1}{3}[111]$	Burgers vector b	
		$\pm\dfrac{a}{3}[\overline{1}12]$	$\pm\dfrac{a}{3}[121]$
$\overline{1}11$	$\pi/3$	$\pm 2/3$	$\pm 2/3$
$11\overline{1}$	$-\pi/3$	$\pm 2/3$	$\pm 2/3$
110	$-2\pi/3$	0	± 1
$0\overline{2}0$	$2\pi/3$	$\pm 2/3$	$\pm 4/3$
200	$-2\pi/3$	$\pm 2/3$	$\pm 2/3$

The disappearance of the contrast from one of the superparticle dislocations bordering the stacking fault was found in the reflex $\underline{g} = 110$ ($gb = 0$) and its contrast was preserved in the reflexes $g = 11\overline{1}, 0\overline{2}0, 200$, and $\overline{1}\overline{1}1$ ($gb = -\dfrac{2}{3}$).

In the Ni_3Al intermetallic compound, occupancy of a vacancy by the <110> dislocation disrupts the stability of this dislocation and may cause its instantaneous splitting, which contributes to the creation of a stacking fault. Perhaps, this is the reason for the absence of the <110> -type dislocations in the investigated alloy. According to the obtained results, the deformation of the Ni_3Al sample in the temperature range of 1200-1250°C is associated with the so-called high-temperature dislocations. Appearance of the high-temperature dislocations is observed under deformation near the melting point, when the number of both structural and thermal vacancies is high, and interaction of the stacking faults leading to new defects is also actively occurring.

The SSFs are always present in the structure of Ni_3Al and Ni_3Al-based alloys after the creep tests or after the high-temperature tests with high percentages of deformation (Baker et al. 1987). The intrinsic stacking faults (SISF) play a special role in the twinning process of Ni_3Al (Vinogradova et al. 2008). The dislocation mechanism of the twin formation in the FCC metals and $L1_2$-type of structures includes a superposition of the intrinsic defects, which appears as a result of the motion of the helical partial twinning dislocations. It is known that plastic straining of crystalline bodies may occur not only due to slips of perfect dislocations or a pair of partial dislocations that produce shifting of atoms to the entire interatomic spacing, but also due to the slips of the special dislocations (the twinning ones). The motion of the latter is accompanied by displacements of atoms over adjacent planes by fractions of the interatomic spacing. The total (resulting) shift is macroscopic. The appearance of twins is usually accompanied by the formation of teeth on the stress-strain curves, because twinning is accompanied by stepped variation of the straining force. On the appearance of the first twin, the twinning stress decreases and numerous other twins arise in a burst-like manner. In the twinning process, the orientation of the twinned layer changes, but the structure or the symmetry of the crystal remains unchanged; the arrangement of atoms in a twin is a mirror reflection of the arrangement of atoms in the matrix crystal. The twinning planes are planes of the densest packing of atoms in the crystal lattice. It is assumed that the twinning process is independent of the temperature, related only to the stress applied (Polukhin et al. 1982). The twinning process occurs much more easily in a crystal with imperfections that serve stress concentrators and in the case of alloying, when the energy of formation of a stacking fault changes. It has been proved experimentally that the lower the energy of formation of a stacking fault in a crystal, the lower is the stress required for the twinning onset (Gol'dshtein et al. 1986). In the FCC metals, twinning occurs as a change in the succession of close-packed planes {111}, over which the twinning dislocations that create intrinsic stacking faults slide in <112> direction. Twinning is not a characteristic strain mode, neither for pure nickel nor for the Ni_3Al-base intermetallic alloys. The energy of formation of a stacking fault in these materials is quite high, i.e., $130 \cdot 10^{-7}$ J/cm^2 for Ni and $180 \cdot 10^{-7}$ J/cm^2 for the antiphase boundary (APB) of Ni_3Al in the (111) plane (for comparison, the energy of formation of a stacking fault in copper is $51 \cdot 10^{-7}$ J/cm^2, Shtremel 1982, Mishin 2004). Twinning in the γ'-phase (Ni_3Al) is associated with the appearance of partial twinning dislocations with the Burgers vector $\frac{a}{3}$<112>(11$\bar{1}$). Twins in Ni_3Al are observed experimentally, when the stress due to the inclusion of twinning dislocations is reduced, which occurs at a low deformation temperature (Travina and Nikitin 1975), at a high degree of strain (about 60-80%) (Dimitrov et al. 2000), or in the high-alloy materials (up to 30% of the alloying elements) (Albert and Gray III 1995).

The special features of the twinning process in the Ni_3Al-based superalloy were studied under the conditions of dynamic loading by Vinogradova et al. (2008). We studied a blade of a cogeneration plant turbine produced from nickel alloy ÉP-800, which had been stricken by fragments of a neighboring blade in operation. The operating temperature of the blade was 850°C, and the service time before the impact was 38 260 hours; the blade had withstood 84 startups of the turbine. We investigated test pieces cut out from the feather in the vicinity of the fractured zone (test piece 1) and the blade root outside the fractured zone (test piece 2). We also studied the alloy in the initial condition (test piece 3) after the standard heat treatment, i.e., 1060°C for 2 hours + 1000°C for 2 hours + 900°C for 2 hours + 850°C for 2 hours, with air cooling after the hold at each temperature. The nickel superalloy ÉP-800 contains 40 vol.% of the hardening intermetallic γ'-phase. The study of the fracture surface of the test piece 1 showed that the fracture occurred not only over the bodies of grains but also over grain boundaries. In the region of the zone of failure (test piece 1), straining covers both the space between the cuboids and the cuboids themselves. A great number of large and small twins are observed in the impact zone. When studying the structure, we discovered that the twins appeared at the boundary of a grain and could end in a crack. Starting at a grain boundary, a twin passes through two phases, i.e., the disordered γ-phase and the ordered γ'-phase. This may be an indication of the favorable orientation of the cuboids with respect to the sliding twinning dislocations in the disordered phase on the one hand and of high stresses arising in the place of the impact on the other hand. Then the twins grow in a step manner, which can be inferred from the presence of characteristic steps on the even boundaries of the twins. It is known that both FCC structures and ordered $L1_2$ superstructures are strained due to the motion of the dislocations of the {111} <110> type. However, the Burgers vector of a perfect dislocation in the ordered $L1_2$ superstructure is a <110>, whereas in the FCC structure, it is $\frac{a}{2}$ <110> (Gol'dshtein et al. 1986). This means that the motion of the $\frac{a}{2}$ <110> dislocations should violate the order in the ordered structure, creating an antiphase boundary. A passage of the second $\frac{a}{2}$ <110> dislocation over the same slip plane will restore this order. Thus, in a binary $\gamma + \gamma'$ structure penetration of dislocations into the ordered γ'-phase occurs in pairs (super-dislocations) from the disordered γ-phase. This requirement creates additional stresses on the boundaries of the phases and thus promotes growth in the creep resistance of the nickel superalloys consisting of two phases, i.e., an ordered (Ni_3Al, $L1_2$) γ'-phase and a disordered (nickel-base FCC solid solution) γ-phase. The presence of an ordered phase in the nickel superalloys hinders the free motion of dislocations in the whole of the material, even in the presence of a coherent γ/γ' interface. It is universally assumed that the dislocation model describing the appearance of twins in the FCC and $L1_2$ structures includes the motion of a partial twinning dislocation. Different models of twinning considered the "sources" of such dislocations (Greenberg et al. 1976, Hirth 1972). During their motion, partial twinning dislocations encounter branching dislocations with the Burgers vector components perpendicular to the plane of sliding and equal to the distance between such planes, coil around them, and form twin layers in the matrix of the crystal. In the FCC metals, twinning dislocations are of the Shockley $\frac{a}{6}$<112> type; such dislocations create a perfect twin in their motion. In the Ni_3Al γ' phase with a $L1_2$ superstructure passage of dislocations of $\frac{a}{3}$ <112> type creates ordered twins that do not require additional "shuffling" of atoms (Albert and Gray III 1995). However, as it has been shown by Gol'dshtein et al. (1986), formation of twins in the γ'-phase with the help of dislocations of the $\frac{a}{3}$ <112> type is observed only at a high degree of the long-range ordering (known as supertwinning). In other cases formation of twins in the γ'-phase can also occur with the help of dislocations of $\frac{a}{6}$<112> type. Such twinning (known as the FCC-twinning; Gol'dshtein et al. 1986, Chakrabortty and Starke, 1975) is possible in the case of the alloyed alloys, for which the formation energy of a complex stacking fault in the intermetallic phase decreases due to the substitution of a part of nickel or aluminum atoms by the atoms of the alloying components. The studied superalloy ÉP-800 contains chromium and cobalt; these alloying elements enter the crystal lattice of the ordered γ'-phase instead of nickel and thus lower the degree of its long-range ordering (Savin et al 1999). Our study has shown that the twins detected in the nickel superalloy ÉP-800 can easily pass through both the ordered γ'-phase and the disordered γ-phase. This is possible only in the case, where the twinning occurs with the help of one type of partial dislocations permitted in both structures, i.e., $\frac{a}{6}$<112> type. Such partial dislocations should move in pairs over one plane, i.e., just like in the case of sliding of perfect dislocations considered above. Twins have been detected only in the impact zone. In the foil region at a distance of about 6 mm and farther from the impact zone, the structure is strongly strained but bears no twins. In this case, we may say that the stresses required for the start of twinning have been created only at the place of the impact and that the twinning process has occurred in a rapidly damping manner despite the fact that the dislocation density in the material was quite high even at a long distance from the fractured zone.

An experimental study of the deformation behavior of the $Co_3(Al,W)$ intermetallic compound was done by Okamoto et al. (2011). The faulted dipoles with the superlattice intrinsic stacking faults (SISF) with the displacement vector $R_F = +\frac{1}{3}[111]$ were observed under a low temperature deformation (77 K) in the structure of the close to single γ'-phase sample of the Co-12 at.% Al-11 at.% W alloys. The Burgers vector of the bounding partial dislocations was $b = \frac{1}{3}[11\overline{2}]$. The density and size of the faulted dipoles decreased with increasing deformation temperature up to 573 K, where the faulted dipoles disappeared (Okamoto et al. 2011).

Defect structure of the $Co_3(Al,W)$ intermetallic compound looks the same as the Ni_3Al structure. Intrinsic stacking faults may have also been observed during aging of $Co_3(Al,W)$-based alloys (Fig. 2.46).

Fig. 2.46 Intrinsic stacking faults in Co-10.8 at.% W-9.3 at.% Al, aging 900°C-1000 hours, the dark-field image $g = (1\overline{1}1)$.

2.1.4   *Anomaly in the temperature dependence of yield stress*

The APB energy γ_{APB} is given by $\gamma_{APB} = K \cdot \dfrac{G \cdot b^2}{2 \cdot \pi \cdot d}$, where K is a constant equal to 1 for screw dislocations and b is the Burgers vector (Marcinkowski et al. 1961). The APB energy in Ni_3Al-based and $Co_3(Al,W)$-base alloys is an important characteristic because it provides a contribution to the precipitation hardening provided by the γ'-phase, together with the coherency strain due to the lattice misfit (Table 2.72).

It is known that the high-temperature strength of alloys directly depends on the concentration of the strengthening phase. In the nickel superalloys, the strengthening phase is γ' (Ni_3Al, Pm-3m, $L1_2$). In the new intermetallic cobalt superalloys , this phase is γ' ($Co_3(Al,W)$, Pm-3m, $L1_2$). The high-temperature and heat-resistant alloys based on Ni_3Al ($L1_2$) and $Co_3(Al,W)$ are interesting, since their yield stress and, correspondingly, the work hardening rate exhibit an anomalous dependence on temperature.

Table. 2.72 Energy of planar defects in $L1_2$-type of structure

$L1_2$-type of structure	Stacking fault energy ΔE_{SISF} (Mottura et al. 2012)	Antiphase boundary energy, ΔE_{APB}
Ni_3Al	53 mJ/m^2	180 mJm^{-2} (Diologent and Caron 2004)
$Co_3(Al,W)$	89-93mJ/m^2	146 mJm^{-2} (Okamoto et al. 2011)

In the Ni_3Al-based alloys, this anomaly persists even if the ordered phase is in the form of precipitates. The yield stress of the Ni_3Al-based alloys grows, rather than decreases, up to some temperatures and it can reach the maximum T_m. The anomaly of the yield stress in both polycrystalline and single-crystalline alloys does not vanish, when the crystallographic orientation of samples changes with respect to the direction of the applied load. However, the lowest and the highest temperatures of the peak are observed at the orientations <111> and <100>, respectively (Thornton et al. 1970). The anomaly takes place only when the degree of deformation increases, while the yield micro stress ($\varepsilon = 10^{-6}$) has a normal temperature trend (Lall et al. 1979). The fracture strength of these alloys also exhibits a normal dependence on temperature (Aoki and Izumi 1978). However, the microhardness of the alloys, which is linearly related to the yield stress, has the same anomaly (Westbrook 1957). It is remarkable that the temperature dependence of plasticity in nickel superalloys exhibits a special behavior, such that the minimum of this dependence exactly fits T_m (Aoki and Izumi 1978). Thus, it may be stated that the increase in hardness, rather than strengthening, as erroneously conjectured by some researchers, takes place in these alloys. The anomalous behavior of the temperature dependence of the yield stress in Ni_3Al usually is related to a special mechanism of dislocation movements in the $L1_2$ structure. It is assumed that thermally activated transformations of glissile superdislocations to sessile configurations (Kear-Wilsdorf locks) represent processes in control of the plastic deformation of the aforementioned alloys. The model, which assumes formation of the locks, adequately describes the deformation behavior of the alloys (Greenberg and Ivanov 2002). Suzuki and Oya (1981) showed that a high anisotropy of the APB energy (namely, the APB energy in the cube plane is much smaller than the APB energy in the octahedron plane) is responsible for the temperature anomaly of the yield stress in the $L1_2$ structure. The APB energy in the (111) plane changes from 226 mJ/m^2 at 0 K to 193 ± 31 mJ/m^2 at 900 K. The APB energy in the (100) plane changes from 35 mJ/m^2 at 0 K to 44 ÷ 57 mJ/m^2 at 900 K. These APB energies approach one and the other at room temperature: 150 ± 15 mJ/m^2 in the (111) plane and 120 ± 15 mJ/m^2 in the (001) plane (Skinner et al. 1995).

Yield anomaly in Co$_3$(Al,W) was found only in a limited temperature range between 950 and 1100 K (Okamoto et al. 2011, Suzuki and Pollock 2008) with an onset temperature higher than in Ni$_3$(Al,X). The reason for the high onset temperature for anomalous strengthening stays is not clarified.

Temperature dependence of the 0.2% flow stress was studied in compression in polycrystalline and single crystal γ/γ' two phases Co-9 at.% Al-9 at.% W alloy (Suzuki and Pollock 2008), Co-9.4 at.% Al-10.7 at.% W alloy (Suzuki and Pollock 2008), and close to single phase (γ') Co-12 at.% Al-11 at.% W (Okamoto et al. 2011). It was found that the flow stress exhibited a strong, positive dependence on temperature above 873 K. Suzuki (2008) showed that, in comparison with the two-phase γ/γ' Ni$_3$Al-based alloys, where the anomalous increase in the 0.2% yield stress was slight, the two phase Co$_3$(Al,W)-based alloys exhibited larger flow stress anomalies, with a 50 MPa increase, regardless of the varying volume fractions of the γ' precipitates. These alloys showed a flow stress of about 520 MPa at room temperature. With increasing temperature, polycrystalline Co-9.4 Al-10.7 W alloy showed a minimum of 420 MPa at 873 K, and then showed a maximum of about 460 MPa at 973 K, followed by a rapid decrease with temperature (Suzuki and Pollock 2008). The single-crystal Co-9.4 Al-10.7 W showed strength, which is equivalent to the polycrystalline alloy up to 973 K, but the anomalous second stage was extended up to 1073 K. The flow stress at the peak temperature was 530 MPa. Above the peak, the flow stress also decreased rapidly (Suzuki and Pollock 2008). Results obtained by Suzuki and Pollock (2008) also show that the 0.2% flow stresses of the Co$_3$(Al,W)-based two phase alloys are superior to the conventional Ni-base sheet alloy (IN617) and Co-base wrought alloy. Unlike two phase γ/γ' alloys, for Co-12 at.% Al-11 at.% W (close to single phase γ') the high-temperature strength of the present $L1_2$ compound, Co$_3$(Al,W), was not so high when compared to the Ni$_3$Al-based $L1_2$ compounds (Okamoto et al. 2011). The yield stress of the Co-12 at.% Al-11 at.% W (close to single phase γ') exhibited a rapid decrease at low temperatures (down to room temperature) followed by a plateau (up to 950 K), then it increased anomalously with a temperature increase in a narrow temperature range of 950-1100 K, followed again by a rapid decrease at higher temperatures (Okamoto et al. 2011). As pointed out by Miura et al. (2007), such a temperature dependence differs from that reported for the Ni_3Al-based $L1_2$ compounds.

As it was shown by Vamsi and Karthikeyan (2017), the activation energy for a cross-slip is significantly higher in $Co_3(Al,W)$ than in Ni_3Al, due to the higher shear modulus in the former compound, which results in a higher peak temperature. The anomalous yield stress increase observed in $L1_2$ -$Co_3(Al,W)$ was consistent with the thermally activated cross-slip of the APB-coupled dislocations from (111) to (010). The dislocation locking in the deformation region 950-1100 K was associated with the complete or near complete Kear–Wilsdorf locks. As it was shown by Okamoto et al. (2011), this behavior is entirely similar to the microstructural signature of the positive temperature dependence of the yield stress in the Ni_3Al- and Co_3Ti-based $L1_2$ compounds. A dislocation slip in $Co_3(Al,W)$ was observed to occur exclusively on the (111) octahedral planes for deformations in the temperature region $T = 77$-1273 K, with no preference of the crystal orientations (Okamoto et al. 2011). However, unlike many other Ni_3Al- and Co_3Ti-based $L1_2$ compounds, where high deformation temperature proceeds by a slip on (010), the rapid decrease in the yield stress at high temperatures of the present $L1_2$ compound was considered to be due to the transformation from γ' to γ- phases (Okamoto et al. 2011).

2.1.5 Dependence of the phase composition of the Co-Al-W alloys on casting conditions

The Co-Al-W alloys refer to the materials with large differences in their melting temperatures and in the specific weight of the alloying components. Therefore, the method of casting may have a significant impact on the structure and phase composition of the alloys, because of the non-uniform distribution of the chemical elements throughout the ingot. Usually, different mechanical processing (rolling, forging, hydrostatic extrusion) combined with the temperature aging can be used to improve the homogeneity of the samples.

The Co-19 at.% Al-6 at.% W alloy was prepared by two casting methods. We used arc melting under an argon atmosphere with casting into a copper water-cooled casting mold and induction melting furnace with casting into a ceramic Al_2O_3 mold. The resulting cylindrical ingots with a diameter of 18 mm were tested in four conditions: (1) cast, (2) cast + annealing at 1200°C-10 hours followed by water quenching, (3) cast + forging at 1200°C, (4) cast + forging + annealing at 1200°C-10 hours followed by furnace cooling. The X-ray diffraction patterns of the samples show that the phase composition depends on the casting conditions and heat treatments. Kazantseva et al. (2015a,b) presented experimental results on the influence of casting conditions on the structure and phase composition of the Co-Al-W alloy (Table 2.73).

Table 2.73 Phase composition of the alloys obtained by different ways

Casting condition	Phase composition
Arc melting with casting into a cupper water cooling mold (Sample 1)	γ + B2 (CoAl) + Co$_3$W (D0$_{19}$)
Arc melting with casting into a cupper water cooling mold + annealing at 1200°C-10 hours followed by water quenching (Sample 2)	γ + B2 (CoAl)
Melted in an induction furnace + forging at 1200°C followed by furnace cooling (Sample 3)	γ + B2 (CoAl) + μ(Co$_7$W$_6$)
Melted in an induction furnace + forging + annealing at 1200°C-10 hours followed by furnace cooling (Sample 4)	γ + B2 (CoAl) + μ(Co$_7$W$_6$)

The structure of Sample 1 consists of the alternating areas of the solid solution (γ-phase) and the Co(Al,W) (B2) phase. Thin plates of the Co$_3$(Al,W) (D0$_{19}$) phase were found inside the areas of the solid solution. After the heat treatment (HT) at 1200°C-10 hours followed by water quenching, the D0$_{19}$-phase plates were dissolved. The microstructure of the forged sample (Sample 3) contains the small-sized μ-phase particles, whose numbers increase after the heat treatment followed by a slow furnace cooling (Sample 4). Average chemical compositions of the samples obtained by statistical processing of five measurements are presented in Table 2.74. As can be seen from Table 2.74, the average chemical compositions of the alloys obtained by the different methods of casting differ from each other. Cast sample (Sample 1) obtained by arc melting followed by fast quenching has considerable heterogeneity of the chemical composition. High-temperature homogenization of this alloy (Sample 2) equalizes the chemical composition of cobalt and aluminum throughout the ingot. We have used a precooked Co-Al alloy for melting of the samples by induction furnace and obtained more uniform chemical composition of the ingots.

The chemical composition of the Co(Al,W) (B2) phase is similar to the value obtained by Dmitrieva et al. (2005) for the Co-18.5 at.% Al-6.3 at.% W alloy, where the B2 phase contained 26.9 at.% Al and 4.5 at.% W. The appearance of the intermetallic Co$_3$W phase in the alloy obtained by arc melting (Sample 1, Fig. 2.47) can be attributed to the cooling rate of the ingot. Fast quenching allows the sample to pass through the high temperature area of the equilibrium phase diagram and bypass the low temperature area. Chemical compositions of the Samples 2 and 3 are presented in Tables 2.75-2.76.

Table 2.74 Chemical composition of the samples, at.%, SEM (EDS)

Element	Initial batch	Sample 1 cast	Sample 2 cast + HT	Sample 3 forged	Sample 4 forged + HT
CoK	75	77.05	73.15	72.41	72.95
AlK	19	18.46	22.45	21.52	21.29
Wl	6	4.50	4.40	6.07	5.76

Table 2.75 Chemical composition of Sample 2, at.%, SEM (EDS)

Element	γ-phase (FCC)	Co(Al,W) (B2)
CoK	74.89	72.51
AlK	20.63	23.02
Wl	4.48	4.46

Our results show that the studied part of the phase diagram contains the high-temperature area of the $D0_{19}$ phase and the low-temperature area of the μ-phase. This explains the discrepancy between the experimental results on the Co-Al-W phase diagram obtained by different authors. For example, a vertical slice of the equilibrium phase diagram in (Dmitrieva 2006) contains the μ-phase area, but the same temperature part of the phase diagram presented by Sato et al. (2006) contains the $D0_{19}$ phase.

Table 2.76 Chemical composition of Sample 3, at.%, SEM (EDS)

Element	γ-phase (FCC)	Co(Al,W) (B2)	Co_7W_6 (μ-phase)
CoK	79.31	67.25	48.96
AlK	13.63	27.58	3.78
Wl	7.06	5.17	47.17

Thus, the casting method (especially the rate of cooling) followed by a thermo-mechanical treatment, significantly affects the chemical homogeneity and phase composition of the Co-Al-W alloys, because of limited solubility of the alloying components.

A structure stability of the high temperature composite materials is usually determined by many factors: the misfit in the crystal lattices, the rate of diffusion of the different elements, the solvus temperature of the strengthening phases, and the long-range order of the intermetallic phases. In $Co_3(Al,W)$-based alloys both factors are important, since the γ'-phase is stable only when the composition is similar to $Co_3(Al_{0.5}W_{0.5})$ (Joshi et al. 2014).

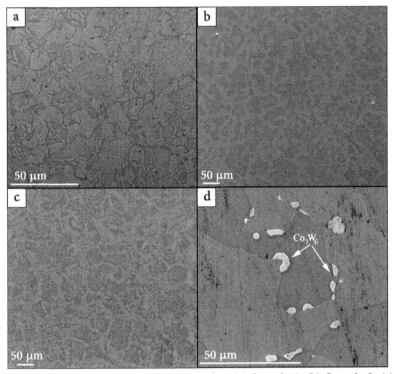

Fig. 2.47 The SEM results of the samples: (a) Sample 1; (b) Sample 2; (c) Sample 3; (d) Sample 4.

Alloying affects the phase stability, thus it needs to take into account the structural parameters such as site preference, for designing alloys with the desired misfit and optimal volume fraction of γ'-phase (Joshi et al. 2014). The site preference of quaternary elements in γ' precipitates was determined by (Meher et al. 2012). The authors indicated that Ni and Co occupy one sub-lattice, while Al, W and Ta occupy another sub-lattice in the ordered $L1_2$ structure.

The parameters of the structure stability have been studied in detail for the nickel superalloys because nickel superalloys are the natural composites with unique properties. The main feature of the γ/γ'-phases nickel superalloys is a low coherency, namely, the absence of the elastic strain fields on the boundary between the disordered gamma matrix and gamma prime precipitate. The morphology of a coherent γ' particle is determined by minimizing the sum of the elastic strain energy of the particle, the surface energy of the particle, and the elastic interaction energy between particles ((Doi et al., 1984). The precipitate morphology (namely, the shape and distribution of γ' precipitate particles) greatly influences the properties of nickel-base superalloys. Mismatch is a governing factor determining the precipitate morphology, as well as the coarsening and creep strength at high temperature for these alloys. Shape of the precipitates can change from a sphere to a cuboid, needle, and plate with an increase of the lattice mismatch degree (Doi et al., 1984). The average magnitude of the lattice mismatch is determined as follow:

$$\delta = \frac{a1 - a2}{0.5(a1 + a2)},$$

where $a1$, and $a2$ are the lattice parameters of the γ' and γ phases, respectively. Spherical precipitates generally were observed when the $\Delta^* = \delta / \gamma_s$, ratio of γ / γ' lattice misfit δ to surface energy density γ_s, ranged from 0.0 to 0.2 % , i.e. when the elastic interaction is weak. The cubic shape develops for $0.2 < \Delta^* < 0.4$, when the elastic interaction becomes stronger $(0.4 < \Delta^*)$ the cuboids are split up, and the plate-like structure is formed (Doi et al., 1984). In nickel superalloys, the different shape of the γ' precipitation can be obtained by alloying, forging, or special heat treatment, when the chemical composition of the alloy can change, and variations in lattice parameters can occur. The lattice mismatch creates the elastic coherent strains. A large mismatch in the lattice parameter may be beneficial for some alloys under certain testing conditions. However, in the alloys with a large mismatch, the precipitates can lose coherency with the matrix.

The sign of the lattice mismatch can be either a plus or a minus, depending on the size of atoms in the crystal lattices of the matrix and strengthening phase. Unlike Ni base superalloys, $Co_3(Al,W)$ based superalloys have a positive mismatch, namely, $a_{\gamma'} > a_\gamma$ by ~0.5%. Sato et al. (2006) showed that the lattice parameter mismatch in the two phase γ / γ' Co-9.2 at.% Al-9 at.% W alloy was 0.53%. The lattice parameters of the γ and γ' phases of the alloy heat-treated at 1173 K were determined by the room temperature X-ray diffraction at 0.3580 and 0.3599 nm, respectively. The alloy had a cuboidal γ'-phase structure. Positive mismatch can result in high energy semi-coherent γ / γ' interfaces, with a tendency to coarsen rapidly (Joshi et al. 2014). Variations in the lattice parameter may occur from both local changes in chemical composition and from the constraints of coherency. Because of this, the alloying elements that increase the lattice parameter of γ-phase and decrease the lattice parameter of γ'-phase may be attractive for optimal mechanical properties. The example of the mismatch change in $Co_3(Al,W)$-based alloy is presented in Fig. 2.48. The Co-9.7 at.% Al-10.8 at.% W alloy was prepared by induction melting under an argon atmosphere. The cast samples were cold rolled and heat treated at 1350°C for 24 hours followed by water quenching. After homogenization, the samples were subjected to heat treatments at 900°C in a vacuum of 10^{-3} Pa.

Decreasing the lattice parameters of the γ- and γ'-phases with an increase in the annealing time indicate a depletion of theses phases of the alloying elements, namely tungsten (Table 2.77). The μ-phase (Co_7W_6) diffraction reflexes are present in X-ray diffraction patterns of the alloy.

Table 2.77 The crystal parameters of γ- and γ' phases and lattice mismatch in the Co-9.7 at.%Al-10.8 at.% W alloy

Annealing, h	a_{γ}, ±0.0001 nm	$a_{\gamma'}$, ±0.0001 nm	δ,%	$V_{\gamma'}$, %
4	0.35949	0.36034	0.24	64
72	0.35796	0.36011	0.6	71

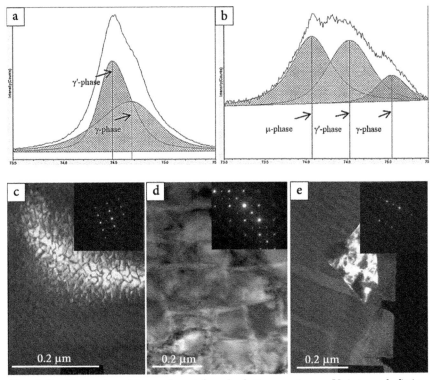

Fig. 2.48 Example of the mismatch calculations using a Voigt peak-fitting for the separation of experimental diffraction line $(220)\gamma/\gamma'$: (a)-(c) annealing at 900°C for 4 hours, SAED pattern to (c), zone axis $(011)\gamma'$; (b)-(d) annealing at 900°C for 72 hours, SAED pattern to (d), zone axis $(100)\gamma'$; (e)- the μ-phase precipitation in the sample after annealing, the dark-field image in $(024)\mu$, SAED pattern to (e), zone axis $(071)\mu$.

Volume fractions of the γ' phase in the alloy after annealing were calculated from the ratio of the intensities of the separated X-ray reflexes as follows (Petrushin et al. 2015):

$$V_{\gamma'} = \frac{I_{\gamma'}}{I_\gamma + I_{\gamma'}}$$

2.1.6 Mechanical properties of the Co₃(Al,W)-base alloys

Strengthened $Co_3(Al,W)$ based alloys are considered as a new class of high-strength materials, because of the $Co_3(Al,W)$ hardening intermetallic compound (γ'-phase, $L1_2$, structural type Cu_3Au). The mechanical properties of the Co-Al-W alloys were investigated in a number of papers and compared to the well-known refractory commercial alloys based on Ni_3Al. Such a comparison makes sense, because the Co-Al-W alloys are mainly considered in the literature as promising hot-strength materials. The cobalt alloys are less strong, but they possess better wear resistance than the nickel heat-resistant strong alloys (Suzuki and Pollock, 2008). The cobalt alloys, just as the nickel alloys, exhibit an anomalous positive temperature dependence of 0.2% flow stress above 600°C (Suzuki and Pollock, 2008, Ishida 2008). However, the cobalt alloys, unlike the nickel alloys, have good casting properties (Petrushin et al. 2015). The cobalt alloys show excellent mechanical properties at a temperatures of up to 807°C (1077 K), but their strength rapidly decreases with increasing temperature. This is assumed to be related to low values of the temperature of the γ' solvus, compared to the nickel alloys, in which the temperature of the γ' solvus reaches 1340°C (1613 K) (Ooshima et al. 2010, Pyczak et al. 2015). Influence of the tungsten content on the mechanical properties of Co-Al-W alloys was investigated by Pyczak et al. (2015). The authors have shown that in the Co-8.5 at.% Al-7.4 at.% W, Co-8.5 at.% Al-8.7 at.% W, and Co-8.9 at.% Al-10.3 at.% W alloys, an increase in the tungsten content leads to a growing volume fraction of the γ' phase and of the solvus temperature of the γ' phase. As a consequence, the creep resistance increases at 850 and 900°C. The creep rate at 850°C in the alloy with 11 at % W was 4.8×10^{-8} s^{-1} (Pyczak et al. 2015.). The creep resistance of cobalt alloys was significantly improved and the γ' solvus temperature was increased after alloying with Ta, Ti, Ni, Nb, and V (Suzuki and Pollock 2008, Ooshima et al. 2010, Pyczak et al. 2015, Petrushin et al. 2015, Bauer et al. 2010).

Elastic constants are the important parameters that describe responses to applied macroscopic stresses. The elastic constants for a $Co_3(Al,W)$ single crystal (Co-10 at.% Al-11 at.% W) were experimentally determined by Tanaka et al. (2007) by resonance ultrasound spectroscopy at liquid helium temperatures and theoretical determined by Yao et al. (2006). Table 2.78 presents the elastic constants for this intermetallic compound in both states, in comparison with those for Ni_3Al (Yao et al. 2006).

Table 2.78 Elastic constants for $Co_3(Al,W)$ and Ni_3Al

Alloy	C_{11}, GPa	C_{12}, GPa	C_{44}, GPa
$Co_3(Al,W)$, experiment	271	172	162
$Co_3(Al,W)$, theory	363.39	189.92	211.64
Ni_3Al, theory	227	148	120

Using the single-crystal elastic modulus obtained by Tanaka et al. (2007), the other parameters of the stoichiometric "ideal" $Co(Al_{0.5}W_{0.5})$ composition were calculated (Table 2.79). The Debye temperature was derived from the data on the elastic constants for the $Co_3(Al,W)$ stoichiometric single crystal and density temperature was derived using equations obtained by Chen and Sundman (2001): $\Theta_D = \dfrac{h}{k_B}\left(\dfrac{3n}{4\pi V_a}\right)^{\frac{1}{3}} V_m$,

where V_m the Debye sound velocity and V_a is the volume of the solid.

For an elastically isotropic medium, the Debye sound velocity is (Fatmi et al. 2011):

$$V_m^{-3} \frac{1}{2}\left(\frac{1}{V_L^3}+\frac{1}{V_T^3}\right),$$

where V_L and V_T are longitudinal and transverse sound velocities, related to the shear modulus (G), bulk modulus (B), and the density (ρ) by

$$\rho V_L^2 = B+\frac{4}{3}G, \rho V_T^2 = G.$$

For cubic systems, the isotropic bulk and shear moduli are given as follows: $G = \dfrac{G_V + G_R}{2}$, $B = \dfrac{(C_{11}+2C_{12})}{3}$.

The bounds on the shear modulus are (Fatmi et al. 2011):

$$G_V = \frac{(C_{11}-C_{12}+4C_{44})}{5}, G_R = \frac{5(C_{11}-C_{12})C_{44}}{4C_{44}+3C_{11}-3C_{12}}.$$

Table 2.79 Calculated parameters for stoichiometric $Co_3(Al,W)$

Density, ρ, g/cm^2	Longitudinal velocity V_L, ms^{-1}	Transverse velocity V_T, ms^{-1}	Average sound velocity V_m, ms^{-1}	Debye temperatures $\Theta_{D,K}$
10.071	6858.59	4100.68	4135.96	343

In comparison with the density of the Ni_3Al intermetallic compound (7.7 g/cm^2), the calculated density of the $Co_3(Al,W)$ intermetallic compound is higher, because of the "heavy" tungsten. The calculated Debye temperature for $Co_3(Al,W)$ is similar to that for Ni_3Al (360 K).

An experimental study of the mechanical and magnetic properties of Co-X at.% Al-Y at.% W (where X= 9; 7.9; 8.2; 8.7; 8.2, Y = 4.6; 6.8; 8.5;

10; 12.6) polycrystalline alloys has been carried out depending on the tungsten content by Kazantseva et al. (2017a). The alloys were obtained by the Bridgman method; the rate of the directional solidification was 0.83 mm/min, and the temperature gradient at the solidification front was 80 K/cm.

There are three clearly pronounced peaks in the DSC curves of the heating of the samples, which correspond to the temperature ranges of 760-780, 1000-1050, and 1250-1270°C. With allowance for the data from the high-temperature X-ray diffraction study obtained by Kazantseva et al. (2016a), it can be assumed that the low-temperature range probably corresponds to the region of the formation of the $D0_{19}$ (Co_3W) phase, the second range is linked with the dissolution of the γ' phase, and the third range corresponds to the region of existence of the μ phase (Co_7W_6).

However, according to the magnetic measurements, a magnetic transition was found near the temperature of 776°C (1049 K); thus, there are probably two transformations in this range that are superimposed, namely, a weak diffusion-controlled transition (formation of a phase with a $D0_{19}$ structure) and a magnetic transition (formation of the γ' phase). There is a weak anomaly in the DSC curves in the region of 900°C (1173 K). Comparing the DSC data and the data of magnetic measurements, it can be concluded that the weak anomalies in the DSC curves in this temperature range are related to the magnetic transitions in the γ' phase. The weak thermal effect may indicate the fast reaction rate. As can be seen from the DSC curves, the most clearly pronounced is the order–disorder ($\gamma' \rightarrow \gamma$) phase transition that occurs with heat absorption. Results of the specific heat estimation of the $\gamma' \rightarrow \gamma$ phase transition, determined from the integrated area of the DSC peak are presented in Table 2.80. As can be seen, the highest amount of heat was absorbed upon the phase transition in Sample 5 with 12.6 at.% W, which contains the highest content of tungsten. The results indicate the diffusional character of the transition, the rate of which is controlled by the diffusion of the slowest component (tungsten) of the alloy.

Table 2.80 Results of differential scanning calorimetry

Co-Al-X alloy, (X = at.% W)	4.6	6.8	8.5	10	12.6
Specific heat of transition, J/g (exo)	141	172	176	179	304

Besides the heat effect, we note an increase in the temperature range of the phase transition and the deceleration of the rate of the $\gamma' \rightarrow \gamma$ reaction, together with an increase in the tungsten content in the alloy. The decrease in the onset of the $\gamma' \rightarrow \gamma$ phase transition at low and high contents of tungsten in the alloy may indicate the presence of a low-temperature dome-like γ' region (miscibility gap); a possible mechanism of its appearance is the spinodal decomposition. The latter was observed by the method of atomic-probe tomography in a commercial nickel superalloy after rapid quenching from 1300°C (Tan et al. 2014). In this alloy, particles of the γ' phase with size of 30-50 nm with a diffuse non-equilibrium interphase boundary were revealed, the appearance of which Tan et al. (2014) ascribed to the processes of ordering and spinodal decomposition. The assumption on the presence of spinodal decomposition in the Co-Al-W alloys can explain the variation of the temperatures of the onset of the $\gamma' \rightarrow \gamma$ phase transition reported in the literature, depending on the changes in the chemical composition of the alloy. In this case, the formation of a characteristic cuboid-like γ/γ' structure in a temperature range of 900-1000°C is a result of the γ'-phase decomposition.

The γ' solvus temperature and the range of the $\gamma' \rightarrow \gamma$ transformation were determined by Pyczak et al. (2015). The shift of the peaks toward higher temperatures at the onset of the transformation and a narrower range of the transformation in the DSC curves may be related with the heating rate of the samples. For example, Petrushin et al. (2015) performed the heating at a rate of 273 K/min and, Pyczak et al. (2015) made the heating and cooling rates equal to 20 K/min.

The temperature of the γ' solvus was determined based on the DSC data presented in a number of researches (see Table 2.81).

Table 2.81 Summary results of the differential scanning calorimetry

Composition of alloy, at %	Temperature of γ' solvus, °C	Reference
Co-9Al-10W	990	Sato et al. 2006
Co-9.2Al-9W	1000	Petrushin et al. 2015
Co-9.4Al-10.7W	1000	Suzuki and Pollock 2008
Co-9Al-9W	963	Bauer et al. 2010
Co-9Al-6W	948	Ooshima et al. 2010
Co-9Al-7W	964	Ooshima et al. 2010
Co-9Al-9W	985	Ooshima et al. 2010
Co-7Al-7W	854	Yan et al. 2014a
Co-8.5Al-7.4W	960	Pyczak et al. 2015
Co-8.5Al-8.7W	985	Pyczak et al. 2015
Co-8.9Al-10.3W	1036	Pyczak et al. 2015
Co-8Al-4.6W	1009	Kazantseva et al. 2017a
Co-7.9Al-6.9W	1000	/-/
Co-8.2Al-8.5W	1014	/-/
Co-8.7Al-10W	1021	/-/
Co-8.2Al-12.6W	1044	/-/

Kazantseva et al. (2017a) used the heating rate that was 4.6 K/min, which fully explains the low temperatures of the onset of the $\gamma' \rightarrow \gamma$ transition and the wider range of transition observed in our alloys compared to the data presented by Petrushin et al. (2015) and Pyczak et al. (2015) (Table 2.81).

The increased values of the γ' solvus published in the literature compared to Co-8.7Al-10W alloy investigated by Kazantseva et al. (2017a) (see also Table 2.81) may be related to the changes in the morphology of the γ' phase. The alloys investigated by Suzuki and Pollock (2008) were preliminarily aged at 900°C for 150 hours. Bauer et al. (2010) aged them for 200 hours and Ooshima et al. (2010) aged the alloys at 850°C for 96 hours, which led to a coarsening of the particles of the γ' phase and to the possible delay of its dissolution.

The review of mechanical properties of the $Co_3(Al,W)$-based alloys compared with nickel alloys was presented by Sato et al. (2006). Unlike the polycrystalline Ni_3Al samples, which possess an enhanced brittleness up to the temperatures close to the melting point, the cobalt polycrystalline samples are relatively ductile both at room temperature and at high temperatures (Suzuki and Pollock 2008). Mechanical properties of the Co-Al-W alloys can be estimated by Young's moduli E. It was found that Young's moduli of Co-Al-W alloys exceed the E values determined under the same conditions for the commercial nickel-based superalloys (Romanov et al. 2017). For example, acoustic experiments by Rinkevich et al. (2013) showed that, for the ZhS-36 alloy (with the volume fraction of the γ' phase of 64%), E = 246 GPa, and for the VKNA-4U alloy (with the volume fraction of the γ' phase of 90%), E = 249 GPa (this value agrees with the results of our nanoindentation experiments). Young's moduli E and hardnesses H of the ZhS-32 and VKNA-4U nickel superalloys were determined by nanoindentation (F = 1.5 N). The experimental investigation of the elastic characteristics of a single-crystal sample of the $Co_3(Al,W)$ alloy showed the presence of a stronger anisotropy of Young's modulus as compared to the $Ni_3(Al,Ta)$ alloy. The elastic moduli were determined by the method of resonant ultrasonic spectroscopy at the temperature of liquid helium (Table 2.82) (Ooshima et al. 2010, Tanaka et al. 2007). As was shown by Ooshima et al. (2010), the elastic moduli of the alloys based on the $Co_3(Al,W)$ intermetallic compound are by 15-25% higher than those for the alloys based on $Ni_3(Al,Ta)$. In cobalt alloys alloyed with nickel, at cryogenic temperatures the $Co_3(Al,W)$ intermetallic compound has the tensile strength that is approximately 20% larger than that of Ni_3Al. However, it was shown by Yan et al. (2014b) that the cobalt alloys are softer than the nickel alloys at room temperature. This disagreement may be related to the differences in the methods of investigation.

Table 2.82 Mechanical properties of the Co$_3$(Al,W)-based alloys
and Ni superalloys

Alloys	E, GPa	H, MPa	Reference
Co$_3$(Al,W)	260	-	Tanaka et al. 2007
Ni$_3$(Al,Ta)	216	320	Ooshima et al. 2010, Booth-Morrison et al. 2008
Waspaloy	211	200	Ooshima et al. 2010
Co-9Al-10W	-	280	Ooshima et al. 2010
Co-9.2Al-9W	-	400	Ishida 2008
Co-9Al-9W	242 (γ)	-	Povstugar et al. 2014
	245 (γ')	-	/-/
Co-7Al-7W	-	468	Yan et al. 2014a
ZhS-32	244	466	Romanov et al. 2017
VKNA-4U	244	410	Romanov et al. 2017

As has been shown in this study, the microhardness (H) and Young's modulus (E) of the alloys investigated in this work depend not only on the tungsten content in the alloy, but also on the value of the load applied upon the nanoindentation. Note that the dependence of the elastic modulus on the force applied to the indenter is much higher than the dependence of the microhardness, which is caused by the transition from elastic to plastic deformation on an increase in the indentation force. To obtain reliable results, the nanoindentation of nickel alloys with different contact loads was performed. The results obtained were compared with the literature data on the ultrasonic determination of Young's moduli of the same nickel alloys. According to the data of acoustic experiments (Rinkevich et al. 2008), for the VKNA-4U alloy (volume fraction of the γ'-phase of 90%), the modulus is E = 249 GPa. We determined that under the maximum load F = 1.5 N, Young's modulus of the VKNA-4U alloy was E = 249 GPa. It may be assumed that the use of maximum load F = 1.5 N under nanoidentation also produces most reliable values of the elastic moduli in the case of the cobalt alloys. Results of the investigation of the mechanical properties of alloys carried out by Kazantseva et al. (2017a) are presented in Table 2.83.

Table 2.83 Mechanical properties of Co$_3$(Al,W)-based alloys
investigated by the nanoindentation (F = 1.5 N)

Alloy	E, GPa	H, MPa
Co-9 at.% Al-4.6 at.% W	240	290
Co-7.9 at.% Al-6.8 at.% W	269	414
Co-8.2 at.% Al-8.2 at.% W	248	400
Co-8.7 at.% Al-10 at.% W	259	413

As can be seen from this table, the value of the microhardness obtained for the alloy 3 (Co-8.2Al-8.5W) under the maximum indentation load of

1.5 N is 400 MPa and Young's modulus is equal to $E = 248$ GPa, which is close to the value obtained for the Co-9Al-9W alloy (Table 2.83).

Thus, we conclude that the $Co_3(Al,W)$-based intermetallic alloys are more rigid and have higher hardness at room temperature than the commercial nickel superalloys. On the other hand, the cobalt intermetallic alloys are more ductile and are easily deformed at room temperature, unlike nickel alloys.

2.2 Magnetic Properties of the $Co_3(Al,W)$-Based Alloys

2.2.1 Effect of Al and W alloying on magnetic properties of Co-based alloys

Cobalt has the highest Curie temperature (1121°C) among all known ferromagnets. This allows one to create cobalt-based materials with desired magnetic properties, such as the soft or permanent magnets (Mishin 1981). Cobalt has wide industrial applications; it enters into the composition of many steels, heat-resistant alloys, and coatings, and is also used widely in the production of various magnetic materials. The magnetic properties of heat-resistant cobalt-containing alloys can be significantly affected by alloying. For example, alloying with tungsten or aluminum decreases the Curie temperature of the solid solution based on α(FCC) cobalt. In the heat-resistant cobalt alloys, besides the ferromagnetic solid solution based on α(FCC) cobalt (the solid solution in the Co-Al-W system based on the FCC phase is called the γ phase), various intermetallic phases can be present both ferromagnetic and paramagnetic.

Phase diagrams of the binary Co-W system have been studied by Rama (1986). It is known that tungsten reduces the Curie temperature of the cobalt solid solution down to 865°C at 6 at.% W. The Co-W binary equilibrium diagram demonstrates both phase and magnetic transformations. Magnetically-induced phase separation associated with the solubility of tungsten in the various phases was confirmed by Sato et al. (2005) and Okamoto (2008).

Except the heat resistant superalloys, the magnetic properties of electrodeposited Co-W alloys are of interest for the recording media. These materials may be interesting for the magnetic MEMS devices, such as microactuators, frictionless microgears, sensors, or micromotors which are constituted of both magnetically hard and soft electrodeposited components. It was found that, depending on the electrodeposition conditions, Co-W coatings may exhibit either hard or soft ferromagnetic properties. The saturation magnetization, M_S, and coercivity, H_C, decrease monotonically as the amount of W increases (Tsyntsaru et al. 2013). The authors found that H_C increases from 170 Oe for pure Co films to 470 Oe for films containing 2-3 at.% W. At higher tungsten contents the coatings are magnetically softer, and the electrodeposits become non-ferromagnetic beyond ~30 at.% W. Formation of the hard-magnetic HCP Co_3W clusters might be partially responsible for this semi-hard ferromagnetic behavior. The Curie temperatures of the Co_7W_6 intermetallic phase (μ phase) and of the Co_3W phase in the ternary alloy Co-19 at.% Al-6 at.% W were determined by Kazantseva et al (2015b) to be equal to 327°C (600 K) and 527°C (800 K), respectively.

The Co-Al system has the ($B2$-CoAl) phase, whose magnetic state depends on the chemical composition and ferromagnetic cobalt solid solution. Aluminum reduces the Curie temperature of the alloy down to 840°C at 7 at.% Al (Lyakishev 1996). In the Co-Al system, the intermetallic compound CoAl ($B2$-phase) is paramagnetic, when the cobalt content in the alloy is less than 50 at.%; it is ferromagnetic with a cobalt content of more than 50 at.%. Thus, changes in the magnetic properties of weakly magnetic alloys based on the intermetallic compound CoAl can serve as an indicator of the degree of the imperfection of its crystal lattice. The Curie temperature of the intermetallic phase CoAl in the alloy with a content of cobalt equal to 57.9 at.% is ~153°C (Kobayashi 2009). The Curie temperature of the α phase decreases to 840°C with an increase in the aluminum content in the alloy, remaining constant beyond 8 at.%, according to Rama (1986). According to other data (Mishin 1981), it decreases to 800°C and remains constant above 10 at.% Al. The defect state of the crystal lattice of the $\beta'(B2)$ phase significantly affects the physical properties of the CoAl intermetallic compound. The investigation of the defect state and of the magnetic properties of CoAl in the binary alloy performed by Tamminga (1973) and Kulikov et al. (1999) showed that, with the content of cobalt of less than 50 at.%, the intermetallic compound CoAl (β' phase ($B2$)) is paramagnetic; with a greater content (>50.4 at.% Co), CoAl is ferromagnetic. The alloy Co-44% Al is ferromagnetic; with a Curie temperature T_C = 180 K (Tamminga 1973) and increasing cobalt content in the alloy, the Curie temperature increases. In the alloy with a cobalt content equal to 57.9 at.%, T_C = 426 K (Wachtel et al. 1973).

Magnetic properties of the different phases in Co-Al system were studied by Kazantseva et. al (2016b). Studies were provided with a *Lake Shore 7407* vibrating sample magnetometer in magnetic fields to 17 kOe, in a temperature range from the room temperature to 1000°C, in an inert atmosphere at the frequency of vibrations of 82 Hz and their amplitude of 1.5 mm.

A comparison of the DSC data of the studied Co-9 at.% Al alloy with the results of thermo-XRD (Kazantseva et. al. 2016b) and the literature data (McAllster 1989, Lyakishev 1996) indicates that the temperature of 280°C corresponds to the $\alpha + \varepsilon \rightarrow B2 + \alpha$ phase transition in that alloy.

The temperature of 950°C can be related to the presence of a mixing zone of the paramagnetic and ferromagnetic α phases with different aluminum contents (magnetically induced miscibility gap observed in many binary alloys of cobalt (Co-W, Co-Cr, Co-Mo) and given by analogy in the Co-Al phase diagram (Ishida 2008).

In our case, the sample is in the ferromagnetic state to 853°C (1126 K), retaining the value of the specific magnetization $\sigma = 13.6$ A·m^2/kg. Note the existence of bends in the graph of the temperature dependence of the specific magnetization σ of the initial sample, which correspond to three values of the Curie temperatures, which are well visible in the differential curves of the temperature dependence dσ/dt: 144°C (417 K), 254°C (527 K), and 853°C (1126 K). The results agree well with the literature data for the Curie temperatures of the $B2$ phase of a nonstoichiometric composition, the ε phase ($2H$, HCP), and α phase (FCC). Furthermore, in the curve of the differential dependence dσ/dt, a small broadening was found near 254°C (527 K), which indicates the presence of one more Curie temperature of ~288°C (561 K). We assume that it may be related to an unknown phase, lines of which were revealed in the X-ray diffraction pattern of the initial alloy. The difference in the slope of the curves of the specific magnetization of the sample in the initial state (1) and after heating (2) to 1000°C (1273 K) indicates the different magnetic state of the alloy. In both cases, the samples have high specific saturation magnetization $\sigma = 140$ A·m^2/kg. The coercive force decreases by a factor of two after heating and cooling the alloy.

As it is known, the X-ray diffraction analysis is a surface method, making it possible to reveal phases with contents greater than ~5%. The magnetic studies give the opportunity to reveal magnetic phases, whose content in the volume of the alloy is at least 1%. The transmission electron microscopy also makes it possible to reveal low percentages of phases (2-3%); however, this method also refers to surface local methods of study and depends on the selected region of a sample. Therefore, the absence of the $B2$ phase in the X-ray patterns of the initial sample and in the TEM images, and its detection by magnetic measurements can indicate its extremely low content. The appearance of lines of the $B2$ phase in the XRD pattern on heating to 550°C can mean that its percentage reached the level of the sensitivity of the instrument (5%).

Thus, only the application of a complex analysis, which includes structural and magnetic methods of study, makes it possible to obtain true information about the state of the magnetic material.

Magnetic properties of the $B2$(CoAl) phase were also studied by Kazantseva et al. (2017a) in the Co-Al-Si alloys. Measurements have shown that the magnetization of the alloys increases linearly with increasing magnetic field and that the magnitude of the magnetic susceptibility does not exceed 10^{-5}. No anomalies were observed in the temperature dependences of the magnetization and AC susceptibility. Therefore, it can be concluded, that these compounds are paramagnetic in the temperature range under investigation (77-360 K). According to Bester et al. (1999) and Tamminga (1973), deviations from the stoichiometric composition of CoAl lead to changes in the character of the occupation of the crystal lattice of the $\beta'(B2)$ phase by different atoms and in its magnetic properties. In our case, taking into account that silicon, which is a substitutional element, enters the aluminum sublattice (Bozzolo et al. 1998), all three alloys have an excess of cobalt. According to the literature data, they should exhibit an antisite character of the replacement of atomic positions and be weak ferromagnets (Tamminga 1973, Wachtel et al. 1973). The paramagnetic state of the alloys, investigated by Kazantseva et al. (2017a), may testify for the presence of defects of the crystal lattice of the intermetallic compound CoAl.

2.2.2 Magnetic properties of the phases in $Co_3(Al,W)$-based alloys

Cobalt-based intermetallic alloys are comparable in strength to nickel superalloys (Sato et al. 2006, Suzuki and Pollock 2008, Kazantseva et al. 2005, Stepanova et al. 2011), which are paramagnetic (Stepanova et al. 2011). In contrast, the cobalt-based heat-resistant alloys have properties of the ferromagnetic materials.

Magnetic properties of a set of $Co_3(Al,W)$-based alloys with different tungsten composition were studied by Kazantseva et. al. (2017a, 2017c). The $Co_3(Al,W)$-based alloys were melted by the Bridgman method. The ingots were subjected to a homogenizing annealing at 1250°C for 24 hours with a subsequent slow cooling in a furnace. The alloy compositions were determined by the spectral analysis. The magnetization was measured using a vibrating-sample magnetometer in magnetic fields with strength of up to 18 kOe in a temperature range of 77-320 K. The *ac-* magnetic susceptibility was measured using the method of a compensated transformer in a sinusoidal magnetic field with an amplitude of 5 Oe and a frequency of 80 Hz, in a temperature range of 77-360 K. The magnetic domain structure was studied by the method of a Magneto-Optical Indicator Film (MOIF) (Nikitenko et al. 2003). The film was superimposed on the sample and the image of the domain structure was created due to the Faraday rotation of the polarization of light in the stray fields of the sample magnetization.

All of the investigated alloys are in the ferromagnetic state at temperatures of up to ~900°C (1173 K). The curve of the temperature dependence of the specific magnetization σ of the alloys 1 and 5 is shown in Fig. 2.49. There are kinks in the graphs of the temperature dependence of the specific saturation magnetization σ_s (149-151°C (422-424 K), 752-776°C (1025-1049 K), and 905-911°C (1178-1184 K)), which correspond to three values of Curie temperatures and are clearly seen in the differential curves of the temperature dependence $d\sigma/dt$. These results may be related to the magnetic transitions in three different phases, whose Curie temperatures depend on the tungsten content in the alloy (Fig. 3.13). The first value, 149-151°C (422-424 K), is in a good agreement with the literature data for the Curie temperatures of the $B2$ phase (Wachtel et al. 1973, Kazantseva et al. 2016b, Kazantseva et al. 2016a) with a small content, hence not found by the X-ray diffraction analysis and transmission electron microscopy (Kazantseva et al. 2016a).

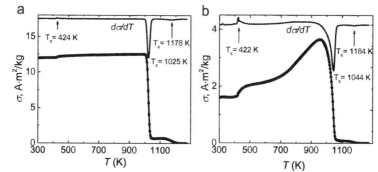

Fig. 2.49 Temperature dependence of specific saturation magnetization, $H = 100$ Oe, rate of measurement was 2 K/min: (a) Co-9 at.% Al-4.6 at.% W alloy; and (b) Co-8.2 at.% Al- 12.6 at.% W alloy.

It is possible that a small amount of this phase is retained in the dendrite structure of the alloy. The presence of the $B2$ phase in the dendritic regions was observed in cast homogenized Co-10Al-12W and Co-10Al-13W alloys (Miura et al. 2007). The second value of the Curie temperature, 752-776°C (1025-1049 K), is related to the magnetic transitions in the γ' phase, whereas the third value, 905-911°C (1178-1184 K), is attributed to the magnetic transitions in the γ phase. The dependence of the Curie temperature of the γ' phase on the content of tungsten in the alloy is presented in Table 2.84.

Table 2.84 Magnetic properties of the Co$_3$(Al,W)-based alloys

Tungsten content, at.%	4.6	6.8	8.5	10	12.6
Curie temperature of γ' phase (Co$_3$(Al,W)), K	1025	1052	1050	1049	1044
Curie temperature of γ' phase, K	1178	1182	1182	1184	1184
Specific saturation magnetization σ_s, emu/g	86	61	38	31	12
Coercive force H$_c$, Oe	1.5	2.5	140	220	500
Size of the γ'-precipitations	3-5	10	20	25	50

The Curie temperature of the γ' phase increases sharply with an increase of the tungsten content from 4.6 to 6.8 at.%; beyond, it changes only slightly with increasing tungsten content to 12.6 at.%. An increase in the tungsten content in ternary alloys leads to a decrease of the specific saturation magnetization, which changes from 85 A m^2/kg in the alloy with 4.6 at.% W to 14 A m^2/kg in the alloy with 12.6 at.% W. In contrast, the coercive force increases with the tungsten content in the alloy (Fig. 2.50). It is known that the coercivity of soft magnetic materials is in a range of 8-800 A/m (0.1^{-10} Oe) (Buschow and de Boer 2003). Thus, changes of the chemical composition of the Co$_3$(Al,W)-based alloys, especially the tungsten content, leads to the magnetic state of the alloys changing from soft to hard magnet (Table 2.84, Fig. 2.50).

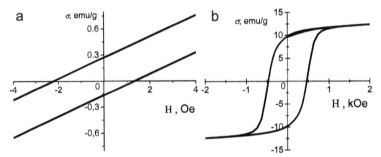

Fig. 2.50 Room-temperature hysteresis loops for the ferromagnetic materials: (a) Co-9Al-4.5W; (b) Co-8.2Al-12.6W.

As shown in Table 2.84, the changes in the size of the intermetallic γ'-precipitations and tungsten content effect on the magnetic properties of the alloys. It is known that specific saturation magnetization is not a structure-sensitive property of materials. This property is determined by the material composition and it does not depend on the size of the precipitation, the level of macro- or micro- stresses, and the changes in dislocation density (Geller 1983).

Specific saturation magnetization depends on the nature of the chemical bond and the distance between atoms in the crystal lattice; for ordered alloys saturation magnetization depends on the degree of long-range order (Geller 1983). During phase transitions in alloys the magnitude of the saturation magnetization may vary noticeably; formation of new phases or/and changes of the quantitative ratio between the magnetic phases, which are present in the alloy, influence the saturation magnetization (Moniruzzaman et al. 2012). An abrupt change in the saturation magnetization was also observed during the transition of the alloy to nano-scale state (Andrievski and Ragulya 2005). The Curie temperature is the same magnetic characteristic of the materials. Unlike the Curie temperature and saturation magnetization, coercive force depends on the structural changes in the materials (Geller 1983). Structural defects and grain size increase the value of the coercive force of soft magnets. Grain size dependence of the magnetic properties of various types of soft magnets was especially studied by (Herzer 1996). A maximum of the coercive force (10 A/m) in the industrial soft magnets was found in the nano-scale structure with grain size of 100 nm; a minimum was found in nano-scale structure with 10 nm or in single crystals with a grain size of 1 mm (Herzer 1996).

Our study shows that the alloys can be seen as two-phase composite ferromagnetic materials; magnetic properties of these materials depend on the tungsten content and size of the intermetallic γ'-phase ($Co_3(Al,W)$). The best result for these materials as soft magnetic materials may be achieved in alloys with a very small size of the $Co_3(Al,W)$ intermetallic precipitations (Fig. 2.50).

Influence of the tungsten content on the specific saturation magnetization and Curie temperature (Table 2.84) in the intermetallic $Co_3(Al,W)$-based alloys may be explained by taking into account the long- range order of the $Co_3(Al,W)$ intermetallic compound. It is known that the changes in the long-range order of the non-stoichiometric Ni_3Al intermetallic compound with the $L1_2$-type of the crystal lattice is explained by the anti-structural bridge mechanism (ASB) which allows existence of Al_{Ni} antisite (anti-structure) atoms or Ni_{Al} antisite atoms. The anti-structure bridge (ASB) mechanism consists in the sequences of atomic jumps when the anti-structure Al atom and the Ni vacancy exchange their positions (Moniruzzaman et al. 2012). It may be suggested that the $L1_2$ crystal lattice of the $Co_3(Al,W)$ intermetallic compound also admits the existence of the anti-site atoms. It is known that the long-range order is determined by the measurements of the difference between the intensity of the structural and super structural reflexes of the ordered phase. Suggestion about anti-site exchanges in the $Co_3(Al,W)$ crystal lattice supported by the changes in the intensity of the γ'-super structural reflexes in the SAED patterns with changes of tungsten content in the alloy. In non-stoichiometric composition of the alloys, the chemical bond between the atoms in the center of the face (Co-position) and in the corners of the crystal lattice (Al or W-position) will be different in different directions.

Results of the investigation of the magnetic domain structure of the $Co_3(Al,W)$-based alloy in two different states are presented in Fig. 2.51. For study, the alloy with 10.8 at.% W was homogenized at 1350°C for 24 hours, then aged at 900°C for 1000 hours, followed by a water quench. Figure 2.51 shows the surface of the ferromagnetic samples with the crystal grains divided into several domains, parallel to its "easy" axis of magnetization. Stripe domains are common phenomenon for many different physical systems, for example, for pure cobalt or cobalt films (Yan et al. 2006). Existence of stripe domains in ferromagnetic samples is based on energy minimization consideration, since the formation of stripe domains can reduce the surface charge (Yan et al. 2007).

One can see that the stripe domains start from the grain boundary and run through the entire grain regardless of the γ'-phase region size. Our alloy has a two-phased structure such as γ-phase (cobalt solid solution) and γ'-phase ($Co_3(Al,W)$). Both phases are ferromagnetic, however we do not see any γ'-phase magnetic domains. It means that the composite material "works" as a whole one; this fact can be explained by the orientation between two crystal lattices. Orientation relationship between the two phases is considered as follows: <001> directions of $L1_2$-type ordered crystal lattice of the γ'-phase are oriented parallel to the <001> FCC directions of the γ-phase (cobalt solid solution) (Sato et al. 2006). Thus, presence of the stripe domains in the structure means also the existence of the special configurations and orientation of the magnetic domains in the $Co_3(Al,W)$-based alloys.

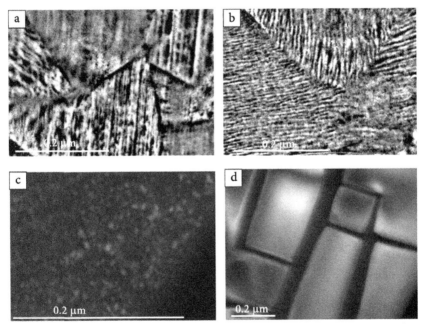

Fig. 2.51 Magnetic domain structure of the alloy with 10.8% W treated in different ways. The magnetic field was applied parallel to the sample plane and horizontally, and then turned off: a, c- initial state, a- stripe magnetic domains, H = 0 Oe, c- TEM; b, d- aged state, b- stripe domain structure of the remaining magnetization, H = 0 Oe, d- TEM.

2.2.3 Effect of heat treatment and plastic deformation on the structure and phase composition of $Co_3(Al,W)$-based alloys

Transition from experimental to industrial materials includes the deep understanding of the structure and chemical features of the alloys.

Multistage heat treatments are the usual procedure for the technological process of obtaining materials with specified physical or mechanical properties. Kazantseva et al. (2016a) showed that the structure of cast single crystalline ingot of the $Co_3(Al,W)$-based alloys had a cellular-dendritic structure. Moreover, we found a *liquation* through whole *length* of the ingot; the lower part of the ingots was enriched with tungsten. Cellular-dendritic structure is non-uniform and provides the homogenization process. *Liquation* is more a serious process which is difficult to reduce by heat treatments only, because of that we also used plastic deformation.

In this section we present the regimes of heat treatments including plastic deformation for obtaining the uniform γ/γ' structure in the $Co_3(Al,W)$-based alloy. These regimes are new and have never been used earlier for $Co_3(Al,W)$-based alloys.

Morphology of the γ'-precipitations in the Co-7.9 at.% Al-6.8 at.% W alloy depending on heat treatments and deformation was investigated by Davydov et al. (2018). Single crystalline alloy Co-7.9 at.%Al-6.8 at.% W was melted by the Bridgman method. The ingot was subjected to a homogenizing annealing at 1250°C for 24 hours with a subsequent slow cooling in a furnace. Then samples were annealed at 1050°C for 40 minutes, followed by water quenching and cut in two parts: the first part was subjected by many steps of heat treatments (annealing was done for 6 hours followed by water quenching). The second part was deformed by compression up to 10% and then it was subjected by many steps of heat treatments, followed by water quenching. The regimes of heat treatments were chosen according to the DSC result.

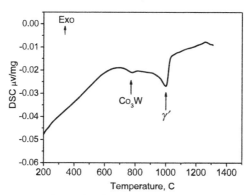

Fig. 2.52 DSC result of the alloy after homogenization.

Differential scanning calorimetry (DSC). As can be seen from the DSC curve, the most clearly pronounced is the order-disorder ($\gamma' \rightarrow \gamma$) phase transition that occurs with heat absorption (Fig. 2.52). The γ' solvus temperature calculated form the DSC curve is 1000°C; the temperature

900°C may be defined as the beginning temperature of the $\gamma' \to \gamma$ transition (dissolution of the ordered γ' phase); the temperature 670°C is the temperature of the beginning of the Co_3W (DO_{19}) intermetallic phase formation (Fig. 2.52).

According to the isothermal section of the ternary Al-Co-W phase diagram suggested by (Liu et al. 2016), the Co-7.9Al-6.8W alloy should be in a three-phase $\gamma + \gamma' + Co_3W$ region at 600°C, however the concentration boundaries of this three-phase region by (Liu et al. 2016) were based on theoretical calculations. Our experimental result correlates with the high temperature X-ray result obtained by Kazantseva et al. (2016a) and DSC results obtained by Xue et al. (2012). We used two temperatures for the high and low temperature steps of the heat treatments, such as 1050°C (γ-phase region) and 650°C (lower than the temperature of the Co_3W formation). Also we used room temperature plastic deformation by compress of 10% after the high temperature step of annealing.

X-ray results. The diffraction lines of γ'-phase were found in the X-ray pattern after annealing of the sample at 650°C. Co_3W diffraction lines in the X-ray patterns were found in the sample after annealing at 800°C. After annealing of the sample at 850°C, the Co_7W_6 diffraction lines were also found in the X-ray diffraction pattern (Table 2.85).

Table 2.85 X-ray results of the studied samples

Heat treating	1050°C	650°C	650°C+800°C	650°C+850°C
Phase composition	γ	$\gamma + \gamma'$	$\gamma + \gamma' + Co_3W$	$\gamma + \gamma' + Co_3W + Co_7W_6$
Heat treating and deformation	650°C	650°C + 800°C	650°C + 850°C	650°C + 900°C
Phase composition	$\gamma + \gamma' + Co_3W$	$\gamma + \gamma' + Co_3W$	$\gamma + \gamma' + Co_3W$	$\gamma + \gamma' + Co_3W$

Very weak traces of this phase were found in the initial homogenized sample by magnetic measurement, because the homogenization was done on the boundary of the temperature region of this phase (Xue et al. 2012, Kazantseva et al 2016a). Usually, the defect structure increases the possibility of the phase transition. In our case, the deformation of the sample after high temperature annealing (1050°C) promotes the Co_3W formation at lower temperatures than those found in the samples without plastic deformation and it suppresses the Co_7W_6 growth at the high temperature (Table 2.85).

TEM study. Microstructural studies show that the homogenized sample may be considered as the composite material that consists of cobalt solid solution matrix and precipitations of γ'-phase (Fig. 2.53). However, as can be seen from the Fig. 2.53b, the structure of the quenched sample after annealing at the temperature just higher than

γ' solvus does not show any presence of the γ'-phase. Compared to the sample after annealing at 1050°C, the structure of the homogenized sample shows the γ'-phase round precipitations of about 5 nm in size (Fig. 2.53a).

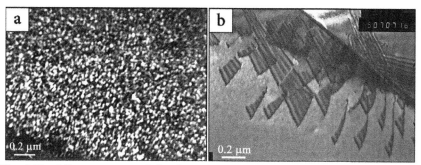

Fig. 2.53 Microstructure of the Co-8Al-6.8W alloy, the dark-field image taken with γ'-reflex, TEM: (a) homogenized at 1250°C-24h; (b) 1050°C-40 minutes, water quenching.

The TEM pictures of the microstructure of the samples after annealing at 1050°C for 40 minutes and an addition annealing at 650°C, and 800°C for 6 hours shows the change of the morphology and size of the γ'-phase precipitations (Fig. 2.54a). The SAED patterns of the studied samples after annealing in all cases show the γ'-phase reflexes (Fig. 2.54a). These results correlate well with a diffusion study provided by (Chang et al. 2015), where increasing values of and interdiffusion coefficients were found with temperature in FCC-ternary Co-Al-W system. Analysis of the size distribution of the γ'-precipitation in the samples shows the normal law of the distribution, which is characteristic of the normal growth of the particles with closely related sizes (Fig. 2.53a-b). This experimental fact is supported by the experimental results obtained by (Pyczak et al. 2015) relating to dissolution of γ'-phase at 1000°C. Also, it may lend support to the homogeneous formation mechanism (like spinodal) suggested by Liu et al. (2016), which takes place under deep overcooling creating the concentration gradient in the sample. Spinodal decomposition was observed by the method of atomic-probe tomography in a commercial nickel superalloy after rapid quenching from 1300°C (Tan et al. 2014). In (Tan et al. 2014), particles of the γ' phase with sizes of 30-50 nm with a diffuse non equilibrium interphase boundary were revealed, the appearance of which the authors (Tan et al. 2014) ascribed to the processes of ordering and spinodal decomposition.

Plastic deformation make a great impact on the process of precipitation of the γ'-phase. One can see in Figs. 2.54b and 2.54d that cuboids of γ'-phase in the sample annealed at 850°C have the close size (about 25-30 nm), however the volume fraction of γ'-phase in the sample with plastic deformation is lower than that in the alloy without deformation.

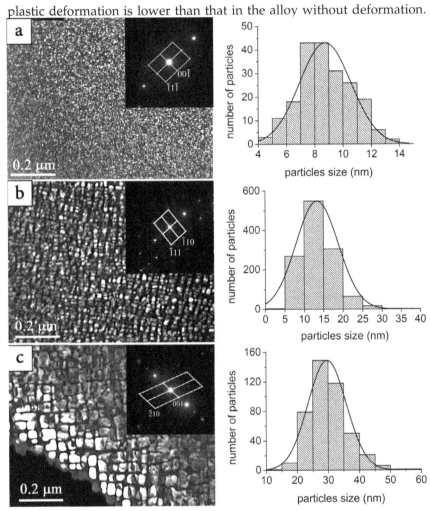

Fig. 2.54 Morphology and size dependence of the γ'-phase particles: a-b -heat treatments, (a) 650°C + 800°C; (b) 650°C + 850°C; c-d -heat treatments including plastic deformation, (c) 650°C + 800°C; (d) 650°C + 850°C.

Size distribution of the γ'-particles after the heat treatments including plastic deformation testifies for the abnormal law of the growth. It suggests the existence of very heterogeneous size distribution, when large particles serve as embryos for growth. Such a type of process is characteristic for the coarsening of the particles (Fig. 2.54c). Delay of the growth of the particles may be due to texture or the release of a second phase. In our case, this delay is probably associated with Co_3W formation, which decreases the tungsten enrichment of the γ-phase.

As can be seen from Fig. 2.55, the normal low of distribution of the γ' particles is observed in the sample after increasing the temperature of the heat treatment to 900°C. Structure of the sample presents the uniform distribution of γ/γ' phase (Fig. 2.55).

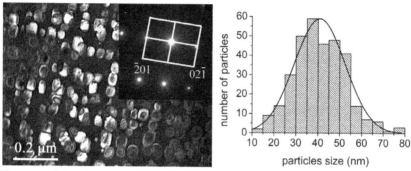

Fig. 2.55 Morphology and size dependence of the γ'-phase particles on heat treatments including plastic deformation, multistep annealing at 650°C + 900°C.

Magnetic measurements. Structure studies are supported by the magnetic measurements (Tables 2.86).

Table 2.86 Coercive force, size and volume fraction of γ'-phase particles in dependence on annealing temperature in the samples without/with deformation

Heat treatments/+def.	650°C	650°C+800°C	650°C+850°C	650°C+900°C
Hc, Oe	4.35/7.7	10.4/27.9	117.9/35	/29.6
Size of γ'-phase particles, nm	9/7	13/15	30/25	/40
Vol. fraction of γ'-phase, %	50/52	58/50	87/34	/47

It is known that the coercive force is a structurally censored characteristic of the magnetic materials. Structural defects increase the value of coercive force of soft magnets (Herzer 1996). Grain size also affects the value of coercive force. Maximum coercive force (10 A/m) in the industrial soft magnets was found in the nanoscale structure with grain size of 100 nm. A minimum of the coercive force was found in nanoscale structure with 10 nm or in single crystals with the grain size of 1 mm (Herzer 1996). As can be seen from the Table 2.86, in the case of our ferromagnetic material coercive force depends on both the volume fraction and γ'-precipitation size (Table 2.86).

At the annealing temperature of 850°C, when the active growth of the γ'-phase precipitations associated with increasing of tungsten diffusion occurs, magnetic measurements allow us to get the precise results of the structure and phase composition changes in the sample. Coercive force of the studied alloy is found to be decreased in about of four times after heat treatments including plastic deformation (Table 2.85). Because the coercive force change is proportional to the change of the tensions in the materials (Herzer 1996), we can suggest that the increasing of the coercive force depends on the size of γ'-precipitations, but drastic increasing of the coercive force in the sample may be associated with the increasing of the tension in the material at the Co_7W_6 formation (Tables 2.84-2.85). Decreasing of the coercive force in the sample after heat treatments including plastic deformation and annealing at 900°C, may point to the approach to an equilibrium state. It may be suggested that this latter processing mode that is optimal for obtaining a homogeneous alloy structure without undesirable precipitations of a topologically close packed phase Co_7W_6.

Thus, our studies have shown that the most successful mode of heat treatment for obtaining a uniform γ/γ' structure of the $Co_3(Al,W)$-based alloys should be multistage and include plastic deformation.

CONCLUSION AND FUTURE OF THE $CO_3(AL,W)$-BASED SUPERALLOYS

In contrast to the Ni_3Al intermetallic compound, which is formed directly from the liquid, Co (Al,W) is formed due to transformations in the solid state; because of this, it belongs to the category of unstable chemical compounds. The temperature range of this intermetallic compound does not exceed 1000°C. However, the high plasticity of this intermetallic compound, in comparison with Ni_3Al, makes it possible to reliably consider Co_3 (Al,W)-based alloys as the promising ones for manufacturing products operating under conditions of elevated temperatures and stresses.

Because $Co_3(Al,W)$ hardened alloys have a much higher creep strength than conventional carbide-hardened Co-based superalloys, we can say that the discovery of a new intermetallic hardening γ' phase in the Co-Al-W system initiated the development of new cobalt high-temperature alloys. This development is aimed to increase the operating temperatures and strength of products. A number of studies focused on the influence of doping on the solvus temperature of the γ' phase (Co_3Al,W)) (Sato et al. 2006, Suzuki and Pollock 2008, Bauer et al. 2010, 2012). Xue et al. (2013) showed that alloying with titanium and tantalum increases the γ'-phase stability in $Co_3(Al,W)$-based alloys. It has been shown that alloying of the $Co_3(Al,W)$-based alloy with tantalum and molybdenum allows one to get a 91% γ'-volume fraction in the alloy (Bauer et al. 2010). The greatest effect on the liquidus, solidus, and γ'-solvus is found to be due to doping with tantalum, iridium, or titanium. In the Co-9Al-8W-2Ta-2Ir alloy, the γ'-solvus temperature reaches 1140°C; the liquidus temperature of the alloy is within 1460°C

(Bauer et al. 2010). The same situation is observed in Co-9Al-8W-2Ta-2Ti with a 92% γ'-volume fraction, where the γ'-solvus temperature is found to be 1153°C; the liquidus temperature of the alloy is 1411°C (Bauer et al. 2012). The room-temperature mechanical properties of the Co-9Al-9W-2Ta-2Mo-0.02B alloy was estimated by Feng et al. (2016), giving $\delta = 13.6\%$ and $\sigma_{0.2} = 755$ MPa. Co$_3$(Al,W)-based alloys, alloyed with tantalum, were obtained by the slow high-gradient direct solidification (Petrushin et al. 2015). Results obtained by Petrushin et al. (2015) showed that segregation of alloying elements along the length of the Co$_3$(Al,W)-based ingot at slow high-gradient directed crystallization was much lower, in comparison with nickel superalloys. That means that this method may be successfully used for manufacturing the parts made from γ' strengthened cobalt superalloys of the new generation.

The next direction in the development of new Co$_3$(Al,W)-based superalloys is the addition of nickel (Tituts et al. 2012, Yan et al. 2014b). Ni additions increased the solvus, but reduced the observable γ' fraction (Yan et al. 2014b). The Co-29.2Ni-9.8Al-6.3W-2.4Ta-6.4Cr was found to have creep properties comparable to those of the first-generation nickel-based superalloys (Titus et al. 2012). The Co-32Ni-12Cr-6Al-3W-2.5Ta-2.5Ti (Zr, Si, Hf, B, C) alloys were claimed for a patent (Bauer et al. 2016). These alloys showed a similar yield strength and higher creep strength at 750°C, compared to commercial Ni-base superalloys Waspaloy and Udimet 720Li (Neumeier et al. 2016).

Thus, new cobalt γ' strengthened superalloys developed for elevated temperatures are in progress. However it is difficult to say if these alloys could replace nickel superalloys in the future, however, there is no doubt that these alloys find their application as a new class of high-temperature alloys.

Chapter Three

Iron Superalloys

Iron-based superalloys contain iron as the base metal and relate to the grades of high-temperature and corrosion resistance steels. Steels and alloys of austenitic and austenitic-ferritic types are widely applied in modern industries, constituting a significant part of all used materials (approaching 20%, according to many experts). Although austenitic steels are significantly more expensive than conventional structural steels, they have a number of advantages. The main advantages are their high heat-resistance and, at the same time, resistance to corrosion, high plastic properties, in combination with mechanical strength. The required levels of technological and operational properties in most cases are ensured by controlling the Content of the Ferritic Phase (CFP) at the production stage of austenitic material.

This chapter discusses the main methods and means for controlling the properties and phase composition of high-temperature austenitic steels (iron superalloys) with an emphasis on the results produced by the author. The chapter gives a brief overview of iron superalloys used in Russia and other countries for the production of modern parts and structures with increased requirements on heat and corrosion resistance. The main emphasis is made on the development of new methods of nondestructive testing of the phase composition of materials, because the phase composition largely determines operational properties. Although non-destructive testing and control of two-phase steels is relatively well done in different countries, non-destructive testing of the three-phase austenitic steels has not been developed yet for industrial applications. In this chapter, we present our method for the phase control of three-phase austenitic steels by non-destructive testing.

We developed new methods for phase control of iron superalloys using both the traditional measurements of magnetic properties and the electrical measurements. As will be shown below, the value of the electrical resistivity ρ is directly related to the phase composition and structure of steels. In most plants in Russia, however, there is practically no instrumentation base for nondestructive testing of ρ values. The author's development of small-scale means for control of the specific electrical resistance makes it possible to solve this problem to a great extent.

Particular attention is paid to the development for local measurement of the relative magnetic permeability (μ). Despite the fact that most of the parts from austenitic steels operate at temperatures above the magnetic transition to the paramagnetic state (Curie temperature), dangerous structural transformations and the process of nucleation and development of cracks can be detected at early stages by local measurements of the magnetic permeability at room temperature. Such measurements can be done at the stages of scheduled shutdowns of mechanisms.

At the end of this chapter, we present the instrumental developments at large Russian enterprises to ensure product quality, as well as to optimize production process.

3.1 Austenitic Steels: Physical and Mechanical Properties and Area of their Application

Austenitic steels (chromium-nickel, nickel, manganese, and others; Table 3.87) have an unusual range of performance properties: a relatively low yield strength is combined with high ductility, fracture toughness, increased corrosion resistance, and high manufacturability (Gol'dshtein et al. 1999, Filipov et al. 1988). Increase in plasticity and resistance to brittle fracture and stress corrosion is achieved by obtaining fine austenite grains forming a homogeneous substructure and reducing amounts of nonmetallic inclusions. Doping with chromium and nickel significantly increases steel hardness and strength. Alloying of these elements is also used when manufacturing products operating in corrosive environments at temperatures up to 800°C. In addition to having valuable temperature properties, austenitic chromium-nickel steels are also widely used at cryogenic temperatures (down to helium temperatures) due to high plastic properties with a viscous character of failure (Kurdumov et al. 1977, Ulianin 1994).

3.1.1 Chemical composition of the main grades of corrosion-resistant steels

The austenitic class includes high-alloy steels that predominantly form a single-phase austenite structure of γ-Fe with a Face-Centered

Crystal lattice (FCC) during the crystallization process, retaining this crystal structure on cooling to cryogenic temperatures. The amount of the other phase, namely a high-alloyed $\alpha(\delta)$-Fe ferrite with a Body-Centered Crystal lattice (BCC), varies from 0 to 10%. The studied steels contain from 18 to 25% Cr, which provides the heat and corrosion resistance, and from 8 to 35% Ni, which stabilizes the austenite structure, increases the strength and ductility at high temperatures, and also increases the manufacturability of steels over a wide temperature range. These properties allow us the use the austenitic steels, on one hand, as corrosion- and heat-resistant and, on the other hand, as cryogenic structural materials in chemical, thermal, power, and nuclear plants. In this connection, austenitic steels are divided into two main groups: high-temperature and corrosion-resistant steels. The main grades of the steels mostly used in Russia and their compliance with similar steels in the USA (ASTM- American Society for Testing and Materials, AISI- American Iron and Steel Institute), Germany (DIN- Deutsche Industrie Norm) and France (AFNOR- Association Francaise de Normalisation), in accordance with technological and national standards, are given in Tables 3.87-3.88.

Table 3.87   The main grades of stainless steel, operating up to 750°C

DIN	GOST	EN	AISI	AFNOR
X5CrNi1810	05X18H10	1.4301	304	Z6CN1809
X5CrNi18 12	05X18H12	1.4303	305	Z8CN1812
X10CrNiS189	10X18H9	1.4305	303	Z10CNF1809
X2CrNi1911	02X19H11	1.4306	304 L	Z3CN1810
X12CrNi177	12X17H7	1.4310	301	Z11CN1808
X2CrNiN181002	X18H10A	1.4311	304 LN	Z3CN1810Az
X1CrNi2521	01X25H21	1.4335	310 L	Z1CN2520
X6CrNiTi1810	0618H10T	1.4541	321	Z6CNT1810

Table 3.88   The main grades of stainless steel operated at elevated temperatures (600°-1200°C)

DIN	GOST	EN	AISI	AFNOR
X10CrAl7	10X17Yu3	1.4713	-	Z8CA7
X10CrSiAl13	10X1Yu3	1.4724	-	Z13C13
X10CrAI18	10X18Yu3	1.4742	442	Z12CAS18
X18CrN28	18X28A	1.4749	446	Z18C25
X10CrAlSi24	10X24AC	1.4762	-	Z12CAS25
X20CrNiSi254	20 25H4C	1.4821	327	Z20CNS2504
X6CrNi2213	06X22H13	1.4833	309 (S)	Z15CN2413
X12CrNi2521	12X25H21	1.4845	310 (S)	Z8CN2520
X12CrNiTi189	12X18H9A	1.4878	321 H	Z6CNT1812

Where the first letter in the formula indicates the carbon content, divided by 100, and the next letter shows the main alloying additives and their percentage. The most common group includes the corrosion-resistant steels. In this group, the steel X5CrNi1810 has 0.05% chromium-18% nickel-10% carbon. According to the European standard (EN), it is designated as the steel 1.4301, according to the *Russian* national standard (GOST), it is the steel 05X18H10.

Along with chromium and nickel, other alloying elements can also be present in the solid solution or excess phases of the austenitic steels. Those elements are divided in austenitizers (carbon, nitrogen, manganese) and ferritizers (titanium, niobium, molybdenum, tungsten, silicon, vanadium). Like chromium and nickel, they improve the indicated service properties and act on the stability of the austenitic structure. Ferritizers contribute to the formation of high-alloyed ferrite $\alpha(\delta)$-Fe with a BCC crystal lattice; austenitizers stabilize the austenite structure γ-Fe with FCC crystal lattice.

The combined effect of the alloying elements on the final structure is evaluated using the Scheffler structural diagram and the ratio Cr_{eq} / Ni_{eq} called the chromium-nickel equivalent. This structural diagram describes the structures obtained after the crystallization of the metal, most often the structure of the weld seam. The equivalent content of chromium and nickel in the Schaeffer diagram (in percentage is calculated by the following relationships, Maslenkov 1983, Kershenbaum 1995):

$$Cr_{eq} = \% \ Cr + 2 \ (\% \ Mo + \% \ Nb + \% \ Al) + 0.5 \cdot (\% \ Si + \% \ W) + 5 \cdot \% \ Ti + 1 \cdot \% V;$$
$$Ni_{eq} = \% \ Ni + 0.5 \cdot \% \ Mn + 30 \cdot (\% \ C + \% \ N).$$

Knowing the equivalent of chromium and nickel, Cr_{eq}/Ni_{eq}, it is possible to determine the equilibrium phase composition in austenitic steel with a known chemical composition.

3.1.2 *Effect of ferrite on the operating properties of austenitic steels*

The formation of ferrite occurs during the manufacturing of austenitic steel, depending on its chemical composition. Ferrite is located along the boundaries of austenite grains; it forms a continuous or partially continuous rim or lies along the boundaries of austenite grains in the form of separate particles that do not form a continuous rim. In some cases, ferrite (or α-phase) may be practically absent in austenitic steel. Austenite has plasticity with a relative elongation of 40-50% and possess more hardness than ferrite (Kershenbaum 1995, Shcherbinin and Gorkunov 1996, Bida et al. 2001). The presence or absence of ferrite determines the technological, mechanical, and corrosion properties of steels.

In practice, the Content of the Ferritic Phase (CFP) is regulated, and often the requirements for the content of the CFF are stricter than those for the chemical composition of austenitic steel (Kershenbaum 1995, Kliyuev 2004). Thus, according to the Russian standard OST 95 503-2006, the ferrite phase should not exceed 8%. For example, for a welding wire made of S-08X16N8M2, the regulated amount of ferrite is no more than 2-6%, and for a welding wire made to the grade SV-04X19N11M3, the regulated amount of ferrite is no more than 3-8%. Thus, the specified CFP at the stage of obtaining austenitic materials ensures required levels of technological and operational properties of the steels. Accordingly, strict requirements are also imposed on the methods of controlling the manufacture of steel.

3.1.3 Application of destructive methods for phase control of austenitic steels

Current methods for controlling the phase composition of austenitic steels may be divided into the magnetic (electromagnetic) and nonmagnetic ones. Nonmagnetic methods are mainly used in conjunction with the destruction of an object under study and special preparation of controlled samples. These include X-ray diffraction, micro-X-ray spectroscopy, metallographic method, electron microscopy, and some non-destructive methods (Kliyuev 2004, Apaev 1976). Below, we briefly discuss the advantages and limitations of these methods.

Non-magnetic methods for controlling the phase composition of austenitic steels. In X-ray structural analyses, the crystal lattice parameters are selected as the phase characteristics, which are indirectly determined from the diffraction pattern (Kurdiumov et al. 1977, Mirkin 1978). To obtain the X-ray diffraction patterns of steel, the monochromatized $K\alpha$ radiation of chromium is routinely used. Each phase is represented in X-ray diffraction patterns by a series of diffraction lines. The phase percentage is determined from intensities of the diffraction lines. An X-ray diffraction pattern is usually obtained from the surface area of the sample, sized between several square millimeters and one square centimeter, with a depth of 0.001 to 0.01 mm. With the X-ray diffraction patterns, it is possible to determine the phase composition, density, crystallographic structure of the studied phase, provided that the phase content in the sample is above 4%. The X-ray diffraction methods analyze only thin surface layers of samples, not providing objective information on the content of ferromagnetic inclusions in the entire volume of the material. This fact may lead to inconsistencies between the X-ray diffraction data and the measurements obtained by other methods of controlling the amount of the ferromagnetic phase. Essential shortcomings of the X-ray methods of phase analysis are also the strong dependence of the measurement results on the metal temperature, stress state of the phases, texture, and coarse-grained samples. Also, the X-ray methods are of no use for timely monitoring the material directly during its operation (cutting of samples is required), and because of the need to protect personnel from the ionizing radiation (Mirkin 1978, Litovchenko et al. 2006).

1. *Metallography.* This method involves an evaluation of the microstructure in thin surface layers, using optical microscopes and preliminary etching of the sample surface. It should be noted that the metallographic (optical) method is the main method for determining the structure and phase composition, with which results of the other methods are compared. For the metallography methods, cutting of the control samples is required; this fact often makes these methods unsuitable for use in the manufacture and operation of the product (Shcherbinin and Gorkunov 1996, Apaev 1976).

2. *Transmission and scanning electron microscopy.* To study the phase composition, electronic scanning and transmission microscopy methods based on the use of a high-energy electron beam and focusing through a system of magnetic lenses for image formation of a sample structure, with a maximum magnification of up to 10^6 times, are widely used. However, these methods require a special preparation of samples (for example, the production of polished foils with a thickness of ≤ 0.4 mm for transmission microscopy) and creation of a high-vacuum environment. With these techniques, a small area of the investigated material is controlled; the method is characterized by the instability of the samples under the electron beam, which limits the application of these techniques in the industrial conditions (Utevski 1973).

3. *Methods of scanning probe microscopy.* Atomic force, magnetic force, and electro sensor microscopy are also used to visualize the structural-phase composition of studied materials. These methods are realized on the basis of modern scanning probe microscopes (Schwartz and Wiesendanger 2008, Femenia et al. 2003, Jacobs et al. 1998, Nonnenmacher et al. 1991, Melitz et al. 2011). The work of the atomic force microscope is based on the force interaction (van der Waals forces) between the probe and the sample surface - the probe experiences attraction from the sample side at large distances and repulsion at small distances. The methods of magnetic force microscopy make it possible to obtain images of the spatial distribution of the magnetic interaction forces of a probe coated by a ferromagnetic material with a sample at a submicron scale. Presently, techniques employing electrical

interaction between a probe having a conductive coating and a sample located on a conductive substrate are used. Such methods include scanning electric power and capacitive microscopy, the Kelvin probe method. These methods make it possible to estimate the change in the electrical parameters of different phase components, inhomogeneity, and inclusion in a selected region with submicron resolutions (better than 10 nm). It allows description of the changes in the electrical parameters of different phase components during deformation of two-phase austenitic steels (Nonnenmacher et al. 1991, Melitz et al. 2011). The methods of scanning probe microscopy provide additional information, at a submicron scale, on the relief, magnetic, and electrical properties of the local surface area of the material under study. However, the need for special preparation (cutting) of samples, limited scanning field (usually no more than 100x100 µm), and a long duration of the scanning process make it difficult to widely apply these techniques in non-destructive testing.

4. *Electron backscatter diffraction analysis (EBSD) microanalysis.* To evaluate the phase composition, an analysis of diffraction patterns of backscattered electrons (EBSD) with the indication of Kikuchi patterns is also used (Örnek and Engelberg 2015, Sathirachinda et al. 2009, Danilenko et al. 2012). In this method of microstructural analysis, a sample is inclined at 70° and is placed in a scanning electron microscope; the surface under investigation is scanned step-by-step ("point-to-point"). Back-scattered diffracted electrons form a diffraction Kikuchi-pattern. These diffraction patterns are digitized and indicated. On the basis of the scanned data, information on the grain structure, grain misorientation, and the texture is derived. However, this method also requires special preparation of samples, which limits its practical applications outside of laboratory conditions.

3.1.4 Determination of the phase composition of austenitic steels by methods of nondestructive testing

Method for measuring electrical resistivity. The relationship between electrical properties and phase transformations in austenitic steels has not been sufficiently studied yet. It was established by Apaev (1976) and Gorkunov et al. (2012) that changes in the electrical resistivity during the transformation of austenite to martensite occurred at earlier stages than the changes in magnetic induction were done. This fact indicates preparation processes in austenite before the actual transformation. Specific resistance can be affected by structural factors (concentration, sizes, and shapes of inclusion of other phases, dispersity), chemical factors (interaction of solid solution components), and a number of other factors, which may lead to large errors in derived electrical resistivity.

The thermoelectric method. This method is based on the fact that changes in the phase composition of an alloy entails changes in the thermoelectromotive force (E) measured with respect to some element of comparison (most often, a copper electrode). For austenitic steels studied by Kuznetsov and Okunev (1993), and Gorkunov et al. (1998), the changes in the thermopower as a function of the phase composition has the same character, as the saturation magnetization. A complicated dependence of the thermoelectric power on several phases, the degree of doping, the minor impurities, and the arrangement and shape of the crystallites make it difficult to use this relatively simple method in practice.

The eddy-currents testing. This method is based on the excitation of eddy currents in electrically conducting objects, such as metals, alloys, semiconductors, etc. Studies of phase transformations using the eddy-current method were done by Gerasimov et al. (1992), Sandovski et al. (2000, 2001). An investigation of the electromagnetic properties of pipe samples made of a chromium-nickel austenitic steel showed a unique correlation between the magnitude of the introduced reactance of the feed-through and eddy-current transducers and amount of the deformation martensite, formed during cold rolling above 25% of deformation. The presence of such correlation allows us to use it for determining the percentage of deformation martensite by eddy-current testing.

However, this method requires a careful determination of the optimum operating conditions of eddy-current converters for tuning interfering parameters and obtaining sufficient sensitivity. In addition, the control zone depth in the eddy-current method is small.

Method of acoustic emission. The appearance of acoustic emission signals was recorded in austenitic steel parts, which were operated at cryogenic temperature and experienced friction during operation, which led to the appearance of deformation martensite in the steel. The authors noted the relationship between the parameters of the acoustic emission signals and the dynamics of the martensitic transformation. However, the acoustic emission signals are also initiated by microplastic deformation of the metal, the formation and movement of dislocations, the growth of micro cracks. Because of this , it is difficult to distinguish the signals from phase transformations and the appearance of martensite deformation against a background of interference (Aßmus and Hübner 2004, Moore et al. 2005).

The ferromagnetic phases formed in austenitic steels are highly dispersed and chaotically located. This fact makes it difficult to reliably determine them using local methods, such as the X-ray diffraction, electron microscopy, etc. At the same time, the appearance of a highly dispersed ferromagnetic phase in the bulk of the paramagnetic material significantly changes the magnetic properties of the austenitic steel. Thus, it is possible to determine the percentage of ferrite and (or) deformation martensite by nondestructive evaluation using various magnetic parameters.

Numerous studies, related to the control of the ferrite or/and martensite amount, convincingly show that the magnetic method is preferable and more accurate than the above-mentioned non-magnetic methods. To assess the changes in the phase composition of steels, magnetic parameters such as the saturation magnetization (M_s), the Curie temperature (T_c), and the constant of the crystallographic magnetic anisotropy (K) are used. To determine the amounts of magnetic phases in a paramagnetic steel material, the method of magnetization before saturation period is used (International Standard ISO 8249-198544, GOST RF11878).

Measurement of the content of the ferrite phase by the magnetic saturation method. It was shown by Rigmant et al. (2000, 2015) that, in order to estimate the content of a number of ferromagnetic phases in austenitic chromium-nickel steels, one can successfully apply monitoring methods based on measuring the saturation magnetization of the controlled material. The magnitude of the specific saturation magnetization is used in the magnetic phase analysis as the most reliable measure of the content of the ferromagnetic component of any degree of dispersion in a complex-doped paramagnetic matrix (Merinov et al 1978, Elmer and Eagar 1990, Stalmasek 1986, Rigmant and Gorkunov 1996).

The method of magnetic ferritometry is widely used in nondestructive testing due to its efficiency, contactlessness, high productivity, and the ability to determine the content of the ferrite phase directly in finished products (equipment, pipelines, welded joints). The method of magnetic saturation is stipulated by the National Standards of Russia (State Standard System, GOST) for the certification of standard samples of the ferrite content in austenitic steels (GOST no. 26364-90 1985, 8.518-84, 1984, 11878, 1967, Merinov and Mazepa 1997). In Russia, the term "percentage of ferrite" is used; it is defined as one-hundredth of the value of the specific saturation magnetization of the ferrite phase in the studied steel. In other countries, the control of the content of the ferritic phase (CFP) arose as an engineering problem and was developed by the methods and tools of magnetic thickness measurement (ASME Code ISO 8249-56:2000, ASTME562-08, AWSA4.2M:2006). These methods, based on the measurement of the detachment force of a permanent magnet from the surface of a sample, can also give reproducible results, however, it is difficult to correctly express the measurement result in percentage. Therefore, it is necessary to use conditional quantities, for example, a

Ferrite Number (FN). Currently, the project of a single international standard considered by the International Welding Institute provides for an exact relationship between the ferrite number and the percentage of ferrite in austenitic steels: FN ~ 1.80 CFP (%).

Magnetization to the technical saturation of the controlled area or sample cut from the part occurs in the method of magnetic ferritometry. In this case, the following expressions are used:

$$\%F = \left(\frac{J^F_{samp}}{J^F_{100\%}} \right) \cdot 100\%. \tag{1}$$

where $\%F$ is the percentage of ferrite in the investigated material; J^F_{samp} is the saturation magnetization of the steel that contains ferrite in the amount of $x\%$; $J^F_{100\%}$ is the satu/ration magnetization of the steel that contains the maximum possible (100%) amount of ferrite, at a given chemical composition. Bosort (1956) suggested an empirical approach, developed by Mikheev and Gorkunov (1985), Deriagin et al. (2007). This approach showed that the $J^F_{100\%}$ value can be obtained with high accuracy, when the chemical composition of the studied steel was determined with the formula:

$$J^F_{100\%} = 1720 - 21.9(\% \text{ Cr}) - 26.3(\% \text{ Ni}) - 22.3(\% \text{ Mn}) - 48.6(\% \text{ Si}) - 20.7(\% \text{ Mo}) - 53.3(\% \text{ Ti}) - 50.2(\% \text{ V}) - 8(\% \text{ Cu}) - 39.8(\% \text{ P}), \tag{2}$$

where the percentage of chemical elements in the ferrite phase of the steel is indicated in parentheses. This formula is sufficiently clear from the physical point of view. The value of 1720 Gs is the saturation magnetization of pure iron (Belenkova and Miheev 1967). The presence of alloying elements decreases this value. Formula (2) was obtained by generalizing the results of the studies provided by Elmer and Eagar (1990), who performed a multiple regression analysis of the dependence of the saturation magnetization on chemical composition of the single-phase ferromagnetic iron-chromium-nickel alloys with an FCC solid solution structure and different chemical compositions. It should be mentioned that formula (2) is not applicable to all steels, it may be used only for austenitic steels present in the Scheffler diagram.

Industrial devices for determining the ferrite phase percentage were developed in Russia together with the system of state standards (Himchenko and Bobrov 1978, Stalmasek 1986). In Russia, the devices include two main types: ferritometers for measuring the volume content of the ferrite phase, giving an objective idea of the average of the CFP in a sufficiently large volume of the metal, and the devices for local measurements. The results of measurements with ferritometers using the volumetric or local methods mutually complement each other. The volumetric measurements of the CFP characterize the material (wire, ingot, etc.) as a whole. In this case, witness samples with a length of 60 mm and a specified diameter (usually with diameters of 5 or 7 mm) are cut out for the control. In the USA and Europe, the ferritometers with volumetric measurements are not common, used, as a rule, by agreement with a customer. Advantages of applications of ferritometers using local measurements are in the usage of measuring transducers of various designs. This allows us to identify the dangerous areas in parts of complex shape, and even to explore heterogeneous welded joints.

The development of the ferritometry methods includes expansion of their functional capabilities, miniaturization, and improvement of their metrological characteristics. However, despite the declared high accuracy of measurements for the ferrite phase, results of such measurements on the same samples of metal in different laboratories using magnetic devices and devices of different designs differ from each other (Merinov et al. 1978, Stalmasek 1986). The results of measurements of the ferromagnetic phase may depend on the design features of the measuring device. First of all, insufficient intensities of magnetic fields may be used in the devices for magnetizing the products.

Particularities of measuring the martensite phase by magnetic methods. The control methods, based on the magnetic saturation measurements of the controlled material, can be successfully used to estimate the amount of martensite deformation in austenitic chromium-nickel steels (Merinov and Masepa 1997). However, the main problem is that, unlike ferrite, the martensite of deformation is a more magnetically rigid ferromagnetic phase. Under insufficient magnetic fields, saturation magnetization of materials is not a single-valued function of its phase composition. It depends on the magnetic background of the sample, external uncontrolled magnetic fields, structural factors, morphology of the ferromagnetic phase, and degree of interaction between phases (Apaev 1976). Most modern instruments, manufactured in Russia or Europe, have local primary transducers, in which magnetization during the ferrite phase measurements is carried out to magnetic fields of the order of 100-150 A/cm. To control ferrite, these are quite sufficient fields. The ferrite phase has small residual magnetization (J_F) below 1-2% of the saturation magnetization (J_S) (Bosort 1956). According to GOST no. 26364-90, the magnetic field must be at least 250 A/cm in the operated area of the measuring transducer. Such value is not sufficient to magnetize the martensitic phase to saturation. The residual magnetization of austenitic steels containing a martensitic phase can be 10-15% or more of the saturation magnetization.

At the end of this section, we would like to emphasize limitations of the method of saturation magnetization measurements. Determination of the saturation magnetization of steels gives information only on the presence or absence of ferromagnetic inclusion, enabling also estimates of their total percentage. From the measured value of the saturation magnetization, it is impossible to establish whether the ferromagnetic phase is a ferrite or a deformation martensite. Moreover, it is impossible to determine homogeneity of this phase, i.e., whether it consists, for example, entirely of ferrite or partly of martensite. It also does not allow us to measure the amounts of the ferromagnetic phases.

To summarize, it may be said that, for nondestructive testing, it is necessary to take into account the different behavior of the ferrite and martensite phases during magnetization of a three-phase austenitic steel. It is necessary to search for new quantitative parameters to reliably determine the type and amount of the ferromagnetic phase contained in a three-phase austenitic steel.

Modern means of ferritometry. Magnetic control methods may be distinguished by the way the primary information is obtained: the Ferro-Probe (FP) method, the Hall Effect (EH) method, the induction (I) method, the ponderomotive (PM) method, and the magnetoresistive method (MP). The Ferro-Probe (FP) method is based on measurements of magnetic field strength by ferro-probes. The Hall Effect (EH) is based on the recording of magnetic fields by Hall sensors. The induction (I) method is based on recording the magnetic scattering fields by the magnitude or phase of the induced EMF. The ponderomotive (PM) method is based on recording the pull-off force (attraction) of a permanent magnet or the core of an electromagnet from a controlled object. The magnetoresistive method (MR) is based on the detection of magnetic scattering fields by magnetoresistors (from the electrical resistance changes in the applied magnetic field).

To determine the content of the ferrite phase, ferritometers with eddy-current method are used. In the process of control, the sample is magnetized by a special impulse field, then the output signal is fixed and the conclusion about the magnetization of the material is done. The use of the eddy-current method greatly expands the functionality of these devices; they can also be used as the thickness gauges, electrical and thermal conductivity meters, and magnetic permeability or flaw detectors. Examples of such ferritometers are the universal ferritometer MF 510, as well as the ferritometer MF-51NC ("AKA-SCAN" Moscow), designed to measure the content of the ferritic phase (CFP) in cast bucket samples under melting of corrosion-resistant stainless chromium-nickel austenitic steels.

The ferritometer MVP-2M (KROPUS, Moscow, Samara) was designed to implement various tasks of controlling materials with the eddy-current method. This device measures the content of the ferrite phase in parts made on austenitic and perlite-class steels, the conductivity of materials, the thickness of protective and decorative coatings applied to a conductive material, as well as it can determine sizes of defects by the eddy-current method. The MVP-2M ferritometer can be used as the ferritometer, conductivity meter, and coating thickness gauge by connecting various sensors.

The universal ferritometer MK-1.2F and the Ferritometer MK-1F (Introscope, Yekaterinburg) were designed to measure the local and volumetric content of the ferrite phase between 0.8 to 20% in welds and products made on stainless steels of austenitic and austenitic-ferromagnetic classes.

Feritscope MP30 and Feritscope MP30E-S (Germany) are also used for measuring the ferrite content in Ferrite Numbers (FN) according to the ASME code.

It should be emphasized once again that the eddy-current method and induction magnetization, realized in many ferritometers, allows one to obtain reliable results in most cases. However, it must be emphasized that such a control does not allow measurements in strong magnetic fields for the implementation of the "magnetic saturation" method. Special requirements for eddy-current testing are imposed on the quality of the controlled surface, which are very difficult to fulfill under operating conditions. Moreover, not only the electrical properties can change during operation due to heating and deformation. In the case of thermal deformation, decomposition of austenite may occur with the formation of a magnetic-hard martensite phase. Because of this, the eddy-current and induction control methods for the phase composition of austenitic steels in the case of martensite formation cannot be used.

Thus, we can note the necessity and huge benefits of the ferritometers in the manufacture and operation of products made of austenitic steels. At the same time, it is necessary to improve the devices in controlled depth, as well as to significantly increase the magnitude of the created magnetic field in the controlled facility or within the controlled area. The problem of local phase control remains unresolved, when the magnetically hard martensite phase appears in the structure.

3.2 Magnetic Control of the Phase Composition of Two-Phase Austenitic-Ferritic and Austenite-Martensitic Steels

This section presents the author's development for monitoring the phase composition in two-phase austenitic steels contained austenite + ferrite or austenite + deformation martensite.

The controlling method of the phase composition of the two-phase austenitic-ferritic or austenite-martensitic steels, based on the measurement of saturation magnetization, often requires sufficiently large magnetizing devices to create magnetic fields, which are necessary for the realization of this method. Because of this, an application of this method in the industrial conditions is difficult or impossible.

We proposed methods for creating large magnetic fields, necessary for the phase control of products from austenitic steels. The methods use the local high-energy permanent magnets that ensure magnetization of the portion of the controlled material to the state of technical saturation. Presently, these methods are introduced at many Russian industrial enterprises. Figure 3.56 shows a schematic diagram of the device operated with this method. At the place of contact of the local magnet with the surface of the controlled article, the magnetic field is at least 3.5-4.5 kA/cm (Fig. 3.56). The scattering field from the magnetized section is measured by miniature flux-gate transducers, which are located on the neutral axis of the attached magnet. This arrangement of the magnetic sensors allows detachment from the magnetic field of the magnet itself. The signal measured by the FE1 and FE2 ferro-probes arises due to the magnetic properties of the monitored product; this signal is related to the amount of ferromagnetic inclusion in the austenitic base of the material.

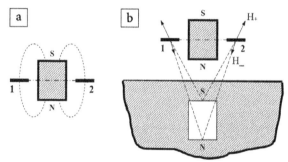

Fig. 3.56 Effects of the primary converter on the ferro-elements: (a) by a permanent magnet field; (b) by the "reflected" magnet field inside the seam material.

Ferritometers with FM-3 IMP ferro-probe transducers and Hall transducers FK-3 IMP are described in (Rigmant and Gorkunov 1996, Deriagin et al. 2007, Merinov et al. 1977). These devices are successfully implemented at a number of large enterprises in Russia. In addition to successfully controlling the ferrite content in austenitic steel, they also allow measurement of the amount of martensite phase of deformation in austenitic martensitic steels to 35-40%.

Monitoring the phase content of deformation martensite in austenitic steels is very important, because it is associated with safe usage of complex structures and products. To study the effect of deformation on phase transformations, three compositions of deformation-metastable manganese and chromium-manganese steels with a two-phase $(\gamma + \varepsilon)$

structure were selected by (Rigmant and Gorkunov 1996). The chemical and phase composition of the investigated steels after the initial quenching in water from 1050°C is shown in Table 3.89. Due to the different chemical content of the steels, such as carbon, nitrogen, silicon, and chromium, which have a significant effect on M_s and M_f (temperatures at the beginning and the end of the martensitic transformation), the phase composition and deformation stability of the steels were significantly different.

The determination of the amount of the ferromagnetic phase in the investigated steels was carried out at room temperature with a specially designed low-magnetic setup, using a uniaxial tensioning scheme with a force of up to 50 kN; this setup was equipped with a device for recording the deformation curve. The amount of the ferromagnetic α-phase was measured in the volume of the stretched sample until the time of the localized flow in the neck and destruction of the sample by using the built-in ferritometer (the α-phase meter).

Table 3.89 Chemical composition of the studied steels

Grade	Chemical composition, mass %							Phase content, %	
	C	Mn	Cr	Si	N	S	P	γ	ε or δ
1-05G20C20	0.05	19.7	-	1.88	-	0.008	0.006	45	55 (ε)
2*-03G21X13	0.03	21.66	13.22	0.14	-	0.006	0.012	96	4 (δ)
3-07G21AX13	0.07	19.28	13.71	0.29	0.15	0.020	0.010	100	0

*steel contained 4% of δ-ferrite

In addition, the phase composition of metastable steels in the fracture zone of studied samples was estimated by X-ray diffraction analysis using a DRON diffractometer with a Fe Kα radiation; the integral intensities of the $(111)_\gamma$, $(110)_\alpha$, and $(101)_\varepsilon$ diffraction lines and the texture were taken into account. Figure 3.57 shows results of the control of the effect of deformation on the process of formation of the martensite phase of deformation in various manganese steels with the FM-3 IMP device.

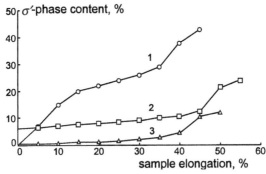

Fig. 3.57 The effect of deformation on the α phase formation in the different steels: (1) 05G20C20; (2) 03G21X13; (3) 07G21AX13.

One can see that in steels with austenitic and two-phase ($\gamma + \varepsilon$) structures, $\gamma \to \varepsilon \to \alpha$ martensitic transformation successively develops under elastic-plastic deformation. It was found that the formation of the first portions of the deformation α-martensite with amounts of 0.5-2.5% occurs in the elastic region at stress values close to the conditional yield strength $\sigma_{0.2}$.

Mechanical tensile tests have shown that, depending on the composition and structure of the steels, the experimental tensile curves considerably differ from each other. It was found that the maximum strength is realized in the nitrogen-containing steel 3; low plasticity was found in the two-phase ($\gamma + \varepsilon$) steel 1. A general feature of the metastable steels is the ability to delocalize the plastic flow, which manifests itself in a high degree of uniform elongation of the samples. An analysis of the deformation kinetics of α-martensite formation in sample 1 steel showed that the greatest increase in the amount of the ferromagnetic phase at the stage of uniform deformation was up to 15% of relative elongation. The intensity of the α'- phase formation remains constant until the moment of a neck formation in the sample. In the austenitic steels 2 and 3 containing Mn and Mn + Cr, a more linear dependence of the α'-martensite formation on the relative elongation was observed, in comparison with the steel 1. The transition from the uniform to the concentrated deformation for all the studied steels causes a sharp activation of the ferromagnetic α'-phase formation. One can see from Fig. 3.57, a sharp increase in the content of the ferromagnetic phase is observed at certain elongation of the samples of all investigated steels. After unload of the test samples, the content of the ferromagnetic phase remained the same as under the action of the maximum load. Thus, amounts of α'-phase can serve as a reliable parameter of nondestructive testing to assess the level of acting stresses.

In general, it may be noted that the experimental data obtained by the method of precision magnetic analysis are characterized by high accuracy, repeatability, and direct character of measurements. These results are also in good agreement with the data obtained by the method of magnetometric analysis with intermediate unloading. Additional measurements of the magnetic phase in the unloaded samples showed no differences in the phase composition of the steels. Such differences could be associated with a possible decrease in the α'-martensite content due to the possible development of the elastic-reversibility of the $\gamma \leftrightarrow \varepsilon$ transformation.

However, the successful use of the FM-3 IMP, FX-3 IMP and IMDS-1 ferritometers to control the ferrite or deformation martensite phases of two-phase austenitic steels does not solve the problem of controlling the phase composition of three-phase austenitic steels (austenite, ferrite, and martensite) (Rigmant et al. 2005). The next paragraph is devoted to solving this complex problem.

3.2.1 Magnetic phase control of three-phase austenitic-ferritic-martensitic steels

In this section, we present a method for controlling the phase composition of austenitic three-phase steels. This method, based on both the static magnetic characteristics (M_S saturation magnetization, remaining magnetization M_R, coercive force H_C), and on the dynamics of changes of the differential magnetic susceptibility (χ_{dif}) with the magnetic field during the magnetization reversal (Rigmant et al. 2005, 2012, 2015, Korkh et al. 2015a).

At the first stage of the development of the methodology, a set of standard two-phase samples containing either austenite and ferrite, or austenite and martensite was produced (Rigmant et al. 2005). At the second stage, the model was corrected with samples of three-phase steels. For this purpose, the initially two-phase (austenitic-ferritic composition) samples were previously pre-cooled and subjected by cold rolling to form the deformation martensite phase. Metallographic, magnetic, and other methods confirmed the simultaneous presence of three phases in the material. The developed method of phase control of three-phase steels was successfully confirmed on samples of austenitic steels of various compositions (Rigmant et al. 2012, Korkh et al. 2015b, Rigmant et al. 2015).

As mentioned above, in addition to the main austenite phase (γ-phase), the steel sample may contain the initial ferrite (α-phase). During the operation, design or repair, under the influence of high temperatures and plastic deformations, a deformation martensite phase (α'-phase) can be formed, while the content of ferrite remains unchanged.

The method of determining martensite (M) in two-phase austenitic steels was used by (Merinov and Mazepa 1997). The martensitic percentage was calculated similar to the method for calculating the ferrite content in ferrite-austenite steels:

$$\%M = (J_x^m / J_{100\%}^m) \cdot 100\%,\qquad(3)$$

where J_x^m is the saturation magnetization of the studied steel that contains martensite in the amount of $x\%$; and $J_{100\%}^m$ is the saturation magnetization of the studied steel that contains the maximum possible (100%) amount of ferrite at a given chemical composition. This value can be obtained experimentally by quenching for martensite formation, as well as by calculation, knowing the complete chemical composition of the steel. The value of the magnetization of saturation of martensite in chromium-nickel steels is estimated from the generalization of a large amount of empirical material consisted of specially fused single-phase ferromagnetic iron-chromium-nickel alloys with a solid solution structure and a different chemical composition:

$$J_{100\%}^m = 1720 - 21.9(\% \text{ Cr}) - 26.3(\% \text{ Ni}) - 22.3(\% \text{ Mn}) - 48.6(\% \text{ Si})$$
$$- 20.7(\% \text{ Mo}) - 53.3(\% \text{ Ti}) - 50.2(\% \text{ V}) - 8(\% \text{ Cu}) - 39.8(\% \text{ P}). \quad (4)$$

As can be seen, the formulas (2) and (4) are almost identical, the difference being that the magnetization of saturation in steels containing magnetic ferritic phase is determined by (1), and the magnetization of saturation of steels containing deformation martensite phase is determined by (4). Both iron phases are the ferromagnetic ones; however, the saturation magnetization of the steels containing ferritic or martensitic phases depends on the chemical composition in practically the same way (Merinov and Mazepa 1997, Rigmant et al. 2005, 2012, 2015, Korkh et al. 2015b).

With the accurate data on the chemical composition of the martensite, the relative error in determining the martensite content in steels usually does not exceed 3-5%.

We can conclude that it is possible to determine the phase composition of the two-phase austenitic chromium-nickel steels, containing, in addition to austenite, either ferrite or martensite. In practice, there are cases, when it is unknown, which of the ferromagnetic phases (martensite or ferrite) is contained in the austenitic steel. Formulas (2) and (4) are applicable simultaneously to steels with ferrites and to steels with martensite, because the saturation magnetization for ferrite and martensite is similar. However, the operating properties of steels with the same content of different magnetic phases vary significantly. In addition, the implementation of the method for determining the content of ferromagnetic phases (martensite) in austenitic steels has a number of practical difficulties. In particular, it is possible to determine the saturation magnetization of the steels only with a stationary setup that provides large magnetic fields (at least 5000-6000 A/cm). The second difficulty arises when using local control devices for samples or controlled areas on finished products. It is associated with a mandatory magnetization of samples or controlled parts to strong magnetic fields, to ensure that residual magnetization would not affect the accuracy of measurements of the content of ferromagnetic masses in the product. Most of modern devices have local primary converters, in which magnetization during the measurement of the ferromagnetic phase is carried out to magnetic fields of the order of 150-200 A/cm. Such fields are sufficient for ferrites, since the ferrite phase has small residual magnetization values (J_r), below 1-2% of the saturation magnetization (J_s). For austenitic steels containing a martensitic phase, the residual magnetization may be 10-15% or more of the saturation magnetization. Such a magnetic background can lead to large errors when measuring the martensite content M (%).

A more difficult problem arises for three-phase steels, in which both ferrite and deformation martensite phases can be present simultaneously.

Below, we present the magnetic properties of austenitic chromium-nickel steels, investigated in order to develop a magnetic method for a separate control of the phase composition (austenite, ferrite, and martensite) in three-phase austenitic steels.

Samples, measurement techniques, and setup. To measure the magnetic properties of austenitic chromium-nickel steels, two sets of cylindrical specimens with a length of 60 mm and a diameter of 7 mm were used. The first set of 10 samples was a State Standard Samples set manufactured by the method of centrifugal casting (Merinov et al. 1978). This set consisted of samples of austenitic steel grade 12X18H10T with a content of ferritic phase within 0.5-15%.

A second set of 16 samples was produced of the chromium-nickel steel grade 09X17H5Yu-VI. The samples contained, in addition to the austenite, either a ferrite phase or a martensite phase (quenched martensite or deformation martensite). The manufacturing technology for the samples included the sequence of a series of thermal and thermo-mechanical processing, in order to eliminate the chemical segregation of the initial cast state. The ferromagnetic component in the steel structure was presented by the phases of delta-ferrite (up to 10%), deformation martensite (from 10 to 40%), and quenched martensite (from 40 to 80%) (Rigmant et al. 2005, 2012).

To control the magnetic phase content in three-phase steels, we used Sets 1 and 2, including two-phase, cylindrically-shaped samples with a diameter of 7 mm and a length of 60 mm. Set 1 consisted on the austenitic-ferritic samples, made of steels graded 06X18H11, 09X17H5Yu-VI, with various contents of the α-phase. Set 2 consisted of the austenitic-martensitic samples with different contents of α'-phase, made of steel graded 09X17H5Yu-VI. This steel, initially containing austenite, was subjected to plastic deformation for obtaining deformation martensite. The samples were studied with the transmission electron microscope JEM 200CX. Such studies are described below. They allow us to reasonably determine the phase composition and structure of the samples.

For the production of three-phase ($\gamma + \alpha + \alpha'$) samples from Sets 3 and 4, two-phase austenitic-ferritic steels of grades 06X18H11 (Set 3) and 08X21H6M2T (Set 4) were used. To obtain a three-phase structure, the samples from these sets were deformed by rolling with a preliminary exposure in liquid nitrogen, which was supposed to promote a better transition $\gamma \rightarrow \alpha'$. The degree of deformation was fixed by the change in the cross section of the sample relative to its initial value (see Table 3.90).

Austenitic-ferritic samples in all sets were fabricated by the centrifugal casting method to distribute the ferritic phase uniformly throughout the entire volume of the steel ingot. Samples were cut from the ingots by the method of electric spark cutting, which allowed us to avoid their overheating and deformation. In addition, single samples of austenitic-ferritic steels of grades 05X18H10T and 03X22H6M2 were cut out for comparisons. The percentage of ferromagnetic phases in the samples is given in Table 3.91.

Table 3.90 Degree of deformation of samples from Sets 3 and 4

Sample	Degree of deformation %	Sample	Degree of deformation %	Sample	Degree of deformation %
3-1	0	4-1	0	4-4	15
3-2	5	4-2	3	4-5	30
3-3	15	4-3	10	4-6	45
-	-	-	-	4-7*	15

*sample 4-7 was deformed at 20°C without martensite formation.

Table 3.91 The content of ferrite and martensite deformation in the samples

Steel grade	Set	Sample	Content of the ferromagnetic phase, %	α-phase content, %	α'-phase content, %
06X18H11 09X17H5YuVI	1	1-1	6.35	6.35	0
		1-2	11.5	11.5	0
		1-3	14	14	0
		1-4	20.4	20.4	0
		1-5	25.5	25.5	0
09X17H5YuVI	2	2-1	7.66	0	13.2
		2-2	13.2	0	19.1
		2-3	19.1	0	21.6
		2-4	21.6	0	30.1
06X18H11	3	3-1	5.7	5.7	0
		3-2	14.5	5.7	8.8
		3-3	62.7	5.7	57
08X21H6M2T	4	4-1	18	18	0
		4-2	24	18	6
		4-3	38	18	20
		4-4	46.5	18	28.5
		4-5	57	18	39
		4-6	59.8	18	41.8
		4-7	18	18	0
05X18H10T	-	-	12.8	12.8	0
03X22H6M2	-	-	39	39	0

It was determined using the experimentally obtained values of the saturation magnetization (J_S) by the formulas (2, 4). In Sets 2, 3, and 4, after deformation of samples cooled in nitrogen, the J_S values were increased in comparison with the initial non-deformed material. This fact may be due to the formation of α'-phase in them (an exception is the sample 4-7, deformed at 20°C; its J_S remained the same as in the initial state).

Differences in the J_S values between the initial and deformed steel made it possible to determine the percentage of the resulting α'-phase.

To confirm the phase composition of the samples, Transmission Electron Microscopy (TEM) was used, as well as various methods of scanning probe microscopy (SEM). These studies detected and identified the α' phase formed in deformed austenite-ferrite samples. Tests using various structural methods, as a tool for monitoring the phase composition of chromium-nickel steels were also done.

The magnetic properties and characteristics of the two- and three-phase austenitic steels were studied by their limiting magnetic hysteresis loops built with the vibrational magnetometer *LakeShore 7407*. The results of investigations of the phase composition of samples 4-1 (non-deformed $\alpha + \gamma$ sample, 18% of the α-phase) and sample 4-5 (deformation of 30%, 18% of α-phase, 39% of α'-phase) are presented in the Figs. 3.58 and 3.59. Investigations of the structure by the TEM have shown that, after the cold deformation in a two-phase austenitic-ferrite material in austenite grains, a third phase (deformation martensite) appears. To obtain images of the structure of the studied samples, the following structural methods were used: Atomic Force Microscopy (AFM), Magnetic Force Microscopy (MFM), and the Kelvin Probe Method (KPM).

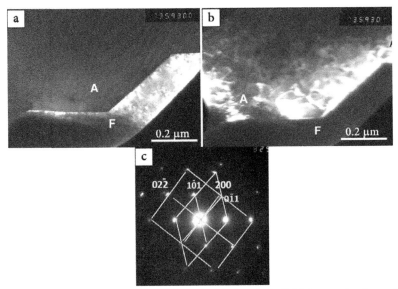

Fig. 3.58 Microstructure of sample 4-1: (a) the dark-field image in the reflex (0$\bar{1}$1) of ferrite; (b) the dark-field image in the austenite reflex (200); (c) SAED pattern to a), b), zone axis $[011]_F \parallel [111]_A$.

With AFM and MFM, the images of the distribution of surface Electric Potentials (EP) of various samples were obtained. This allowed us to evaluate and compare the changes in the magnetic and electrical properties of the investigated material before and after its plastic deformation.

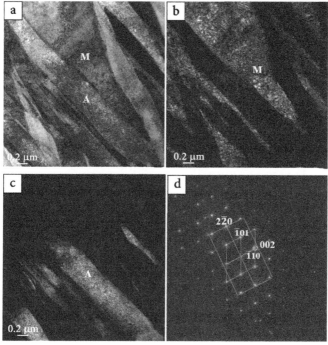

Fig. 3.59 Microstructure of sample 4-5: a- the bright-field image; b- the dark-field image in the reflex ($\bar{1}$10) of the α' martensite (BCT); c- the dark-field image in the austenite reflex (200); d- SAED pattern to (a)-(c), zone axis [111]$_M$ ∥ [110]$_A$.

The results of studies of the structure of austenitic-ferrite steel 08X21H6M2T (Set 4) by AFM and MFM before and after plastic deformation are shown in Fig. 3.60a-d.

As one can see from Fig. 3.60, the structure of both samples, according to the AFM, is two-phase and consisted of the large (light) grains of γ-phase in the environment of the dark α-phase. The MFM of non-deformed sample 4-1 (Fig. 3.60a) also showed a two-phase structure - a dark paramagnetic austenite surrounded by a ferromagnetic ferrite with a complex magnetic structure. In the case of sample 4-3 (Fig. 3.60b), MSM showed the presence of tightly packed elongated magnetic structures - deformation martensite inside the austenite grains. The MFM of other samples from Set 4 showed the deformation martensite of in all deformed samples, except for Sample 4-7.

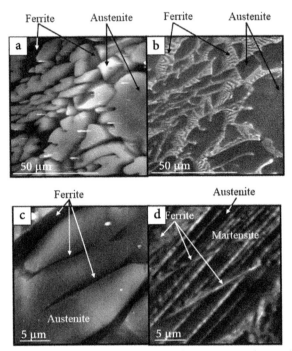

Fig. 3.60 The topography and a magnetic-force image of a surface area of the samples: (a)-(b) 4-1; (c)-(d) 4.3.

The position with Sample 4-7 was expected, since its J_S did not change after deformation. Thus, MFM allowed us to confirm the assumptions about phase changes in deformed austenite-ferrite samples and to obtain images of the structure of the three-phase material.

The samples from Set 4 (the content of the α phase was of 18%) were chosen as objects for the studies with the Kelvin Probe Method (KPM). We used Sample 4-1 (deformation 0%, α' phase 0%); Sample 4-2 (deformation of 3%, α'-phase of 6%); Sample 4-3 (deformation of 10%, α'-phase of 20%), Sample 4-5 (deformation of 30%, α'-phase of 39%); Sample 4-7 (deformation of 15%, α'-phase 0%).

Images obtained with the Kelvin probe method (KPM) for some of the samples are shown in Figs. 3.61-3.64. Each figure contains three images: *a* - Atomic Force Microscopic image (AFM) of a surface area of 35×35 µm; *b* - electro-power image, showing the spatial distribution of Electric Potential (EP), with a brighter color corresponding to a larger value of the surface potential; *c* - profile of the cross section of the EP of the sample along the dotted line between points 1 and 2. The vertical axis displays the EP (mV) along the dashed line, the horizontal axis shows the distance between points 1 and 2 (µm).

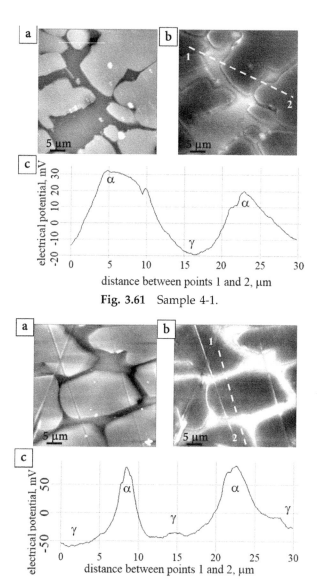

Fig. 3.61 Sample 4-1.

Fig. 3.62 Sample 4-7.

The studies carried out with the Kelvin probe method showed that a high electric potential of α-phase (Figs. 3.61 and 3.62) is characteristic for the two-phase samples 4-1, and 4-7. Grains of γ-phase have a homogeneous and lower electric potential. In the case of a three-phase material, even with small deformation and percentage of α'-phase, the electric potential of the ferrite starts to decrease. The electric potential of the austenite grains loses uniformity and some banded structures appear, with an elevated electric potential resembling the inclusions of the α' phase in the microstructural images (Fig. 3.63b). With increasing deformation degrees and percentages of the α' phase, heterogeneous electric potential in austenite appears as spots (Fig. 3.64b). This is because the α' phase in austenite no longer appears as separate bands, but as irregularly-shaped areas.

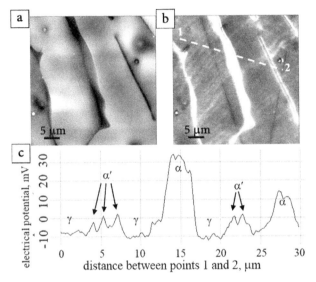

Fig. 3.63 Sample 4-2.

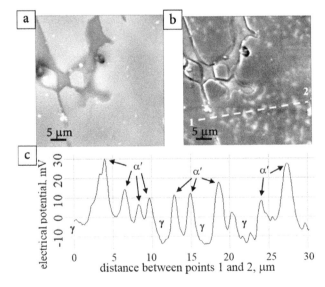

Fig. 3.64 Sample 4-5.

The Kelvin probe method is efficient in studies of the phase structure of the deformed chromium-nickel steels. It is found that the two- and three-phase chromium-nickel high-alloy steels show changes in the electric potential of their constituent phase elements during deformation and formation of the α' phase. The correspondence between the electric potential of the γ- and α-phases in the two-phase samples that do not contain the α' phase is shown in samples 4-1 and 4-7. For the three-phase samples, it was found that, with increasing degrees of deformation and amounts of the deformation martensite, the electric potential of γ and α'-phases also increases, while the electric potential of α-phase decreases to negative values. Results of studies of magnetic properties based on the limiting magnetic hysteresis loops of non-deformed austenite-ferrite samples are shown in Fig. 3.65.

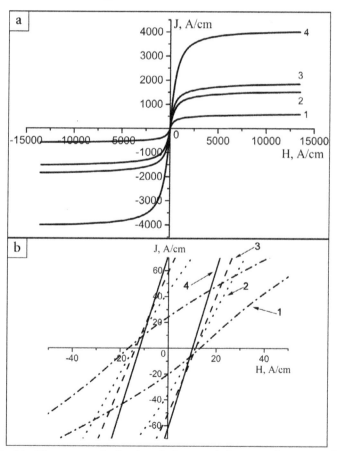

Fig. 3.65 Limiting magnetic hysteresis loops of the two-phase samples with different ferrite content (F): (a) general view of loops; (b) section of loops in fields close to the values of H_C samples; 1-5.7% F; 2-12.8% F; 3-18% F; 4-39% F.

Thus, the proposed microscopic methods allow identification and visualization of the phase components in three-phase samples. However, none of the methods makes it possible to quantify the phase percentage of steels.

Analysis of the magnetic properties of austenite-ferrite samples shows that the values of the coercive force of the two-phase samples are similar to each other, independent of their chemical composition, falling in a relatively small interval 10-15 A/cm.

Figure 3.66 shows the limiting magnetic hysteresis loops for the three-phase samples, taken from the Set 4. Figure 3.66 shows the growth of Js of deformed samples with the increasing of the α'-phase content and its coercive force.

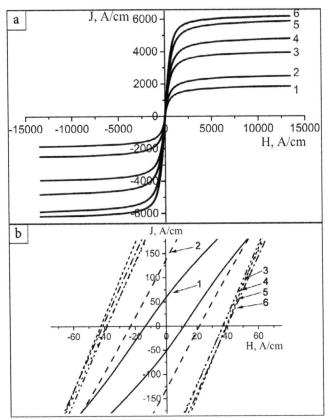

Fig. 3.66 Limiting magnetic hysteresis loops of the three-phase samples from the Set 4: (a) general view of the loops; (b) section of loops in fields close to the values of the coercive force of the samples, (1) undeformed Sample 4-1; (2) Sample 4-2, deformation of 3%; (3) Sample 4-3, deformation of 10%; (4) Sample 4-4, deformation of 15%; (5) Sample 4-5, deformation of 30%; (6) Sample 4-6; deformation of 45%.

It is obvious that the magnetic properties of chromium-nickel steels can serve as an indirect sign of a change in their phase composition. Thus, the increase in the saturation magnetization of the material after its plastic deformation indicates an increase in the content of ferromagnetic inclusions in its composition. These inclusions are nothing more than a deformation martensite phase; the saturation magnetization of ferrite is constant, independent of the deformation degree. Attention must be paid to the increasing coercive force of deformed samples, because this may be a sign of the formation of a more magnetically rigid phase, the deformation martensite.

However, studies have shown that increasing coercive force is only a qualitative characteristic of the changes in the phase composition of austenitic-ferritic steels. Increasing saturation magnetization also does not allow determination of the percentage of each ferromagnetic phase in the steels, if the ferrite content is unknown before deformation. Thus, it is necessary to investigate additional magnetic parameters of three-phase materials to determine their phase composition.

To measure the magnetic properties during magnetization and its reversal in strong magnetic fields, a "Hysteresis" setup was used in which the samples were clamped at the poles of the electromagnet to create a closed magnetic circuit (Rigmant et al. 2012, Korkh et al. 2015). During the experiment, the magnetization curve and hysteresis loops (when present) were measured in the setup. The maximum magnetic field was $H_{max} = 6$ kA/cm, measurements were made in alternating magnetic fields at low frequencies (0.1-1.0 Hz).

Measurements of the χ_{dif} (H) dependence were carried out with magnetization reversal in an alternating magnetic field of low frequency (0.25 Hz). A voltage drop, proportional to the magnitude of the current in the electromagnet came and, consequently, also to the field of the electromagnet, is measured with the Analog-Digital Converter (ADC) of the sample resistance. The waveform was chosen so that:

$$\frac{\partial I}{\partial t} \sim \frac{\partial H_e}{\partial t} = const,$$

where I is the electromagnet current, H_e is the intensity of the magnetizing field, and t is the time. In addition, measurements of the tangential component of the magnetic field on the surface of the samples showed that, for a measuring setup of a given type, $H_e = H_{in}$, where H_{in} is the sample field. Dependences of χ_{dif} (H) for the two-phase samples are shown in Figure 3.67.

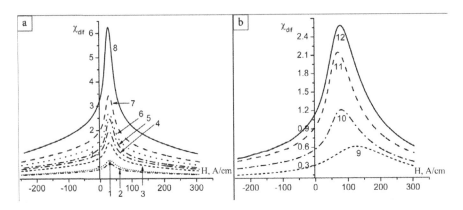

Fig. 3.67 Dependences of χ_{dif} (H) for: (a)—austenitic-ferritic and (b)—austenitic-martensitic samples (Set 2), (1) Sample 3-1 5.7% F; (2) Sample 1-1 6.35% F; (3) Sample 1-2, 11.5% F; (4) Sample 1-3, 14% F; (5) Sample 4-1, 18% F; (6) Sample 1-4, 20.4% F; (7) Sample 1-5, 25.5% F; (8) 03Cr22Ni6M, 39%(F); (9) Sample 2-1, 7.66% M; (10) Sample 2-2, 13.2% M; (11) Sample 2-3, 19.1% M; (12) Sample 2-4, 21.6% M.

Further calculations showed that the electromotive force (EMF), appearing at the terminals of the differential measuring coil, is proportional to the magnitude of the differential magnetic susceptibility: $E \sim c \cdot \chi_{\text{dif}}$, where c is a constant. An analysis of the obtained results showed that all the dependences of χ_{dif} (H) of austenite-ferrite samples have a field maximum of 28 ± 1 A/cm, while the maxima of χ_{dif} (H) for the austenitic-martensitic samples shift towards larger fields, the magnetic stiffness of martensite is higher than that of ferrites. This displacement indirectly suggests that the sample contains the deformation martensite phase.

It was found that the areas under the curves χ_{dif} (H) for two-phase samples are linear functions of the percentage of ferromagnetic phases (see Fig. 3.68).

To simulate a situation, in which two ferromagnetic phases are located in one volume, we chose the dependences χ_{dif} (H) of composite samples simulating a three-phase structure of deformed austenitic-ferritic steels. Two two-phase samples of the same size were placed in the measuring coil: austenitic-ferritic and austenitic-martensitic, and the dependence χ_{dif} (H) of the composite sample was taken. Examples of such dependences are shown in Fig. 3.69.

Figure 3.69 shows that the dependence χ_{dif} (H) for composite three-phase samples is more complex, in comparison with χ_{dif} (H) for the two-phase samples presented above in Fig. 3.67. The appearance of the inflection (Fig. 3.69a) or second maximum (Fig. 3.69b) is caused by different magnetic rigidity of the phases in the composite sample and is a signature of the presence of phase constituents with different magnetic properties. It was also found that the areas under the χ_{dif} (*H*) curves of the composite sample are equal to the sum of the areas of χ_{dif} (*H*) for the two-phase samples of the constituent three-phase sample.

To determine the content of each of the two ferromagnetic phases in three-phase composite samples, the χ_{dif} (*H*) curves having double peaks or kinks were approximated by the Lorentz function. This was done to "restore" χ_{dif} (*H*) of each of the magnetic phases entering into the composite sample (see Fig. 3.70) and determine fractions of these phases from the areas of their χ_{dif} (*H*).

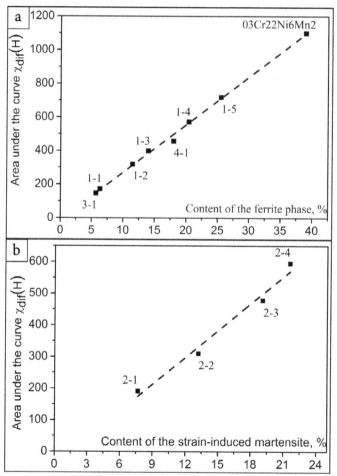

Fig. 3.68 Dependence of the area under the χ_{dif} (*H*) curve on the percentage of: (a) ferrite in austenitic-ferritic samples; (b) deformation martensite in austenitic-martensitic samples from Set 2; dotted line - linear approximation.

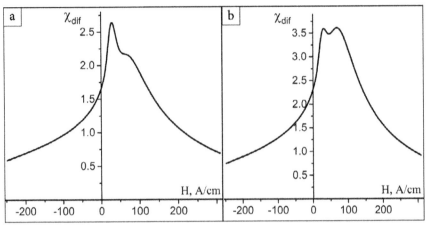

Fig. 3.69 Dependence χ_{dif} (H) for composite samples consisting of: (a) austenitic-ferritic sample 1-3 (α - 14%) and austenitic-martensitic sample 2-2 (α' - 13.2%); (b) austenitic-ferritic sample 1-3 (α - 14%) and austenitic-martensitic sample 2-4 (α' - 21.6%).

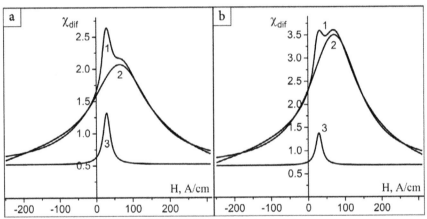

Fig. 3.70 An example of the expansion of χ_{dif} (H) of the composite samples by the Lorentz function: (a) austenitic-ferritic sample 1-3 (α - 14%) and austenite-martensitic sample 2-2 (α' - 13.2%); (b) austenitic-ferritic sample 1-3 (α - 14%) and austenitic-martensitic sample 2-4 (α' - 21.6%).

The digits indicate: *1* - the experimentally obtained χ_{dif} (H) of the composite sample; *2* - χ_{dif} (H) dependence for the martensitic component of the composite sample; *3* - χ_{dif} (H) for the ferrite component of the composite sample.

Such an increase in the "recoverable" areas of χ_{dif} (H) martensitic samples in depending on the ferrite content in the composite sample is illustrated in Fig. 3.71.

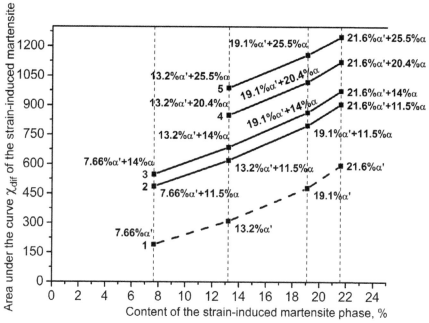

Fig. 3.71 Dependence of the areas under the χ_{dif} (*H*) curves of the martensitic constituents on the martensite fraction.

However, after a "reconstruction" of the χ_{dif} (*H*) curves of the martensitic samples included in the composite three-phase sample (2 in Figs. 3.70a and 3.70b), it was found that the areas under curves are always larger than the χ_{dif} (*H*) areas of single martensitic samples (Fig. 3.70b) by an amount proportional to the content of ferrite in the composite sample. Such an increase in the "recoverable" areas of χ_{dif} (*H*) martensitic samples, as a function of ferrite content in the composite sample is illustrated in Fig. 3.71.

Curve 1 shows a dependence of the area under χ_{dif} (*H*) on the percentage of the α' phase for single samples from Set 2; this dependence is also shown in Fig. 3.68b. Curve 2 presents the areas under χ_{dif} (*H*) of martensitic samples, reconstructed from χ_{dif} (*H*) of composite samples with an α phase content of 11.5%. Curve 3 shows the areas under χ_{dif} (*H*) of the martensitic samples, reconstructed from χ_{dif} (*H*) for the composite samples with an α-phase content of 14.5%. Curve 4 presents the areas under χ_{dif} (*H*) for the martensitic samples, reconstructed from χ_{dif} (*H*) for the composite samples with a phase content of 20.4%. Curve 5 presents the areas under χ_{dif} (*H*) of the martensitic samples, reconstructed from χ_{dif} (*H*) of the composite samples with an α-phase content of 25.5%.

Since the growth of the "reduced" areas of the χ_{dif} (*H*) martensitic component relative to the χ_{dif} (*H*) areas of the two-phase martensitic samples (Fig. 3.71, *1*) is proportional to the increase in the ferrite percentage, this relationship can be expressed as

$$S_M^* - \Delta_\alpha \cdot \alpha_\% = S_M, \qquad (5)$$

where S_M^* is the area under the "restored" martensitic curve χ_{dif} (H), with a physical meaning and the dimension of magnetization (A/m); S_M is the area under the measured χ_{dif} (H) for two-phase samples with martensite; Δ_α is the coefficient of S_M^* increase with respect to S_M with increasing fraction of ferrite ($\alpha_\%$) in the composite sample.

Taking into account that the area under the χ_{dif} (H) curve for the two-phase samples increases linearly with increasing ferromagnetic phase, S_M can be approximated by a straight line

$$S_M = k \cdot \alpha_\%' + b, \qquad (6)$$

where k and b are the coefficients of linear approximation (Fig. 3.68b); $\alpha_\%'$ is the percentage of the martensite phase in the sample.

Then the expression (6) takes the form of an equation with two unknown values, $\alpha_\%$ and $\alpha_\%'$, in a three-phase composite sample

$$S_M^* - \Delta_\alpha \cdot \alpha_\% = k \cdot \alpha_\%' + b. \qquad (7)$$

To determine $\alpha_\%$ and $\alpha_\%'$, it is necessary to compose another equation with these unknown values. Because with traditional methods of ferrometry (i.e., measuring saturation magnetization) it is quite easy to determine the total percentage of ferromagnetic inclusions ($\Sigma_{\alpha+\alpha'}$) in austenitic steels, then expression

$$\alpha_\% + \alpha_\%' = \Sigma_{\alpha+\alpha'}. \qquad (8)$$

The resulting system of equations with the two unknown $\alpha_\%$ and $\alpha_\%'$ has the form:

$$\begin{cases} S_M^* - \Delta_\alpha \cdot \alpha_\% = k \cdot \alpha_\%' + b \\ \alpha_\% + \alpha_\%' = \Sigma_{\alpha+\alpha'} \end{cases}. \qquad (9)$$

Its solution gives fractions of both the ferrite and martensite phases in the three-phase material of austenitic chromium-nickel steels.

However, the study of composite samples gives a simpler picture: the phases with different magnetic stiffness are spatially separated and their influence on each other during magnetization reversal is weakened.

It is necessary to verify the proposed model for the case when all three phases ($\gamma + \alpha + \alpha'$) are present in the volume of a sample. The mutual influence of ferromagnetic phases in that case is much more complicated, and this directly affects the obtained χ_{dif} (H).

Dependences χ_{dif} (H) for the samples from Set 4 (including three-phase samples with different deformation martensite content and with a ferrite content of 18%) are shown in Fig. 3.72.

In contrast to the composite samples, χ_{dif} (H) in Fig. 3.70 do not have an additional peak nor inflection point, as in Fig. 3.69. Consequently, there may be doubts on the presence of the martensite phase in the deformed austenite-ferrite material. However, the presence of the α' phase is confirmed by the growth of H_C for the three-phase material.

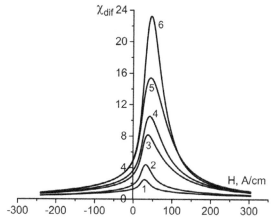

Fig. 3.72 Dependences χ_{dif} (H) for samples with initial ferrite content of 18% (Set 4) subjected to cold deformation: (1) Sample 4-1, undeformed; (2) Sample 4-2, deformation 3%; (3) Sample 4-3, deformation 10%; (4) Sample 4-4, deformation 15%; (5) Sample 4-5, deformation 30%; (6) Sample 4-6, deformation 45%.

When the deformation degree of the samples (hence, the content of the α' phase) increases, both the maximum of χ_{dif} (H) and the area under it increase (Fig. 3.72). It can also be seen that, when the austenitic-ferrite sample additionally exhibits deformation martensite, maxima of the χ_{dif} (H) curves shift relative to the H axis toward larger fields (from 28 A/cm for a non-deformed sample to 46 A/cm for a sample with 45% of deformation). This may serve as an additional evidence (in addition to the H_C growth) of the appearance of α'-phase in the austenite-ferrite material. Apparently, the reason for the displacements is the mutual influence of α- and α'-phases during the reversal of three-phase samples.

The experimentally obtained χ_{dif} (H) for a three-phase sample can be decomposed by Lorentz function approximation into two χ_{dif} (H) curves for the phases of ferrite and martensite entering the sample.

It can be seen from Fig. 3.72, when the degree of deformation of the samples (and hence the content of the α' phase in them) increases, the maximum of the χ_{dif} (H) dependence, as well as the area under it, increase. It can also be seen that when the austenitic-ferrite sample additionally exhibits deformation martensite, the maxima of the χ_{dif} (H) curves shift relative to the H axis toward larger fields (from 28 for an undeformed sample to 46 A/cm for a sample with 45% of deformation).

This fact may serve as an additional sign (in addition to the growth of the H_C) in the appearance of α'-phase in the austenite-ferrite material. The reason for the displacement is, apparently, the mutual influence of the α- and α'-phases on each other during the reversal of three-phase samples.

The experimentally obtained χ_{dif} (H) dependence for a three-phase sample can be decomposed by Lorentz function approximation into two curves χ_{dif} (H) for the phases of ferrite and martensite entering the sample.

Figure 3.73 shows the χ_{dif} (H) curves of the martensitic components calculated from χ_{dif} (H) for the three-phase samples from Set 4.

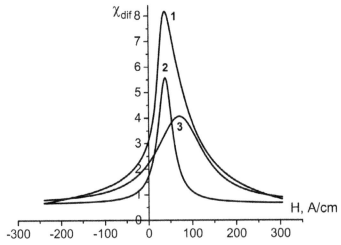

Fig. 3.73   Restored martensitic curves of three-phase deformed samples with an initial ferrite content of 18%.

Having χ_{dif} (H) for the martensitic component of the three-phase samples (Fig. 3.73), it is possible to derive the dependence of the areas under those curves on the fraction of martensite in samples from Set 4. Similar actions can be performed for three-phase samples with a ferrite content of 5.7% from Set 3. The areas under the χ_{dif} (H) curves of martensitic constituents, depending on the percentage of martensite for three-phase samples, as well as for two-phase austenitic-martensitic samples are shown in Fig. 3.74.

As can be seen from Fig. 3.74, when the area under χ_{dif} (H) is plotted against martensite fraction, the resulting dependence shifts upward along the ordinate axis by an amount, proportional to the content of ferrite in the three-phase sample. Therefore, in the case of a three-phase material, it is impossible to directly estimate amounts of deformation martensite from the area under the "restored" curve χ_{dif} (H). However, the content of ferromagnetic phases in a three-phase material (analogous to the definition of ferromagnetic phases in a composite sample) can be determined using the system of equations (9).

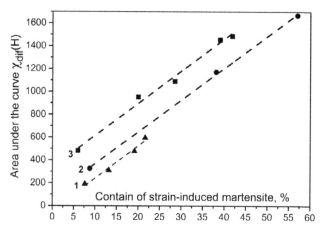

Fig. 3.74 Dependence of the areas under χ_{dif} (H) for the martensite phase on the martensite fraction in the samples: (1) for austenitic-martensitic samples from Set 2; (2) for three-phase samples from Set 3; (3) for three-phase samples from Set 4. The dashed lines show the linear approximations.

This system of two equations with two unknown values ($\alpha_{\%}$, $\alpha'_{\%}$) makes it possible to calculate the phase composition of three-phase chromium-nickel steels.

Thus, in this section we described a magnetic method for estimating the contents of deformation martensite phase and ferrite phase in an austenitic material on the basis of an analysis of the dependences of the differential magnetic susceptibility on the strength of the internal field. This method of estimating the phase composition can be used for chromium-nickel steels with different chemical compositions.

3.2.2 Control of products made from heat-resistant austenitic steels and nickel superalloys having relative magnetic permeability $\mu \leq 1.10$

This section deals with controlling austenitic steels and alloys, in which, according to technical requirements, ferromagnetic inclusions may be contained in very small amounts - less than 1%. The phase composition of such alloys consists of 99% of paramagnetic austenite. Such materials are widely used for the products and parts, in which the magnetic properties within or close to them do not depend on their own magnetic properties. Such properties of the materials are also useful for the watch industry, in the manufacturing of navigation equipment and minesweeper cases, and in the production of the drill pipes for inclined and horizontal drilling. The quality of these materials is determined by their magnetic properties, or rather, by their "almost complete absence", when the magnetic flux density inside the material is less than a hundredth or thousandths to a percent of the magnetic flux in vacuum. This property is called the relative magnetic permeability μ, calculated from the relation $B = \mu_a \cdot \mu \cdot H$, which relates the induction B in material and the magnetic field H, at which B is generated, where μ_a is the vacuum permeability ($\mu_a = 4\pi \cdot 107T \cdot m/A$). Thus, the requirement of "low-magnetism" may be expressed as $\mu \leq 1.01$ or, in some cases, $\mu \leq 1.05$. Such relative magnetic permeability is due to very small ferromagnetic inclusions. However, modern methods and means for measuring magnetic properties enable measurements for even insignificant magnetic inclusions, whose presence can be judged from the relative magnetic permeability μ or the magnetic susceptibility $\chi = \mu - 1$. It is difficult to measure parameter μ, because it is necessary to magnetize products or their controlled areas to sufficiently large magnetic fields. In this case, useful signals from the controlled magnetized product are by orders of magnitude smaller than the magnetizing field itself. The use of local permanent magnets, proposed by Rigmant and Gorkunov (1996), makes it possible to measure in contact small values of $\mu \sim 1.001 \pm 0.0005$, excluding the influence of external strong fields, acting as a disturbance. Using this method, various magnetic permeability meters have been developed (Vedeniov et al. 1993, Rigmant et al. 2005). The use of devices for monitoring the magnetic state of the industrial materials, when $\mu_m \sim 1.001 - 1.100$, made it possible to cull products or local places in products, in which the magnetic permeability exceeds the established levels under production conditions. This helps improve the production quality, as well as avoid emergencies during operation.

To solve such problems, the authors (at the Institute of Metal Physics of the Russian Academy of Sciences) proposed a method for measuring magnetic permeability μ (of susceptibility $\chi = \mu - 1$) by measuring the scattering field from the magnetized controlled part of the product (Rigmant and Gorkunov 1996, and Rigmant et al. 2005). The method features the measurements of μ, carried out by evaluating the scattering field H_{sc} from the locally magnetized area (Fig. 3.75).

The controlled area is locally magnetized by a strong permanent magnet to a state near the technical saturation. The scattering field H_{sc} from the "mirror" reflection (Figs. 3.56 and 3.75) is directly related to the relative magnetic susceptibility χ (related to permeability by $\chi = \mu - 1$). In Fig. 3.75, a Hall transducer was used as a magnetic field meter, in contrast to Fig. 3.56, where a measuring Ferro probe was used. At the same time, high localization of the scattering field was achieved, since the thickness of the Hall transducer does not exceed 0.2 mm and its dimensions are 2 × 2 mm. For comparison, each ferro-probe element, as in Fig. 3.56, is longer than 5 mm.

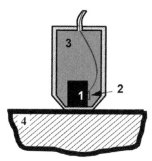

Fig. 3.75 Schematic diagram of the sensor with a meter of the Hall type: (1) permanent magnet built into the sensor, (2) Hall transducer, (3) sensor body made of paramagnetic material, (4) the object of control.

The controlled area is locally magnetized by a strong permanent magnet to a state close to the technical saturation. The scattering field H_{sc} from the "mirror" reflection (Figs. 3.56 and 3.75) is directly related to the relative magnetic susceptibility (it is more convenient to talk about the χ -magnetic susceptibility, which is related to permeability, since $\chi = \mu - 1$). In Fig. 3.75, a Hall transducer is used as a magnetic field meter in contrast to Fig. 3.56 where it is a measuring Ferro probe. At the same time, high localization of the scattering field is achieved, since the thickness of the Hall transducer is not more than 0.2 mm and its dimensions are of 2 × 2 mm. For comparison, the length of each ferro-probe element, as in Fig. 3.56, is not less than 5 mm.

The devices for measuring the relative magnetic permeability for small and super-small ferromagnetic inclusions ($\alpha < 0.01\%$) have been used in some unusual applications, in the case of controlling structural changes in the products of nickel refractory alloys occurring in the material. High-temperature nickel alloys are used for manufacturing turbine blades of stationary Gas Turbine Units (GTU), which operate for a long time, in wide ranges of temperatures and stresses (Stepanova et al. 2011). The main phases of such alloys are a nickel solid solution and a Ni_3Al strengthening phase with a superstructure of the $L1_2$ type (γ'-phase). In addition to the intermetallic γ'-phase, carbides (MC and $M_{23}C_6$) are also present in the superalloy structure. Carbides are released at the boundaries and inside the grains as inclusions.

Magnetic methods of nondestructive testing are widely used in industry to assess the performance of parts, but they have never been used for products made of heat-resistant nickel alloys. The intermetallic compound Ni_3Al is a weak ferromagnet with a Curie temperature of 41 K. All phases of the alloy are in the paramagnetic state, both at room temperature and in the operating temperature range. High-temperature alloys have a low value of magnetic susceptibility ($\sim 10^{-4}$), including the nickel superalloy ChS-70V (40% of γ'-phase).

Presently, active attempts are made to increase capacity and efficiency of gas turbine setup in power engineering. In particular, the experimental turbine GTE-45-3 of the Yakutsk thermal power plant had a capacity four times greater that of the serial turbines. This was achieved by increasing the operating temperature from 800 to 880°C and the rotational speed by 50%. When operating under this regime, the blades made of the alloy ChS-70V experienced extreme conditions, in terms of the temperature and stress levels. As a result, the blades operated for only 11 months, instead of four years. It became necessary to establish the cause of the collapse of the blades of the ChS-70V alloy and to develop a method of nondestructive testing for the early detection and development of cracking processes.

The IMPAS-1 device for local measurements of the magnetic permeability of austenitic steels and alloys allowed us to identify places prone to the initiation and development of cracks. It was found that the cracking process occurred inside the Ni_3Al strengthening phase, not in the zone of ductile austenitic nickel. Thus, a change in the magnetic susceptibility allows detection of a state of pre-destruction that is associated with the appearance of defects within the hardening γ'-phase (Ni_3Al). This allows us to apply the methods of magnetic nondestructive testing to assess the performance of turbine blades. Replacement of the exhausted parts will prevent their emergency destruction.

Similar changes in magnetic properties after deformation were also observed in other high-temperature nickel alloys (EP-800, VKNA-4U, EI-437-V) with different contents of the Ni_3Al hardening intermetallic phase (Stepanova et al. 2011, Kazantseva et al. 2013).

3.3 Relationship between the Specific Electrical Resistance and the Phase Composition of Austenitic Steels

During cold plastic deformations of austenitic chromium-nickel steels, a complex $\gamma \rightarrow \alpha'$ transition can occur (γ-austenite phase, α'-deformations martensite phase). This transition leads to disruptions in the uniform distribution of alloying elements in the solid solution and to changes in the atomic-magnetic austenite states, which determines further operational properties of the austenitic and austenitic-ferritic steels. Austenite is a paramagnetic material, whereas ferrite (α-phase) and deformation martensite are ferromagnetic. In order to determine the contents of ferrite (α) and deformation martensite in austenite material, magnetic methods of phase analysis are widely used (Shcherbinin and Gorkunov 1996, Apaev 1976). Methods for measuring electrical properties for studies of martensitic transformations are also used. For example, the eddy current method for determining deformation martensite in stainless steels is well known (Gerasimov et al. 1992, Sandovski et al. 2001). This method is structurally sensitive; the magnitude of the signal coming from the eddy current transducer is related to the specific electrical resistance and the magnetic characteristics of metals. However, the disadvantage of this method is that it allows investigation of the destruction process (and, as a consequence, the phase composition changes) only on the surface and in the near-surface layers. Pustovoit and Dolgachev (2007) and Netesov and Yaes (1987) investigated the characteristic features of the kinetics of phase transitions of the first and second types by analyzing the temperature dependence of the electrical resistance. Measurements of electrical resistance have high sensitivity; it allows one to estimate the local changes in the positions of atoms in the crystal lattice within a small volume of material. However, in these cases, the study of the martensitic transformation is only qualitative.

Establishment of a correlation between the phase composition and changes in the magnetic or electrical properties of materials under conditions of thermal and deformation effects is warranted. Based on the information obtained from both magnetic and electrical measurements, it would be possible to quickly evaluate materials and predict the degree of their irreversible structural changes. Such predictions are especially important at the initial stages of destruction of structural materials, in order to prevent emergencies. However, the authors of this section have not found any published studies with an established relationship between the amount of formed deformation martensite and electrical properties of austenitic-ferritic heat- and corrosion-resistant steels.

To study the effects of phase composition on the magnetic and electrical properties, we selected samples of austenitic-ferritic steels and chromium-nickel austenitic-ferritic steels with different contents of the ferrite phase in the initial state. Table 3.92 presents the chemical compositions of the experimental alloys examined.

Table 3.92 Chemical composition of studied alloys, mass %

Alloy	C	Ni	Cr	Mn	Ti	Mo	Si
1	0.038	10.01	20.5	1.52	0.64	0.79	1.90
2	0.050	9.06	20.02	1.84	0.39	1.15	2.06
3	0.110	9.24	22.39	1.80	0.40	0.47	3.22
4	0.043	9.78	25.75	1.71	0.52	0.71	1.80
5	0.073	9.66	26.89	2.95	0.72	1.05	1.82
6	0.079	7.08	28.33	1.58	0.76	1.44	2.40
7	0.064	5.18	28.59	1.77	0.52	0.45	3.69

The samples were subjected to plastic deformation by multiple-pass cold rolling, in order to create an additional hard-magnetic phase (deformation martensite) in the initially two-phase austenitic-ferritic ($\gamma + \alpha$) alloys. Prior to each pass, the samples were cooled in liquid nitrogen. Previous research (Korkh et al. 2016) has shown that this kind of treatment promotes the emergence of deformation martensite (α'-phase) in the material structure. Electrical properties were studied using a facility developed at the Institute of Metal Physics (Ekaterinburg, Russia). The facility allows determination of electric resistivity and incorporates a constant current power supply, a seven-digit voltmeter, and a measuring resistance. The operating principle of the facility is based on the four-contact method of measuring electric resistivity. In this scheme, the voltage drop is measured with a voltmeter across a segment of an elongated sample that carries electric current. This drop is proportional to the resistance of this segment of the sample. Electric resistivity was measured along a rolling direction at room temperature. In addition, the resistivity was measured with a "Specialized Micro Ohmmeter" device developed in the Laboratory of Magnetic Structural Analysis at the Institute of Metal Physics (Ekaterinburg, Russia). This device allows one to take measurements of specimens with a thickness of 3-12 mm of both circular and rectangular cross sections. A system of spring-loaded knife contacts ensures firm holding of the examined sample in the transducer. The range of electric resistivity measurements is from 1 $\mu\Omega$ to 20 mΩ.

Magnetic measurements. To offset the effect of the demagnetizing factor, measurements were taken for two groups of samples of different geometry. It was discovered that the results are in agreement with each other, thus enabling a conclusion that the geometry of examined samples (given the right positioning of a rod with the sample in the vibrator unit) has no effect on the results of measurements. Research data are presented in Table 3.93.

Table 3.93 Ferrite (F) and deformation martensite (M) percentages in samples

Alloy	Phase content			
	Initial	Def. 10%	Def. 20%	Def. 30%
1	2.8%F	2.8%F; 13.6%M	2.8%F; 60.0%M	2.8%F; 81.2%M
2	6.7%F	6.7%F; 12.8%M	6.7%F; 27.2%M	6.7%F; 59.6%M
3	12.1%F	12.1%F; 16.0%M	12.1%F; 40,6%M	12.1%F; 47,9%M
4	26.8%F	26.8%F; 10,5%M	26.8%F; 10.6%M	26.8%F; 27.5%M
5	40.9%F	40.9%F; 0.1%M	40.9%F; 14.3%M	40.9%F; 19.3%M
6	67.3%F	67.3%F; 24.1%M	67.3%F; 25,2%M	67.3%F; 32,3%M
7	81.9%F	81.9%F; 18.1%M	-	-

Table 3.93 shows the phase composition of the investigated materials in the initial and deformed states at various stages of deformation, after the preliminary cooling of the samples in liquid nitrogen, calculated from the values of the saturation magnetization.

The resultant saturated hysteresis loops were used to determine the saturation magnetization and coercive force (Fig. 3.77). As research has shown, the coercive force (H_C) in non-deformed samples stays within the range of 5-15 A/cm, values that are characteristic of austenitic-ferritic class steels. The change in the coercive force and saturation magnetization depends on the initial percentage of the ferrite phase in an alloy. Ferrite content in austenitic-ferritic alloys was calculated based on measured saturation magnetization (J_S). The measurement technique was described in detail by Rigmant et al. (2012). The studied alloys contained 2.8-81.9% of ferrite in the initial state (Table 3.93).

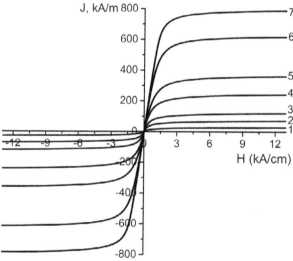

Fig. 3.76 Saturated hysteresis loops of samples with different ferrite percentage in the initial condition: (1) sample of alloy 1, (2) sample of alloy 2, (3) sample of alloy 3, (4) sample of alloy 4, (5) sample of alloy 5, (6) sample of alloy 6, (7) sample of alloy 7.

Magnetic properties of the samples after plastic deformation with pre-cooling in liquid nitrogen changed substantially. Figure 3.76 summarizes results obtained for one of the examined alloys (Alloy 3, Table 3.93). The coercive force H_C in this alloy increased by more than a factor of 3; it was also discovered that the saturation magnetization J_S of the deformed samples increases with the deformation percentage. Similar results were obtained for all the samples. Saturation magnetization is known to be related to the total volume of ferromagnetic phases, whereas amounts of ferrite remain unchanged after deformation (Korkh et al. 2015a). Therefore, the deformation-induced increase of Hc and Js in the initially austenitic-ferritic alloys must occur due to an additional hard-magnetic ferromagnetic phase, namely, α' (strain-induced martensite).

Cracks were detected in the samples 0X32N8 (Alloy 6) and 04X25H5M2 (Alloy 7) containing the highest ferrite value in their initial state. In the sample of Alloy 6, a crack was found after its deformation by 30%. A sample of Alloy 7 collapsed after its deformation by 10%.

The phase composition of the samples was confirmed by Transmission Electron Microscopy (TEM).

Hardness measurements. Deformation martensite in an austenitic-ferritic steel strengthens the material (Gulyaev 1977). Therefore, we also studied, how the hardness of samples changes from their initial condition to that after plastic deformation after pre-cooling in liquid nitrogen. The measured values of hardness are given in Table 3.94.

Table 3.94   Hardness of studied alloys in initial condition and
after deformation with pre-cooling in liquid nitrogen

Alloy	Initial state	Def. 10%	Def. 20%	Def. 30%
1	47	62	67	70
2	45	63	66	68
3	50	66	66	68
4	53	62	65	67
5	58	64	65	68
6	61	64	66	-
7	59	63	-	-

Table 3.94 shows that 10% deformation increases hardness (HRA) in all the samples, with the minimum increase (approximately 5-10%) observed in the samples with the largest ferrite content and the maximum increase in the samples with the ferrite content of 2.8 and 6.7%. However, as shown by the results of magnetic analysis (see Table 3.93), the maximum amount of martensite formed in the sample of Alloy 6, which exhibited the minimum hardness increase. The samples of Alloy 2 (ferrite content of 6.7%) and Alloy 4 (26.8%), in which approximately 10-12% of deformation martensite formed, the deformation-induced changes in HRA were 40 and 17%, respectively. Further plastic deformation (by 20 and 30%) brings no more significant changes and the values of hardness of all the samples remain close. Thus, based on changes in hardness, it is only possible to evaluate the presence of the martensite phase but not to assess the amount of deformation martensite that emerged in the material after plastic deformation. In this case, it is also worth considering that increase in hardness is connected with cold hardening, i.e., with the emergence of multiple structural flaws.

Further plastic deformation (by 20 and 30%) brings no more significant changes and the values of hardness of all the samples remain close. Thus, based on changes in hardness, it is only possible to evaluate the presence of martensite phase but not to assess the amount of deformation martensite that emerged in the material after plastic deformation. In this case, it is also worth considering that increase in hardness is connected with cold hardening, i.e., with the emergence of multiple structural flaws.

Electric resistivity measurements. We measured the electric resistivity of liquid-nitrogen pre-cooled samples after every deformation action as well as the electric resistivity for a group of samples (Alloy 3) that were deformed at room temperature without pre-cooling in liquid nitrogen. The results of electrical measurements are summarized in Fig. 3.77.

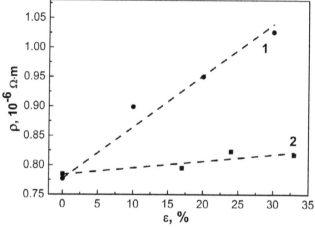

Fig. 3.77 Electric resistivity of samples of 12X21N5T steel (ferrite percentage 12.1%); (1) deformation by rolling at room temperature with pre-cooling in liquid nitrogen; (2) deformation by rolling at room temperature without pre-cooling.

Research of the magnetic properties of the alloy 3 samples deformed at room temperature, without pre-cooling in liquid nitrogen, showed that changes in the saturation magnetization are within the measurement errors; this allowed us to draw the conclusion that there were no phase transitions in the material.

Structural studies performed by Korkh et al. (2016) with 08X21N6M2T steel samples showed that plastic deformation of these samples at room temperature, without pre-cooling in liquid nitrogen, does not lead to the formation of additional hard-magnetic phase of strain-induced martensite. In this research, we additionally measured the electric resistivity of samples made of the 08X21N6M2T steel in the initial non-deformed condition and after a 15% deformation by rolling, without pre-cooling in nitrogen. The measured values were $\rho_{0\%} = (0.79 \pm 0.01) \times 10^{-6}$ and $\rho_{15\%} = (0.81 \times 0.01) \times 10^{-6}$ Ωm, respectively. Together with a comparative analysis of the dependences of the electric resistivity on deformation for the 12X21N5T steel samples, this allowed us to conclude that an appearance of additional hard-magnetic phase (deformation martensite) in the initially two-phase (austenite + ferrite) samples leads to a significant increase in ρ, in comparison to the deformation with no phase transition (Fig.3.78).

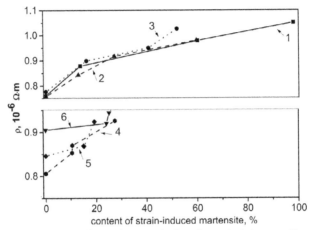

Fig. 3.78 Graph of the dependence of electric resistivity on the amount of martensite phase formed during deformation: (1) Alloy 1, (2) Alloy 2, (3) Alloy 3, (4) Alloy 4, (5) Alloy 5, (6) Alloy 6.

Figure 3.78 provides results on the dependence of electric resistivity on the amount of deformation martensite in the samples of other steel grades (see Table 3.92). Analyzing these data, one can conclude that martensite formed in each of the samples of alloys leads to a significant increase in the electric resistivity. Note that the greatest change in ρ as compared with the initial non-deformed samples is observed in specimens with low ferrite-phase contents. This may be explained by the fact that these specimens contain a considerable percentage of austenite, which decays due to deformation after pre-cooling to form the martensite α'-phase. Interesting results were obtained when studying the samples of Alloys 5 and 6, which contain more than 40% of the ferrite phase. The magnetic and electrical properties of these samples, measured after deformation, indicate that the process of martensite formation in these samples occurs stepwise.

Analysis of the results on electric resistivity in the initial samples showed the existence of a nearly linear dependence of ρ on the ferrite content in the alloy, a fact that is possibly associated with similar structural conditions of the initial samples that were all cut from cast melts obtained under identical conditions.

Thus, our research has shown the possibility for using data on electric resistivity for evaluating the phase composition of austenitic-ferritic steels.

3.4 Instrumental Control of the Specific Resistance of Austenitic Steels at Operated Conditions

At many enterprises producing parts of electric power machines, the Technical Manual (TM) stipulates the requirement for the minimum possible value of the resistivity ρ of the austenitic material of used steels. The resistivity values must not be lower than the preset ρ level, which is related to the optimal phase composition of the products under exploitation. Measurements of ρ are carried out at specialized fixed setups, where a calibrated distance for the samples-witnesses is used to measure the voltage drop, when stable current passes through the sample, using a seven-digit voltmeter. Such setups are available at many enterprises in Russia. However, they are cumbersome and not easily transportable, as well as the expensive equipment for measuring voltage, for creating and measuring current is not always available.

To resolve this problem, small-sized specialized meters of specific electrical resistance were developed at the Institute of Metal Physics (Ekaterinburg, Russia) and introduced at several enterprises of Russian Federation. They are able to measure samples with different transverse dimensions.

Figure 3.80a shows the device "Measuring instrument of the electrical resistivity", which was introduced at LLC "Belenergomash – BZEM" (Belgorod) for the output control of the electrical resistivity of the materials of the manufactured products. The enterprise LLC "Belenergomash - BZEM" (Belgorod) manufactures equipment for the power industry. The most commonly used are two grades of austenitic steels - 30X3N17G2L and 12X18AG18-L with rigorous technical specifications for their electrical properties. For the steel 30X3N17G2L, the electrical resistivity ρ should be at least 0.75 $\Omega \cdot mm^2/m$ (10^{-6} Ω); for the steel 12X18AГ18, it must be not below than 0.65 $\Omega \cdot mm^2/m$ (10^{-6} Ω).

The device developed by the author makes it possible to measure the sample resistivity for different cross-sections, between 2 and 12 mm in the lateral section. A feature of the proposed device is a system of spring-loaded knife contacts made of a particularly durable material, which allows secure fixation of a controlled rectangular sample in the converter device. The use of spring-loaded contacts makes it possible to vary the dimensions of the measured parts, and in addition to rectangular sections, it is possible to measure the resistance of round specimens, between 3 and 12 mm in diameter.

Our experience in the creation of compact meters of resistivity, acquired with at LLC "Belenergomash – BZEM" (Belgorod), enabled us to solve a more difficult problem, namely the problem of measuring the specific electrical resistance on finished products in the manufacturing process. In particular, for the enterprise Joint Stock Company "Zlatoust Machine-Building Plant" (JSC "Zlatmash") (Zlatoust), the author proposed the device MC-1-IFM – "Milliometer specialized". This device allows measurements of electrical resistance on a finished product in the form of austenitic wire of various diameters (1 to 12 mm). Resistance measurements are carried out over a long length of 1000 mm. In this case, the system of spring-loaded knife contacts also reliably fixes the controlled sample (of various sections) in the converter device (Fig. 3.79b).

The device allowed us to significantly improve the quality of products and significantly accelerate the process of controlling the finished wire to reduce the cost of carrying out the control procedure.

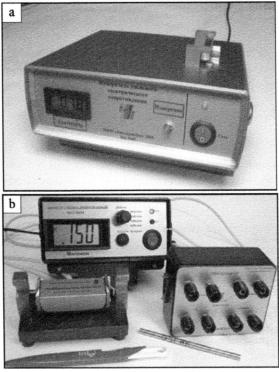

Fig. 3.79 Instruments for measurement of specific electricity resistance in production conditions (IMP, Ekaterinburg).

To simplify the measurements, in which it is necessary to change the length of the working base, the author has developed the instrument "Ommeter specialized MC3-IMP" (Fig. 3.79).

The device has several additional features for easy monitoring of the samples: (1) a large range of measurements of the electrical resistivity from 0.5 to 2 $(\Omega \cdot mm^2)/m$, (2) four contact adjustments, which include copper contacts to create a stable current in the sample, (3) spring-loaded knife contacts to remove Electric Moving Force (EMF) from the controlled sample area, and (4) a remote unit for convenient connection of the transmitter. The stabilized current generator, with errors in current measurements of no more than 0.1%, creates a current of 2A through the copper contacts in the controlled sample. The EMF value is proportional to the voltage drop across a sample portion of different lengths, from 30 to 100 mm. This EMF is fed to a differential instrumentation amplifier with high metrological characteristics, then to a voltmeter and an information display device that is calibrated in units of resistivity.

The system of spring-loaded blade contacts allows fixing the controlled sample in the converter device and in case of deviation in the dimensions of the side or diameter of the circular sample to ± 2 mm. The device MC3-IMP was introduced for monitoring at "Motovilikhinskie Zavody PJSC" (Perm, Russia).

As was shown in the previous section, along with the use of magnetic properties for the phase control of austenitic steels, the knowledge of specific electric conductivity ρ also allows us to derive the phase composition of austenitic steels. The measurement of the resistivity (ρ) is much simpler and cheaper than the measurements of magnetic properties, therefore it seems reasonable that studies of the relationship between the electrical resistivity and the phase composition of austenitic steels may have successful practical solutions.

CONCLUSION

This chapter discusses the effects of phase and chemical composition on the exploitation properties (corrosion resistance, heat resistance, and ductility) of parts made of high-temperature austenitic steels. We present the main modern methods and approaches for the phase control of steel materials, their capabilities and limitations in the production conditions.

The authors did not intend to relate this chapter to all types of the heat-resistant steels used in practice: martensitic-ferritic, pearlitic, austenitic, and martensitic steels. Although it is known that the share of martensitic steels in the total volume of high-temperature materials is very high. For example, the 3X13H7C2 and 4X9C2 steels are successfully used at temperatures of 850-900°C in manufacturing valves of automobile engines; the X5M, 1X12H2VMF, 1X8VF, X6CM, and X5VF steels are widely used for the production of assemblies and various parts working for 1000-10000 hours at temperatures from 500 to 600°C. Pearlitic heat-resistant steels, such as X13H7C2, X7CM, X9C2, X10X2M, X6CM, and X6C of chromium-molybdenum and chromium-silicate compounds are used for parts operating at temperatures up to 650-700°C.

This work emphasizes on austenitic and austenitic-ferritic steels, since they are in great demand. In these steels, the structure is due to the presence of nickel, and the heat resistance is due to the presence of chromium. Such compositions sometimes include insignificant inclusions of niobium, titanium, and carbon (0.01 to 0.1%).

At temperatures of up to 1000° austenitic grades successfully resist the process of scoria formation and at the same time they belong to the group of corrosion resistant steels. Well-known alloys of this group are:

- Dispersive hardening steels with the grades 0X14H28B3TZYR, X12H20TZR, 4X12H8Г8MFB, and 4X14H14B2M are optimal for manufacturing valves of vehicle engines and turbine parts;
- Homogeneous steels with the grades 1X14H16B, X25H20C2, X23H18, X18H10T, X25H16G7AR, X18H12T, and 1X14H18V2B find their application in the field of production of valves and pipes, operating under heavy loads, as well as exhaust system components, ultra-high-pressure units.

In turn, austenitic-ferritic alloys have a very high heat resistance, which is much higher than that of conventional high-chromium materials. This is achieved through the unique stability of their structure. Such steel grades cannot be used for the production of loaded components due to their increased brittleness. But they are perfectly suitable for the manufacture of products, such as pyro metric tubes (steel grade H23N13), or furnace conveyors, pipes, and carburizing tanks (steel grades X20N14C2 and 0X20N14C2) operating at temperatures close to 1150°C.

The chapter also shows the practical significance of measuring the magnetic properties (magnetic permeability μ) of high-temperature steels and alloys at room temperatures; μ may serve as a monitoring parameter for determination of local nucleation sites and developing fracture processes.

Although this chapter focuses mainly on the magnetic properties of high-temperature austenitic steels, considerable attention is paid to the measurements of the specific resistivity ρ. It is convincingly shown that ρ may serve as an additional parameter for the phase control of two- and three-phase high-temperature austenitic steels.

Prospects of Nondestructive Testing Products of High-Temperature Austenitic Steels

Presently, the methods and means of nondestructive testing of materials made of austenitic high-temperature steels are successfully used in Russia and other countries. The main factor determining the quality of the steels is their phase composition. This chapter shows that nondestructive phase control is usually reliably designed only for two-phase materials, namely, for austenitic-ferritic steels. At the same time, there is a lack of methods and means of nondestructive testing of three-phase steels that have austenite, ferrite, and martensite in their composition. Methodological work in this direction is carried out in many countries, although there are no practical developments of means of such control in the literature.

An important direction is the connection between methods of nondestructive testing and practical diagnostics of the working equipment. Early detection of local nucleation sites and the development of fracture processes will reduce the accident rate and, at the same time, increase the reliability of the operation of expensive products. A serious task is monitoring of the state of the parts and mechanisms made of high-temperature austenitic steels. The solution of these problems will be possible with the use of modern robotics and computer processing of the diagnostic information.

References

Introduction and Chapter 1

Aoki, K., K. Ishikawa and T. Masumoto. 1995. Ductilization of Ni_3Al by alloying with boron and substitutional elements. Mater. Sci, Eng. A. 192-193(1): 316–323.

Asoka-Kumar, P., M. Alatalo, V.J. Ghosh, A.C. Kruseman, B. Nielsen and K.G. Lynn. 1996. Increased elemental specificity of positron annihilation spectra. Phys. Rev. Let. 77: 2097–2100.

Bachteeva, N.D., N.I. Vinogradova, S.N. Petrova, V.P. Pilyugin and V.A. Sazonova. 1997. Structure and hardness of Nickel superalloy single crystals after deformation by shear under high pressure. The Physics of Metals and Metallography. 85: 97–104.

Badura-Gergen, K. and H.E. Schaefer. 1997. Thermal formation of atomic vacancies in Ni_3Al. Phys. Rev. B. 56: 3032–3037.

Bagot, P.A.J., O.B.W. Silk, J.O. Douglas, S. Pedrazzini, D.J. Crudden, T.L. Martin, M.C. Hardy, M.P. Moody and R.C. Reeda 2017. An atom probe tomography study of site preference and partitioning in a Nickel-based superalloy. Acta Mater. 125: 156–165.

Bai Lin Lü, Guo Qing Chen, Shen Qu, Hui Su and Wen Long Zhou. 2013. First-principle calculation of yield stress anomaly of Ni_3Al-based alloys. Mater. Sci. Eng. A. 565: 317–320.

Baker, I. and D. Wu 2005. Strain-induced ferromagnetism in $L1_2$ compounds. TSM letters. 2: 57–58.

Basak, A., R. Acharya and S. Das. 2016. Additive manufacturing of single-crystal superalloy CMSX-4 through scanning laser epitaxy: computational modeling, experimental process development, and process parameter optimization. Metall. and Mater. Trans. A. 47: 3845–3859.

Battezzatti, L., M. Baricco and L. Pascale. 1998. High temperature thermal analysis of Ni-Al alloys around the γ' composition. Scripta Mater. 39: 87–93.

Baryshev, E.E., T.K. Kostina, A.A. Ganev, L.G. Savina, A.G. Tyagunov and O.B. Demenok. 1998. Effect of treatment and modification of melt on structure and properties of heat resistant ZhS-6U alloy. Rasplavy. 1: 36–42 (in Russian).

Baryshev, E.E., A.G. Tyagunov and N.N. Stepanova. 2010. The influence of melt structure on the properties of Nickel superalloys in the solid state. Ekaterinburg. Press UB RAS. (in Russian)

Bauer, A., S. Neumeier, F. Pyczak and M. Göken. 2012. Creep strength and microstructure of polycrystalline γ'-strengthened cobalt-base superalloys. pp. 695–703. *In*: E.S. Huron, R.C. Reed, M.C. Hardy, M.J. Mills, R.E. Montero, P.D. Portella, J. Telesman (eds). Proc. Conf. "Superalloys 2012". TMS - The Minerals, Metals & Materials Society. Warrendale, PA, USA.

Bazyleva, O.A., Y.A. Bondarenko, G.I. Morozova and O.B. Timofeeva. 2014. Structure, chemical composition, and phase composition of intermetallic alloy VKNA-1V after high-temperature heat treatment and process heating. Metal Science and Heat Treatment. 56: 229–234.

Bendersky, L.A., F.S. Biancaniello and M. Williams. 1994. Evolution of the two-phase microstructure Ll$_2$ + D0$_{22}$ in near-eutectoid Ni$_3$(Al,V) alloy. J. Mater. Res. 9: 3068–78.

Bhattacharya, B. and R.K. Ray. 2000a. Deformation behavior of a Ni$_3$Al (B, Zr) alloy during cold rolling: Part I. Changes in order and structure. Metal. Trans. A. 31: 3001–3010.

Bhattacharya, B. and R.K. Ray. 2000b. Deformation behavior of a Ni$_3$Al (B, Zr) alloy during cold rolling: Part II. Microstructural and textural changes. Metal. Trans. A. 31: 3011–3021.

Budinovskii, S.A., S.A. Muboyadzhyan, A.M. Gayamov and S.V. Stepanova. 2011. Ion-plasma heat-resistant coatings with composite barrier layer for protecting alloy ZhS-36VI from oxidation. Metal Science and Heat Treatment. 53: 32–38.

Budinovskii, S.A., S.A. Muboyadzhyan, A.M. Gayamov and P.V. Matveev. 2014. Development of ion-plasma refractory metallic layers of heat-insulating coatings for cooled turbine rotor blades. Metal Science and Heat Treatment. 55: 652–657.

Buntushkin, V.P., M.B. Bronfin, O.A. Bazyleva and O.B. Timofeeva. 2004. Effect of alloying and the casting structure on the high-temperature strength of the Ni$_3$Al intermetallic compound at high temperatures. Russian metallurgy (Metally). 2: 194–196.

Cahn, R.W., P.A. Siemers, J.E. Geiger and P. Bardhan. 1987a. The order disorder transformation in Ni$_3$Al and Ni$_3$Al-Fe alloys. I. Determination of the transition temperatures and their relation to ductility. Acta Metal. 35: 2737–2751.

Cahn R.W., P.A. Siemers and E.L. Hall. 1987b. The order-disorder transformation in Ni$_3$Al and Ni$_3$Al-Fe alloys. II. Phase transformations and microstructures. Acta Metal. 35: 2753–2764.

Cahn, R.W. and P. Haasen [ed]. 1996. Physical Metallurgy. Phase transformations in metals and alloys and alloys with special physical properties. 4th Edition. Elsevier.

Campbell, F.C. 2006. Superalloys. pp. 211–272. *In*: Manufacturing Technology for Aerospace Structural Materials. Elsevier Ltd.

Cao, J., J. Zhang, Y. Hua, R. Chen, Zh. Li and Yu. Ye. 2017. Microstructures and isothermal oxidation behaviors of CoCrAlYTaSi coating prepared by plasma spraying on the Ni-based superalloy GH202. Surface and Coatings Technology. 311: 19–26.

Cardanelli, F. 2013. Materials Handbook. A Concise Desktop Reference. 2nd Ed. Springer, New York, London.

Caron, P. and O. Lavigne. 2011. Recent Studies at ONERA on Superalloys for Single Crystal Turbine Blades. J. AerospaceLab. 3: 1–14.

Caron, P., F. Diologent and S. Drawin. 2011. Influence of chemistry on the tensile yield strength of nickel-based single crystal superalloys. Euro Superalloys 2010. Trans. Tech. Publication, Switzerland: Advanced Materials Research. 278: 345–350.

Carona, P. and C. Ramusat. 2014. Optimization of the homogenization and hot isostatic pressing heat treatments of a fourth generation single crystal superalloy. Proc. MATEC Web of Conferences. Publ. by EDP Sciences. 14: 13002.

Carter, L.N., X. Wang, N. Read, R. Khan, M. Aristizabal, K. Essa and M.M. Attallah 2016. Process optimization of selective laser melting using energy density model for nickel based superalloys. Materials Science and Technology. 32: 657–661.

Cermak, J. and V. Rothova. 2003. Concentration dependence of ternary inter-diffusion coefficients in Ni_3Al and Ni_3Al X couples with X = Cr, Fe, Nb and Ti. Acta Mater. 51: 4411–4421.

Chakravorty, S. and C.M. Wayman. 1976. The thermoelastic martensitic transformation in γ' Ni-Al alloys: II. Electron microscopy. Met. Trans. A. 7: 569–582

Christian, R., W. Thomas and K.H. Peter. 2003. TEM investigation of the structure of deformation-induced antiphase boundary faults in Ni_3Al. Physical Review B. 67: 094109 (1–5).

Chowdhury, S.G., R.K. Ray and A.K. Jena. 1995. Structural transformation in $Ni_3Al(B)$ due to cold rolling. Scripta Met. Mater. 32: 1501–1506.

Colombo, L. 1998. A source code for tight binding molecular dynamics simulation. Comp. Mat. Sci. 12: 278–287.

Corey, C.L. and B. Lisowsky. 1967. Electrical resistivity study of Ni_3Al alloys. Trans. Metal. Soc. AIME. A. 239: 239–245.

Courths, R. and S. Löbus. 1999. Electronic band structure of Cu_3Au: an angle resolved photoemission study along the [001] direction. Phys. Rev. B. 60: 8055–8066.

Cozar, R. and A. Pineau. 1973. Morphology of γ' and γ'' precipitates and thermal stability of INCONEL alloy 718 type alloys. Metall. Trans. 4: 47–59.

Crudden, D.J., A. Mottura, N. Warnken, B. Raeisinia and R.C. Reed. 2014. Modelling of the influence of alloy composition on flow stress in high-strength nickel-based superalloys. Acta Mater. 75: 356–370.

Danilov, V.N. and I.N. Ermolov. 2000. Calculations of amplitude-distance-diameter diagrams. Russian Journal of Nondestructive Testing. 36: 491–498.

Danilov, V.N. 2009. Calculation of DGS diagrams of a normal transducer with a rectangular piezoelectric plate. Russian Journal of Nondestructive Testing. 45: 760–774.

Davidov, D.I., N.I. Vinogradova, N.V. Kazantseva and N.N. Stepanova. 2015a. Investigation of the structure of two heat-temperature Nickel-based alloys after high-temperature deformation. The Physics of Metals and Metallography. 116: 200–208.

Davidov, D.I., N.N. Stepanova, N.V. Kazantseva, V.P. Pilyugin and D.A. Shishkin. 2015b. Change of the magnetic properties of Ni₃Al-based alloys after deformation. AIP Conf. Proc. Advanced materials with hierarchical structure for new technologies and reliable structures. P. 020034.

Davidov, D.I., N.N. Stepanova, V.P. Pilyugin and D.A. Shishkin. 2016. Influence of different types of severe deformation on magnetic properties of Nickel-based superalloy. Int. J. Applied Eng. Research. 11: 4197–4201.

Davidov, D.I., N.N. Stepanova, N.V. Kazantseva and M.B. Rigmant. 2017. Application of magnetic methods for study deformed state of nickel-based superalloys. Diagnostics, Resource and Mechanics of materials and structures (DREAM). 1: 6–12.

De Boer, F.R., C.J. Schinkel, J. Biesterbos and S. Proost. 1969. Exchange-enchanced paramagnetism and weak ferromagnetism in the Ni₃Al and Ni₃Ga phase. J. Appl. Phys. 40: 1049–1055.

De Novion, C.H. and J.P. Landesman 1985. Order and disorder in transition metal carbides and nitrides: experimental and theoretical aspects. Pure & Appl. Chem. 57: 1391–1402.

Demura, M., D. Golberg and T. Hirano. 2007a. An athermal deformation model of the yield stress anomaly in Ni₃Al. Intermetallics. 15: 1322–1331.

Demura, M., Y. Xu, K. Kishida and T. Hirano. 2007b. Texture memory effect in heavily cold-rolled Ni₃Al single crystals. Acta Mater. 55: 1779–1789.

Dicson, R.W. and J.B. Wachtman. 1969. Elastic constants of single crystal Ni₃Al from 10° to 850°C. J. Appl. Phys. 40: 2276–2279.

Divinski, S.V., St. Frank, U. Sodervall and Ch. Herzig. 1998. Solute diffusion of Al substitution elements in Ni₃Al and diffusion mechanism of the minority component. Acta Mater. 46: 4369–4380.

Doremus, L., J. Cormier, P. Villechaise, G. Henaff, Y. Nadot and S. Pierret. 2015. Influence of residual stresses on the fatigue crack growth from surface anomalies in a nickel-based superalloy. Mater. Sci. Engin. A 644. P. 234–246.

Druzhkov, A.P., D.A. Perminov and N.L. Pecherkina. 2008. Positron annihilation spectroscopy characterization of effect of intermetallic nanoparticles on accumulation and annealing of vacancy defects in electron-irradiated Fe-Ni-Al alloy. Philos. Mag. 88: 959–976.

Druzhkov, A.P., D.A. Perminov and N.N. Stepanova. 2010. Positron annihilation study of the influence of doping on the 3d electron states in the Ni₃Al intermetallic compound. Physics of the Solid State. 52: 2005–2011.

Duval, S., S. Chambreland, P. Caron and D. Blavette. 1994. Phase composition and chemical order in the single crystal Nickel base superalloy MC2. Acta Metal. Mater. 42: 185–194.

Eldrup, M. and B.N. Singh. 1997. Studies of defects and defect agglomerates by positron annihilation spectroscopy. J. Nucl. Mater. 251: 132–138.

Enomoto, M. and H. Harada. 1989. Analysis of γ'/γ equilibrium in Ni-Al-X alloys by the cluster variation method with the Lennard Jones potential. Metal. Trans. A. 20: 649–664.

Erofeev, V., A. Dergunova, A. Piksaikina, A. Bogatov, E. Kablov, O. Startsev, and A. Matvievskiy 2016. The effectiveness of materials different with regard to increasing the durability. Proc. MATEC Web of Conf. 15 Int. Conf. Topical

Problems of Architecture, Civil Engineering, Energy Efficiency and Ecology. TPACEE - 2016. 73. P. 04021.

Frank, St., U. Södervall and Ch. Herzig. 1995. Self-diffusion of Ni in single and polycrystals of Ni₃Al. A study of SIMS and radiotracer analysis. Physica Status Solidi (b). 191: 45–55.

Fu, C.L. and G.S. Painter. 1997. Point defects and the binding energies of boron near defect sites in Ni₃Al: A first-principles investigation. Acta Mater. 45: 481–488.

Glas, R., M. Jouiad, P. Caron, N. Clement and H.O.K. Kirchner. 1996. Order and mechanical properties of the γ matrix of superalloys. Acta Mater. 44: 4917–4926.

Golberg, D., M. Demura and T. Hirano. 1997. Compressive flow stress of a binary stoichiometric Ni₃Al single crystal. Scripta Mater. 37: 1777–1782.

Golubovskiy, E.R., M.E. Volkov, N.M. Emmausskiy and S.A. Shibayev. 2015. The experimental study of fatigue monocrystals of nickel alloys for GTE turbine blades. Proc. XI All-Russian Congr. on fundamental problems of theoretical and applied mechanics. Kazan. P. 190–191. (in Russian)

Goman'kov, V.I., G.I. Nosova, S.E. Manaenkov, V.V. Rtishchev and L.E. Fykin. 2008. Atomic ordering in the phases of a Nickel-chromium-based superalloy. Russian metallurgy (Metally). 6: 500–505.

Gorsky, W.S. Theorie der elastischen Nachwirkung in ungeordneten Mischkristallen (elastische Nachwirkung zweiter Art). Phys. Zeitschrift Sowjet. 1935. 8: 443.

Greenberg, B.A. and M.A. Ivanov. 2002. Ni₃Al and TiAl intermetallic compounds: microstructure and deformation behavior. Ekaterinburg. (in Russian).

Greenberg, B.A., N.V. Kazantseva, V.P. Pilugin and E.V. Shorokhov. 2006. Phase transformation in intermetallics induced by shock-wave loading. pp. 169–182. In A. Burhanettin [ed]. Severe Plastic Deformation: Toward Bulk Production of Nanostructured Materials. Nova Science Publ. Inc. New York.

Habicht, W., N. Boukis, G. Franz and E. Dinjus. 2004. Investigation of Nickel-based alloys exposed to supercritical water environments. Microchim. Acta. 145: 57–62.

Hancock, G.F. 1971. Diffusion of nickel in alloys based on the intermetallic compound Ni₃Al (γ'). Physica Status Solidi (a). 7: 535–540.

Harrison, W.A. 1989. Electronic Structure and the Properties of Solids: The Physics of the Chemical Bond, Dover Publications, New York.

Hilpert, K., D. Kobertz, V. Venugopal, M. Miller, H. Gerads, F.J. Bremer, and H. Nickel. 1987. Phase diagram studies on Ni-Al system. Z. Naturforsch. A. 42: 1327–1392.

Hino, T., T. Kobayashi, Y. Koizumi, H. Harada and T. Yamagata, 2000. Development of a new single crystal superalloy for industrial gas turbines. Superalloy. Edited by T.M. Pollock, R.D. Kissinger, R.R. Bowman, K.A. Green, M. McLean, S. Olson, and J.J. Schirra , TMS-AIME, Warrendale, PA. 729–736.

Hiraga, K. and M. Hirabayashi. 1979. Modulated structure in metal carbide systems of V/sub 2/C, Nb/sub 2/C and Ta/sub 2/C. AIP Conf. Proc.: Intern. Conf. on modulated structures, Kailua-Kona, HI, USA. 53:1.

Horsfield, A.P., A.M. Bratkovsky, M. Fearn, D.G. Pettifor, and M. Aoki. 1996. Bond order potentials: Theory and implementation. Phys. Rev. B. 53: 12694–12699.

Hoshino, K., S.J. Rothman and R.S. Averback. 1988. Tracer diffusion in pure and boron doped Ni_3Al. Acta Metall. 36: 1271–1279.

Hunziker, O. and W. Kurz. 1997. Solidification microstructure maps in Ni-Al alloys. Acta Mater. 45: 4981–4992.

Idzikowski, B., Young-Hoon Hyun and Y. Kudryavtsev. 2003. Neutron investigation of Ni_3Al itinerant electron system. Experimental report of proposal PHY-02-0397. Hahn-Meitner-Institute (HMI) Berlin.

Iotova, D., N. Kioussis and S.P. Lim. 1996. Electronic structure and elastic properties of the Ni_3X (X = Mn, Al, Ga, Si, Ge) intermetallics. Phys. Review B. 54: 14413–14422.

Jahangiri, M.R. and M. Abedini. 2014. Effect of long time service exposure on microstructure and mechanical properties of gas turbine vanes made of IN939 alloy. Maters & Design. 64: 588–600.

Jozwik, P., W. Polkowski and Z. Bojar. 2015. Applications of Ni_3Al based intermetallic alloys. Current stage and potential perceptivities. Materials. 8: 2537–2568.

Jiang, J., J. Yang, T. Zhang, F.P.E. Dunne and T.B. Britton. 2015. On the mechanistic basis of fatigue crack nucleation in Ni superalloy containing inclusions using high resolution electron backscatter diffraction. Acta Mater. 97: 367–379.

Kablov, E.N., O.A. Bazyleva and M.A. Vorontsov. 2006a. A new base for creating castable high temperature superalloys. Met. Sci. Heat Treat. 48: 348–351.

Kablov, E.N., N.V. Petrushin, M.B. Bronfin and A.A Alekseev. 2006b. Specific features of rhenium-alloyed single-crystal Nickel superalloys. Russian metallurgy (Metally). 5: 406–414.

Kablov, E.N. and N.V. Petrushin. 2008. Designing of high-rhenium single crystal Ni-base superalloy for gas turbine blades. Proc. of the 11 Int. Symp. Superalloys- 2008. Champion. PA. P. 901–908.

Kablov, E.N. and S.A. Muboyadzhyan. 2012. Heat-resistant coatings for the high-pressure turbine blades of promising GTES. Russian metallurgy (Metally). 1: 1–7.

Kablov, E.N., V.I. Titov, N.V. Gundobin, Y.A. Karpov, K.E. Karfidova and G.S. Kudryavtseva. 2015. Rhenium and ruthenium determination in nanostructured high-temperature alloys for aerospace engineering. Inorganic Materials. 51: 1363–1369.

Kablov, E.N., V.V. Sidorov, D.E. Kablov and P.G. Min. 2017a. Metallurgical base for high quality single crystal heat resistant nickel alloys. Aviation materials and technologies. 5: 55–71.

Kablov, E.N., Yu.A. Bondarenko and A.B. Echin. 2017b. Progress of the directional solidification technology of cast heat resistant alloys with variable controlled temperature gradient. Aviation materials and technologies. 5: 24–38.

Kassner, M.E. 2015. Fundamentals of creep in metals and alloys, 3rd ed., pp. 261-273. In: γ/γ' Nickel-based Superalloys. Elsevier Ltd.

Kayser, F.X. and C. Stassis. 1981. The elastic constants of Ni_3Al at 0 and 23.5°C. Phys. Status Solidi. 64: 335–342.

Kazantseva, N.V., B.A. Greenberg, E.V. Shorokhov, A.N. Pirogov, and Yu.A. Dorofeev. 2005. Study of Phase Transformations in Ni_3Al Superalloy after Shock-Wave Loading. The Physics of Metals and Metallography. 99: 535–543.

Kazantseva, N.V., A. Pirogov and M. Rigmant. 2006. Long-period modulated structure in deformed Nickel superalloys. Proc. of Int. Conf. on Advances in Mechanical Engineering (AME-2006). TMS. Fatehgarh Sahib, Punjab, INDIA. M-1: 60–65.

Kazantseva, N.V., N.I. Vinogradova, N.N. Stepanova, A.N. Pirogov and E.O. Golikova. 2009a. Formation of metastable phases in a Ni-9.6 wt.% Al-6.7 wt.% Fe-1 wt.% Cr intermetallic alloy. The Physics of Metals and Metallography. 107: 375–383.

Kazantseva N.V., I.I. Kositsina, N.N. Stepanova and N.I. Vinogradova. 2009b. Modulated structures in complex chromium carbides. Physics of Metals and Advanced Technologies. 31: 1331-1341.

Kazantseva, N.V., A.V. Korolev, D.I. Davydov, N.I. Vinogradova, M.B Rigmant, and N.N. Stepanova. 2013. Concentration heterogeneity and the magnetism in the spatula of heat-resistant Nickel alloy. Materialsovedenie. 4: 18–24. (in Russian)

Kazantseva, N.V., V.P. Pilyugin, V.A. Zavalishin and N.N Stepanova. 2014a. Effect of a nanosized state on the magnetic properties of $Ni_3(Al,Fe)$ and $Ni_3(Al,Co)$. The Physics of Metals and Metallography. 115: 243–247.

Kazantseva, N.V., N.N.,Stepanova, M.B.,Rigmant and D.I. Davidov. 2014b. Magnetic and structural analyses of the deformation failures in Nickel superalloys. Materials Science &Technology Conf. and Exhib. (MS & T-2014). P. 1447–1454.

Kazantseva, N.V., M.B. Rigmant, N.N. Stepanova, D.I. Davydov, D.A. Shishkin, P.B. Terent'ev and N.I. Vinogradova. 2016. Structure and magnetic properties of $Ni_3(Al, Fe, Cr)$ single crystal subjected to high-temperature deformation. The Physics of Metals and Metallography. 117: 451–459.

Keshavarz, S. and S. Ghosh. 2015. A crystal plasticity finite element model for flow stress anomalies in Ni_3Al single crystals. Phil. Mag. 95: 2639-2660.

Kobayashi, S., M. Demura, K. Kishida and T. Hiranoa. 2005. Tensile and bending deformation of Ni_3Al heavily cold-rolled foil. Intermetallies. V. 13(6). P. 608–614.

Kolotukhin, E.V., E.A. Kuleshova, E.E. Baryshev and G.V. Tyagunov. 1995. Structure of high-temperature nickel alloys after time heat treatment of the melt. Metal Science and Heat Treatment. 37: 222–225.

Koneva, N.A., N.A. Popova, E.L. Nikonenko, E.V. Kozlov, M.P. Kalashnikov, M.V. Fedorishcheva, A.D. Pasenova and E.V. Kozlov 2012. Effect of alloying with boron on the phase composition and structure of Ni_3Al. Bulletin of the Russian Academy of Sciences: Physics. 76: 740–743.

Kopelev, S.Z. 1983. Cooled blades of gas turbines: Calculation and profiling. Science, Moskow.

Korznikov, A.V., G. Tram, O. Dimitrov, G.F. Korznikova, S.R. Idrisova and Z. Pakiela. 2001. The mechanism of nanocrystalline structure formation in Ni_3Al during severe plastic deformation. Acta Mater. 49: 663–671.

Kositsyna, I.I., V.V. Sagaradze and O.N. Khakimova. 1997. Specific features of precipitation hardening of austenitic steels with different matrixes: I. Mechanical properties and the effect of primary carbides. The Physics of Metals and Metallography. 84: 80–86.

Kozlov, E.V., A.A. Klopotov, A.S. Tailashev and N.O. Solonitsina. 2006a. The Ni - Al system: crystal-geometrical features. Bulletin of the Russian Academy of Sciences: Physics. 70: 1113–1116.

Kozlov, E.V., N.A. Popova, E.L. Nikonenko, N. Koneva, N.R. Sizonenko, V.P. Buntushkin, and Yu.R. Kolobov. 2006b. Structure of modern superalloys containing topologically close-packed phases. Bulletin of the Russian Academy of Sciences: Physics. 70: 1117–1121.

Kozlov, E.V., A.A. Klopotov, E.L. Nikonenko, M.V. Fedorischeva and V.D. Klopotov. 2011. Structural features of triple equilibrium diagrams of the systems based on Ni-Al. Bulletin of the Russian Academy of Sciences: Physics. 75: 1099–1102.

Kozubski, R., J. Sołtys, M.C. Cadeville, V. Pierron-Bohnes, T.H. Kim, P. Schwander, J.P. Hahn, G. Kostorz and J. Morgiel. 1993. Long range ordering kinetics and ordering energy in Ni_3Al-based γ' alloys. Intermetallics. 1: 139–150.

Kozubski, R. 1997. Long range order kinetics in Ni_3Al based intermetallic compounds with $L1_2$ type superstructure. Progress in Materials Science. 41: 1–59.

Kumar, A., A. Chernatynskiy, M. Hong, S.R. Phillpot and S.B. Sinnott. 2015. An ab initio investigation of the effect of alloying elements on the elastic properties and magnetic behavior of Ni_3Al. Comput. Mater. Sci. 101: 39–46.

Kumari S., D.V.V. Satyanarayana and M. Srinivas. 2014. Failure analysis of gas turbine rotor blades. Eng. Failure Analysis. 45: 234–244.

Kuznetsov, V.P., V.P. Lesnikov, E.V. Moroz and A.S. Koryakovtsev. 2008. Mechanical properties of high-temperature Nickel alloy ZhS-36VI for single-crystal GTE blades. Metal Science and Heat Treatment. 50: 228–231.

Kuznetsov, V.P., V.P. Lesnikov, I.P. Konakova and N.A. Popov. 2011a. Structure, phase composition, and strength properties of single-crystal nickel alloy bearing titanium and rhenium. Metal Science and Heat Treatment. 52(9-10): 499–503.

Kuznetsov, V.P., V.P. Lesnikov, I.P. Konakova, N.V. Petrushin and S.A. Muboyadzhyan. 2011b. Structure and phase composition of single-crystal alloy VZhM-4 with gas-circulation protective coating. Metal Science and Heat Treatment. 53: 131–135.

Kuznetsov, V.P., V.P. Lesnikov, M.S. Khadyev, I.P. Konakova and N.A. Popov. 2012. Structural and phase transformations in single-crystal alloy ZhS36-VI [001] after holding in the range of 1050-1300°C. Metal Science and Heat Treatment. 54: 90–99.

Kuznetsov, V.P., V.P. Lesnikov, I.P. Konakova and N.A. Popov. 2015. Effect of TCP-phases on the tensile fracture behavior of single-crystal nickel alloy ZhS36-VI [001]. Metal Science and Heat Treatment. 56: 495–498.

Lawniczak-Jablonska, K., R. Wojnecki and J. Kachniarz. 2000. The influence of Fe atom location on the electronic structure of $Ni_3Al_{1-x}Fe_x$: LMTO calculation and X-ray spectroscopy. J. Phys.: Condens. Matter. 12: 2333–2350.

Lee, J.H. and J.D. Verhoeven. 1994. Eutectic formation in the Ni-Al system. J. Cryst. Growth. 143: 86–102.

Lepikhin, S.V., G.V. Tyagunov, N.N. Stepanova, Yu.N. Akshentsev and V.A. Sazonova. 2004. Phase and structural transformations in ternary Ni_3Al-based alloys with iron. The Physics of Metals and Metallography. 97: 408–414.

Li, D., K. Kishida, M. Demura and T. Hirano. 2008. Tensile properties and cold rolling of binary Ni-Al γ/γ' two-phase single crystals. Intermetallics. 16: 1317–1324.

Li, M., J. Song and Y. Han. 2012. Effects of boron and carbon contents on long-term aging of Ni_3Al-base single crystal alloy IC6SX. Procedia Eng. 27: 1054–1060.

Lil, Z. and M. Peng. 2007. Microstructural and mechanical characterization of Nb-based in situ composites from Nb–Si–Ti ternary system. Acta Materialia. 55(19): 6573–6585.

Lin, R.Q., C. Fu, M. Liu, H. Jiang, X. Li, Z.M. Ren, A.M Russell and G.H. Cao. 2017. Microstructure and oxidation behavior of Al + Cr co-deposited coatings on nickel-based superalloys. Surface and Coatings Technology. 310: 273–277.

Liu, Li-Li, Xiao-Zhi Wu, R. Wang, Wei-Guo Li and Qing Liu. 2015. Stacking fault energy, yield stress anomaly, and twinnability of Ni_3Al: A first principle study. Chin. Phys. B 24(5): 077102-1 -077102-8.

Liu, R.D., S.M. Jiang, H.J. Yu, J. Gong and C. Sun. 2016. Preparation and hot corrosion behavior of Pt modified AlSiY coating on a Ni-based superalloy. Corrosion Science. 104: 162–172.

Los, V.F., S.P. Repetsky and V.V. Garkusha. 1991. Influence of short- and long-range order on the energy characteristics and the electrical conductivity of the alloy. Metallophysics. 13: 28–39 (in Russian).

Lu, Z.W., S.H. Wei and A. Zunger. 1991. First principles statistical mechanics of structural stability of intermetallic compounds. Phys. Rev. B. 44: 512–544.

Maclachlan, D.W. and D.M. Knowles. 2002. The effect of materials behavior on the analysis of single-crystal turbine blade. Part II – component analysis. Fatigue & Fracture of Engineering Materials & Structures. 25: 399–409.

Maggerramova, L.A. and B.E. Vasiliev. 2011. Influence of the azimuthal orientation of a single crystal on the stress-strain state and strength of the blades of high-temperature gas turbines. Vestnik UGATU, Thermal, Electro-rocket Engines and Power Plants LA. 4 (44): 54–58.

Makoto, T. and K. Yasumasa. 2006. The role of antiphase boundaries in the kinetic process of the $L1_2$-$D0_{22}$ structural change of $Ni_3Al_{0.45}V_{0.50}$ alloy. Acta Mater. 54: 4385–4391.

Masahashi, N., H. Kawazoe, T. Takasugi and O. Izumi. 1987. Phase relation in the section Ni_3Al -Ni_3Fe of the Al-Fe-Ni system. Zs. Metallkunde. 78: 788–794.

Maslenkov, S.B. 1983. Heat-resistant Steels and Alloys. Reference book of Technical Regulations. Moskow: Metallurgy Publ. (in Russian)

Melnikova, G.V., B.F. Shorr, L.A. Magerramova and D.A. Protopopova. 2001. Influence of the crystallographic orientation of a single-crystal and its technological spread on the frequency spectrum of turbine blades. Aviation and space technique and technology. 26: 140–144.

Min, B.I., A.J. Freeman and H.J.F. Jansen. 1988. Magnetism, electron structure and Fermi surface of Ni_3Al. Phys. Rev. B. 37: 6757–6762.

Min, P.G., V.V. Sidorov, D.E. Kablov, V.E. Vadeev and D.V. Zaitsev. 2017. Study of the local distribution of phosphorus and sulfur in the γ- and γ'-phases of the JS36-VI alloy. Proc. of All-Russian Institute of aviation materials. 7(55): 24–35.

Minamino, Y., S.B. Jung, T. Yamane and K. Hirao. 1992. Diffusion of cobalt, chromium, and titanium in Ni_3Al. Metall. Trans. A. 23: 2783–2790.

Minamino, Y., H. Yoshida, S.B. Jung, K. Hirao and T. Yamane. 1997. Diffusion defect forum. 143-147: 257–263.

Mishima, Y., O. Shouichi and T. Suzuki. 1985. Lattice parameters of Ni (γ), Ni3Al (γ') and Ni_3Ga (γ') solid solution with additions of transition and B subgroup elements. Acta Metall. 33: 1161-1169.

Mishin, Y. 2004. Atomistic modeling of the γ and γ'-phases of the Ni-Al system. Acta Mater. 52: 1451–1467.

Mitrokhin, Yu.S., V.P. Belash, I.N. Klimova, N.N. Stepanova, A.B. Rinkevich and Yu.N. Akshentsev. 2005. Effect of alloying on interatomic interaction in the intermetallic compound Ni_3Al. The Physics of Metals and Metallography. 99: 265-271.

Mitrokhin, Yu.S., Ju. Kovneristy and V. Shudegov. 2008a. Molecular dynamic simulation of the melting process in Ni_3Al alloy. J. of Phys.: Conf. Series. 98: 042026.

Mitrokhin, Yu.S, V.P. Belash, N.N. Stepanova and V. Shudegov. 2008b. Theoretical and experimental study of the thermal expansion coefficient of the Ni_3Al alloy. J. of Phys.: Conf. Series. 98: 062036.

Mitrokhin, Yu.S, V.P. Belash, I.N. Klimova and N.N. Stepanova. 2017. Site preference of ternary alloying elements in Ni_3Al-X (X = Co, Nb): a first-principles calculations in combination with XPS study. Mater. Research Express. 4: 25016 (1–8).

Morozova, G.I., O.B. Timofeeva and N.V. Petrushin. 2009. Special features of the structure and phase composition of high-rhenium nickel refractory alloy. Metal Science and Heat Treatment. 51: 62-69.

Mukhopadhyay, J., G. Kaschner and A.K. Mukherjee. 1990. Superplasticity in boron doped Ni_3Al alloy. Scr. Metall. Mater. 24: 857–862.

Muto, S., D. Schryvers, N. Merk and L.E. Tanner. 1993. High-resolution electron microscopy and electron diffraction study of the displacive transformation of the Ni_2Al phase in a $Ni_{65}Al_{35}$ alloy and associated with the martensitic transformation. Acta Metal. Mater. 41: 2377–2383.

Nautiyal, T. and S. Auluck. 1992. Electronic structure and Fermi surface of Ni_3Al. Phys. Rev. B. 45: 13930–13937.

Nozhnitsky, Yu.A. and E.R. Golubovsky. 2006. About strength reliability of single-crystal working blades of high-temperature turbines of advanced gas turbine engines. Aviation and space technique and technology. 9 (35): 117–123.

Ono, K. and R. Stern. 1969. Elastic constants of Ni_3Al between 80° and 600°K. Trans. Met. Soc. AIME. 245: 171–172.

Parlikar, Ch., M.Z. Alam, D. Chatterjee and D.K. Das. 2017. Thickness apropos stoichiometry in Pt-aluminide (PtAl) coating: Implications on the tensile properties of a directionally solidified Ni-base superalloy. Mater. Sci. Eng. A. 682: 518–526.

Petegem, S.V., E.E. Zhurkin, W. Mondelaers, C. Dauwe and D. Segers. 2004. Vacancy concentration in electron irradiated Ni$_3$Al. J. Phys.: Condens. Matter. 16: 591–603.

Petrushin, N.V., A.V. Logunov and A.I. Kovalev. 1977. A method for determining the relative volumetric content of strengthening γ'-phase in heat-resistant Nickel alloys. RF Patent No. 687965.

Petrushin, N.V. and E.R. Cherkasova. 1993. Dependence of the temperature of phase transformations and of the structure of heat-resistant nickel alloys on the heating temperature of melts. Metal Science and Heat Treatment. 35: 34–39.

Petrushin, N.V., I.L. Svetlov, A.I. Samoylov and G.I. Morozova. 2010. Physico-chemical properties and creep strength of a single crystal of Nickel-base superalloy containing rhenium and ruthenium. Int. J. Mater. Research (formerly Zeitschrift für Metallkunde). 101: 594–600.

Petrushin, N.V., E.M. Visik, M.A. Gorbovets and R.M. Nazarkin. 2016. Structure-phase characteristics and the mechanical properties of single-crystal Nickel-based rhenium-containing superalloys with carbide-intermetallic hardening. Russian Metallurgy (Metally). 7: 630–641.

Plaut, R.L., C. Herrera, D.M. Escriba, P.R., Rios and A.F. Padilha. 2007. A Short review on wrought austenitic stainless steels at high temperatures: processing, microstructure, properties and performance. Materials Research, 10(4): 453–460.

Pollock, T.M. 2006. Nickel-based superalloys for advanced turbine engines: chemistry, microstructure, and properties. J. Prop. Power. 22: 361–374.

Pope, D.P. and J.L. Garin. 1977. The temperature dependence of the long-range order parameter of Ni$_3$Al. J. Appl. Cryst. 10: 14–17.

Povarova, K.B., N.K. Kazanskaya, A.A. Drozdov, M.V. Kostina, A.V. Antonova, A.E. Morozov and O.A. Bazyleva. 2011a. Influence of rare-earth metals on the high-temperature strength of Ni$_3$Al-based alloys. Russian Metallurgy (Metally). 1: 47–54.

Povarova, K.B., O.A. Bazyleva, A.A. Drozdov, N.K. Kazanskaya, A.E. Morozov and M.A. Samsonova. 2011b. Structural high-temperature alloys based on Ni$_3$Al: production, structure and properties. Mater. Sci. 4: 39–45.

Povarova, K.B., A.A. Drozdov, A.V. Antonova, A.E. Morozov, Y.A. Bondarenko, O.A. Bazyleva, M.A. Bulahtina, E.G. Arginbaeva, A.V. Antonova, A.E. Morozov and D.G. Nefedov. 2015. Effect of directional solidification on the structure and properties of Ni3Al-based alloy single crystals alloyed with Cr, Mo, W, Ti, Co, Re, and REM. Russian Metallurgy (Metally). 1: 43–50.

Pridorozhny, R.P., A.V. Sheremetev and A.P. Zinkovsky. 2006. Influence of crystallographic orientation on the spectrum of nature frequencies and vibration modes of single-crystal turbine working blades. Herald of aeroengine-building. 2: 42–48.

Pridorozhny, R.P., A.V. Sheremetev and A.P. Zinkovsky. 2008. Influence of crystallographic orientation on the spectrum of natural vibrations and the endurance limit of single-crystal turbine blades. Problems of Strength. 5: 15–27.

Pridorozhny, R.P., A.V. Sheremetev and A.P. Zinkovsky. 2013. Estimation of the influence of the azimuthal orientation on the strength of a single-crystal cooled blade in a perforation system. Herald of aeroengine-building. 1: 53–57.

Protasova, N.A., I.L. Svetlov, M.B. Bronfin and N.V. Petrushin. 2011. Lattice-parameter misfits between the γ and γ' phases in single crystal Nickel superalloys. Russian Metallurgy (Metally). 5: 459–464.

Puska, M.J. and R.M. Nieminen. 1994. Theory of positrons in solids and solid surfaces. Rev. Mod. Phys. 66: 841–897.

Raju, S., E. Mohandas and V.S. Raghunathan. 1996. A study of ternary element site substitution in Ni_3Al using psevdopotential orbital radii based structure maps. Scripra Mater. 34: 1785–1790.

Ramesh, R., R. Vasudevan, B. Pathiraj and B.H. Kolster. 1990. X-ray evidence for structural transformation in Ni_3AI alloys at higher temperatures. Naturwissenschaften. 77: 129–130.

Ramesh, R., R. Vasudevan, B. Pathiraj and B.H. Kolster. 1992. Ordering and structural transformations in Ni_3Al alloys. J. of Mater. Sci. 27: 270–278.

Ramesh, R., B. Pathiraj and B.H. Kolster. 1996. Crystal structure changes in Ni_3Al and its anomalous temperature dependence of strength. J. of Mater. Proc. Technology. 56: 78–87.

Reed, P.A.S., I. Sinclair and X.D. Wu. 2000. Fatigue crack path prediction in UDIMET 720 Nickel-based alloy single crystals. Met. Mat. Trans. A. 31: 109.

Reed, R.C. and C.M.F. Rae. 2014. Physical metallurgy of the nickel-based superalloys. pp. 2215–2290. *In*: D.E. Laughlin and K. Hono (Eds.). Physical Metallurgy. Elsevier.

Ren, P., Sh. Zhu and F. Wang. 2016. Spontaneous reaction formation of $Cr_{23}C_6$ diffusion barrier layer between nanocrystalline MCrAlY coating and Ni-base superalloy at high temperature. Corrosion Science. 99: 219–226.

Rentenberger, Ch., T. Waitz and H.P. Karnthaler. 2003. TEM investigation of the structure of deformation-induced antiphase boundary faults in Ni_3Al. Phys. Rev. B. 67: 094109 (1–5).

Rhee, J.Y., Y.V. Kudryavtsev and Y.P. Lee. 2003. Optical, magneto-optical, and magnetic properties of stoichiometric and off-stoichiometric γ'-phase Ni_3Al alloys. Phys. Rev. B: Condens. Matter Mater. 68: 045104 (1–8).

Rigmant, M.B., V.S. Gorkunov and V.I. Pudov. 2000. A method of measuring ferromagnetic phase in the austenitic steels. RF patent No. 216619.

Rigmant, M.B., A.P. Nichipuruk, B.A. Khudyakov, V.S. Ponomarev, N.A. Tereshchenko and M.K. Korkh. 2005. Instruments for magnetic phase analysis of articles made of austenitic corrosion-resistant steels. Russian Journal of Nondestructive Testing. 41: 701–709.

Rinkevich, A.B., N.N. Stepanova, D.P. Rodionov and V.A. Sazonova. 2003. Elastic waves in single crystals of Nickel-based heat-resistant alloys. The Physics of Metals and Metallography. 96: 227–235.

Rinkevich, A.B., N.N. Stepanova and A.M. Burkhanov. 2006. Acoustical properties of Ni_3Al single crystals alloyed with cobalt and niobium. The Physics of Metals and Metallography. 102: 632–636.

Rinkevich, A.B., N.N. Stepanova and D.P. Rodionov. 2008. Velocity of elastic waves and elasticity moduli of heat-resistant Nickel-based alloys and the 60H21 alloy. The Physics of Metals and Metallography. 105: 509–516.

Rinkevich, A.B., N.N. Stepanova, D.P. Rodionov and D.V. Perov. 2009. Ultrasonic testing of single- and polycrystal articles made of Nickel-based heat-resistant alloys. Russian Journal of Nondestructive Testing. 45: 745–759.

Rinkevich, A.B., N.N. Stepanova, D.P. Rodionov, D.V. Perov and O.V. Nemytova. 2011. Ultrasonic testing of objects made of heat-resistant nickel-based alloys, single and polycrystalline components. Insight: Non-Destructive Testing and Condition Monitoring. 53: 598–602.2.

Rodionov, D.P., D.I. Davydov, N.N. Stepanova, V.G. Pushin, Yu.I. Filippov, N.I. Vinogradova, Yu.N. Akshentsev and V.A. Sazonova. 2010a. Deformation and fracture of single crystals of VKNA type alloys at 1100-1250°C. The Physics of Metals and Metallography. 109: 278–285.

Rodionov, D.P., Yu.I. Filippov, N.I. Vinogradova, N.V. Kazantseva, Yu.N. Akshentsev, N.N. Stepanova and D.I. Davydov. 2010b. High-temperature deformation of single-crystal Ni_3Al-based alloys. Russian metallurgy (Metally). 10: 936–940.

Rossiter, P.L. and B. Bykovec. 1978. The electrical resistivity of Cu_3Au. Phil. Mag. B. 38: 555-565.

Saburov, V.P., G.F. Stasiuk and A.M. Mikitas. 1989. Influence of the complex alloying on the crystallization kinetics of high-resistant alloys. Izvestiya Vuzov. Metallurgy. 8: 92–95. (in Russian)

Sadi, F.A. and Servant C. 2000. Investigation of the ω-phase precipitation in the 0.506 at.%Ti-0.129 at.%Nb-0.365 at.% Al alloy by transmission electron microscopy and anomalous small-angle X-ray scattering. Phil. Mag. A. 80:639-658.

Samoilov, A.I., R.M. Nazarkin, N.V. Petrushin and N.S. Moiseeva. 2011. Misfit as a characteristic of the level of interfacial stresses in single-crystal Nickel superalloys. Russian metallurgy (Metally). 5: 459–464.

Sanati, M., R.C. Albers and F.J. Pinski. 2001. γ'-phase formation in NiAl and Ni_2Al alloys. J. Phys. Condens. Matter. 13: 5387–5398.

Sato, H. and R.S. Toth. 1962. Long period superlattices in alloys. Phys. Rev. 127: 469–484.

Sato, A., Harada H., An-C. Yeh, K. Kawagishi, T. Kobayashi, Y. Koizumi, T. Yokokawa and J-X. Zhang. 2008. A 5th generation SC superalloy with balanced high temperature properties and processability. pp. 131-138. In: K.A. Green, T.M. Pollock and H. Harada (eds.). Superalloys-2008, Pennsylvania USA, Minerals, Metals & Materials Society.

Savin, O.V., B.A. Baum, N.N. Stepanova, Yu.N. Akshentsev, V.A. Sazonova and Yu.E. Turkhan. 1999. Structure and properties of Ni3Al alloyed with a third element: I. Effect of alloying on phase equilibria. The Physics of Metals and Metallography. 88: 377–383.

Savin, O.V., B.A. Baum, E.E. Baryshev, N.N. Stepanova and Yu.N. Akshentsev. 2000a. Structure and properties of Ni_3Al alloyed with a third element: II. Kinetics of ordering. The Physics of Metals and Metallography. 90: 62–67.

Savin, O.V., N.N. Stepanova, D.P. Rodionov, Yu.N. Akshentsev, V.A. Sazonova and Yu.E. Turkhan. 2000b. X-ray diffraction investigation of ordering kinetics in Ni_3Al alloyed with a third element. The Physics of Metals and Metallography. 90: 138–144.

Schoijet, M. and L.A. Girifalco. 1968. Diffusion in order disorder alloys. The face centered AB_3 alloy. Phys. Chem. Solids. 29: 911–912.

Scobie, J.A., R. Teuber, Y.S. Li, C.M. Sangan, M. Wilson and G.D. Lock. 2015. Design of an improved turbine Rim-Seal. J. Eng. Gas turbines power. 138: 022503. Paper No: gtp-15-1266.

Semenov, S.G., A.S. Semenov, L.B. Getsov, N.V. Petrushin, O.G. Ospennikova and A.A. Zhivushkin. 2016. Increasing the lifetime of gas-turbine engine nozzle blades using a new monocrystalline alloy. J. of Machinery Manufacture and Reliability. 45: 316–323.

Semiatin, S.L., D.S. Weaver, R.C. Kramb, P.N. Fagin, M.G. Glavicic, R.L. Goetz, N.D. Frey and M.M. Antony. 2004. Deformation and recrystallization behavior during hot working of a coarse grain, Nickel-base superalloy ingot material. Metall. Trans. A. 35: 679–693.

Shalin, R.E., I.L. Svetlov and E.B. Kachanov. 1997. Single-crystals of nickel refractory alloys. Mechanical Engineering, Moskow.

Shi, Y., G. Frohberg and H. Wever. 1995. Diffusion of ^{63}Ni and ^{114}In in the γ'-phase Ni_3Al. Phys. Status Solidi (a). 152: 361–375.

Shoemaker, C., D. Shoemaker and L.A. Bendersky. 1990. Structure of ω-$Ti_3Al_{2.25}Nb_{0.75}$. Acta Cryst. 46: 374–377.

Siegel, R.W. 1980. Positron annihilation spectroscopy. Annu. Rev. Mater. Sci. 10: 393–425.

Sims, C., N. Stoloff and W. Hagel. 1987. Superalloys II: High Temperature Materials for Aerospace and Industrial Power. John Wiley & Sons.

Sluiter, M.H.F. and Y. Kawazoe. 1995. Site preference of ternary additions in Ni_3Al. Phys. Review B. 51: 4062–4073.

Solly, B. and G. Winquist. 1973. A note on the yield stress behavior of Ni_3Al. Scand. J. Metall. 2: 183–186.

Starostenkov, M.D., N.B. Kholodova, M.B. Kondratenko, D.M. Starostenkov and E.V. Kozlov. 2005. Mechanisms of disordering of two-dimensional crystal of Ni_3Al intermetallic compound. Bulletin of the Russian Academy of Sciences: Physics. 69: 1169–1172.

Stepanova, N.N., V.A. Sazonova and D.P. Rodionov. 1999. Influence of solidification conditions on γ'-phase thermal stability in <001> single crystal of Ni-based superalloys. Scripta Mater. 40: 581-585.

Stepanova, N.N., D.P. Rodionov, V.A. Sazonova and E.N. Khlystov. 2000. Structure formation in <001> single crystals of a Nickel-based superalloy solidified with TiCN powder addition. Mater. Sci. Engin. A. 284: 88–92.

Stepanova, N.N., S.G. Teploukhov, S.F. Dubinin, Yu.N. Akshentsev, D.P. Rodionov and V.D. Parkhomenko. 2003a. Study of the structure of Ni_3Al and $(Ni,Co)_3Al$ crystals grown by the Bridgman method. The Physics of Metals and Metallography. 96: 626–633.

Stepanova, N.N., D.P. Rodionov, Yu.E. Turkhan, V.A. Sazonova and E.N. Khlystov. 2003b. Phase stability of Nickel-base superalloys solidified after a high-temperature treatment of the melt. The Physics of Metals and Metallography. 95: 602–609.

Stepanova, N.N., V.P. Belash, S.V. Lepikhin, O.V. Savin, L.V. Elokhina G.V. Tyagunov and Yu.N. Akshentsev. 2004. Phase and structural transformations in ternary Ni_3Al-based alloys. The Physics of Metals and Metallography. 97: 415–421.

Stepanova, N.N., D.I. Davydov, D.P. Rodionov, Yu.I. Philippov, Yu.N. Akshentsev, N.I. Vinogradova and N.V. Kazantseva. 2011a. Structure and mechanical properties of Ni_3Al single crystal after high-temperature deformation. The Physics of Metals and Metallography. 111: 403–409.

Stepanova, N.N., S.Yu. Mitropolskaya, D.I. Davidov and N.V. Kazantseva. 2011b. Effect of cold rolling on the magnetic behavior of Ni_3Al-based superalloy. Solid State Phenomena. Trends in Magnetism. 168-169: 185–187.

Stepanova, N.N., D.I. Davydov, A.P. Nichipuruk, M.B. Rigmant, N.V. Kazantseva, N.I. Vinogradova, A.N. Pirogov and E.P. Romanov. 2011c. The structure and magnetic properties of a heat-resistant nickel-base alloy after a high-temperature deformation. The Physics of Metals and Metallography. 112: 309–317.

Stepanova, N.N., M.B. Rigmant, N.V. Kazantseva and D.A. Shishkin. 2015. Analyze of magnetic properties and structure in Nickel gas turbine blade after operation at forced regime. Proc.10 Intern. Academic World Conf. on Development in Science and Technology (IACDST-2015), Istanbul, Turkey. P. 1–4.

Stepanova, N.N., N.V. Kazantseva, M.B. Rigmant, D.I. Davidov and V.E. Scherbinin. 2016. Strain-induced magnetism in Nickel-based superalloy after high-temperature deformation. Proc. Academic World 45 International conference on Development in Science and Technology (IACDST-2016). Oxford, United Kingdom. P. 1–5.

Stoeckinger, G.R. and J.P. Neumann. 1970. Determination of the order in the intermetallic phase Ni_3Al as function of temperature. J. Appl. Cryst. 3: 32–38.

Stoloff, N.S. 1989. Physical and mechanical metallurgy of Ni_3Al and its alloys. Int. Mater. Rev. 34: 153–184.

Sun, F., J. Tong, Q. Feng and J. Zhang. 2015. Microstructural evolution and deformation features in gas turbine blades operated in-service. J. Alloys Comp. 618: 728–733.

Sun, J.T. and L. Lin. 1994. Theoretical and positron annihilation study of point defects in intermetallic compound Ni_3Al. Acta Metall. Matter. 42: 195–200.

Suzuki, A., H. Inui and T.M. Pollock. 2015. $L1_2$-Strenthened Cobalt-base Superalloys. Annual Review of Materials Research. 45: 345–368.

Takahashi, S. and K. Ikeda. 1983. The influence of cold-working on magnetic properties in Ni_3Al. J. Phys. F: Met. Phys. 13: 2169–2174.

Thomton, P.H., R.G. Davies and T.L. Johnston. 1970. The temperature dependence of flow stress of the phase based upon Ni_3Al. Met. Trans. 1: 207–218.

Truell, R., Ch. Elbaum and B. Chick. 1969. Ultrasonic methods in solid state physics. N.Y.: Academic Press.

T'sepelev, V.S., B.A. Baum, G.V. Tyagunov, V.V. Vyukhin, V.V. Konashkov and V.Y. Belozerov. 2008. The investigations of the amorphous metal properties in liquid state. J. of Phys.: Conference Series. 98: 072020.

Tsukagoshi, K., A. Muyama, J, Masada, Y. Iwasaki and E. Ito. 2007. Operating status of uprating gas turbines and future trend of gas turbine development. Mitsubishi Heavy Industries. Technical Review. 44: 1–6.

Tyagunov, A.G., E.E. Baryshev, G.V. Tyagunov and V.B. Mikhailov. 2013. Production of heat-resistant EP220 and EP929 alloys by high-temperature treatment of melt. Steel in Translation. 43: 557–560

Tyagunov, A.G., E.E. Baryshev, V.V. V'yukhin, T.K. Kostina and E.V. Kolotukhin. 2014. Increasing the quality of the EP902 alloy using its properties in the liquid and solid states. Russian Metallurgy (Metally). 12: 992–994.

Tyagunov, A.G., V.V. Vyukhin, E.E. Baryshev, G.V. Tyagunov and T.K. Kostina. 2015. Effect of microalloying and the melting temperature on the properties of ChS-70 and ChS-88 alloys in the liquid and solid states. Russian Metallurgy (Metally). 12: 998-1001.

Umakoshi, Y. and H.Y. Yasuda. 2006. Observation of lattice defects and nondestructive evaluation of fatigue life in FeAl and Ni_3Al based alloy by means of magnetic measurement. Materials Science Forum. 512:13–18.

Umakoshi, Y., H.Y. Yasuda and T. Yanai. 2004. Quantitative analysis of γ (gamma) precipitate in cyclically deformed $Ni_3(Al,Ti)$ single crystals using magnetic technique. Proc. MRS Fall Meeting - Boston. USA. 842: S 2.3.1–S 2.3.6.

Van Petegem, S., E.E. Zhurkin, W. Mondelaers, C. Dauwe and D. Segers. 2004. Vacancy concentration in electron irradiated Ni_3Al. J. of Phys.: Condensed Matter. 16: 591–603.

Vinogradova, N.I., N.V. Kazantseva, N.N. Stepanova, E.P. Romanov and A.N. Pirogov. 2008. Twinning and phase transformations in refractory alloy EP-800 after dynamic loading. Metal Science and Heat Treatment. 50: 430–434.

Vorobiov, Yu.S. 1988. Vibration of the turbomachines blading. Nauk. Dumka, Kiev.

Vorobiov, Yu.S., N.Yu. Ovcharova, T.Yu. Berlizova, S.B. Kulishov and A.N. Skritsky. 2014a. Features of the temperature and thermoelastic state of a single-crystal cooled blade of a gas turbine engine. Aviation and space technique and technology. 10 (117): 75–78.

Vorobiov, Iu.S., K.Yu. Dyakonenko, S.B. Kulishov, A.N. Skritskij and R. Zhondkovski. 2014b. Vibration characteristics of cooled single-crystal gas turbine blades. Journal of vibration engineering & technologies. 2(6): 537–541.

Vorobiov, Yu. S., N.Yu. Ovcharova, R. Zhondkovski and T.Yu. Berlizova. 2015. The stress-strain state of the cooled single-crystal blade of a gas turbine engine in the temperature field, taking into account the effect of the orientation of the crystallographic axes. Compressor and power engineering. 4: 15–19.

Vorobiov, Yu.S., N.Yu. Ovcharova, R. Zhondkovski and T.Yu. Berlizova. 2016a. Influence of the azimuthal orientation of the crystallographic axes on the thermoelastic state of the gas turbine blade with the vortex cooling system. Problems of Strength. 3: 15–24.

Vorobiov, Yu.S., N.Yu. Ovcharova, R. Zhondkovski and T.Yu. Berlizova. 2016b. The influence of azimuthal orientation of crystallographic axes on thermoelastic state of a gas turbine engine blade with a vortex cooling system. Strength of Materials. 48(3): 349–356.

Vorobiov, Yu.S., T.Yu. Berlizova and N.Yu Ovcharova. 2017. Influence orientation of crystallographic axes on the thermo-stressed state of the single-crystal blades. Visnyk of Zaporizhzhya National University. Phisical and mathematical sciences. 1: 57–64.

Wang, T.M., M. Shimotomai and M. Doyama. 1984. Study of vacancies in the intermetallic compound Ni_3Al by positron annihilation. J. Phys. F: Met. Phys. V. 14: 37–45.

Wang, X., Sh. Zhu, Z. Gu, M. Chen and F. Wang. 2017. Effect of YSZ-incorporated glass-based composite coating on oxidation behavior of K438G superalloy at 1000°C. J. European Ceramic Society. 37: 1013–1022.

Wang Y., L.-Q. Chen and A.G. Khachaturyan. 1991. Strain-indused modulated structures in two-phase cubic alloys. Scripta Met. Mater. 25: 1969–1974.

Wen, Zh., D. Zhang, S. Li, Zh. Yue and J. Gao. 2017. Anisotropic creep damage and fracture mechanism of nickel-base single crystal superalloy under multiaxial stress. J. Alloys Comp. 692: 301–312.

Wolverton, C. and D. de Fontaine. 1994. Site substitution of ternary additions to Ni_3Al (γ') from electronic structure calculations. Phys. Rev. B. 49: 12351–12354.

Wu, Q. and Sh. Li. 2012. Alloying element additions to Ni_3Al: Site preferences and effects on elastic properties from first-principles calculations. Computational Materials Science. 53: 436–443.

Wurschum, R., K. Badura-Gergen, E.A. Kummerle, C. Grupp and H.-E. Schaefer. 1996. Characterization of radiation-induced lattice vacancies in intermetallic compounds by means of positron-lifetime studies. Phys. Rev. B. 54: 849–856.

Xu, Z., Z. Wang, J. Niu, J. Dai, L. He and R. Mu. 2016. Phase structure, morphology evolution and protective behaviors of chemical vapor deposited (Ni, Pt) Al coatings. J. Alloys Comp. 676: 231–238.

Yamaguchi, M. and Y. Umakoshi. 1990. The deformation behavior of intermetallic superlattice compounds. Progress in Materials Science. 34: 60–73.

Yoo, M.H. 1987. High temperature ordered intermetallic alloys. Proc. MRS Symposia. Materials Res. Soc. Pittsburgh. 81: 207–211.

Zavodov, A.V., N.V. Petrushin and D.V. Zaitsev. 2017. Microstructure and phase composition of a superalloy ZhS-32 after selective laser melting, vacuum heat treatment and hot isostatic pressing. The Physics of Metals and Metallography. 7:111–116.

Zeng, Q. and I. Baker. 2007. The effect of local versus bulk disorder on the magnetic behavior of stoichiometric Ni_3Al. Intermetallics. 15: 419–427.

Zhang, G.P. and Z.G. Wang. 1998. Deformation and fracture behavior in Ni_3Al alloy single crystals. J. Mater. Sci. Let. 17: 61–64.

Zhang, P., Q. Zhu, G. Chen and C. Wang. 2015. Review on Thermo-Mechanical Fatigue Behavior of Nickel-Base Superalloys. Mater. Trans. 56(12):1930-1939.

Zhilyaev, V.A. and E.I. Patrakov. 2016a. Regularities of reactions of titanium carbonitrides and oxycarbides with nickel. Russian Journal of Non-Ferrous Metals. 57: 69–74.

Zhilyaev, V.A. and E.I. Patrakov. 2016b. Influence of alloying titanium carbonitride by transition metals of groups IV-VI on the interaction with the nickel melt. Russian Journal of Non-Ferrous Metals. 57: 141–147.

Zhivushkin, A.A., L.B. Getsov, A.I. Rybnikov and E.A. Tikhomirova. 2011. Study of the properties of heat-resistant Nickel single-crystal superalloys. Vestnik of Samara University. Aerospace engineering, technology and engineering. Part 3. 3: 43–49 (in Russian).

Zhong, J., J. Liu, X. Zhou, S. Li, M. Yu and Z. Xu. 2016. Thermal cyclic oxidation and interdiffusion of NiCoCrAlYHf coating on a Ni-based single crystal superalloy. J. Alloys Comp. 657: 616–625.

Zhondkovski, R., Yu.S. Vorobiov, N.Yu. Ovcharova and T.Yu. Evchenko. 2013. Analysis of the thermoelastic state of a cooled single-crystal blade of gas turbine engine. Mechanic and machine-building. 1: 23–28.

Zulina, N.P., E.V. Bolberova and I.M. Razumovskii. 1997. Nickel self-diffusion along grain boundaries in Ni_3Al-base intermetallic alloys: diffusion and defect data. Part A. Defect and Diffusion Forum. 143–147: 1453–1456.

Chapter 2

Albert, D.E. and G.T. Gray III. 1995. Dislocation reactions responsible for the formation of a twin in the ordered intermetallic alloy Ni-20A1-30Fe. Phil. Mag. 71(7):473–487.

Aoki, K. and O. Izumi. 1978. On the Ductility of the Intermetallic Compound Ni_3Al. Transactions of the Japan Inst. 19(4): 203–210.

Ashbrook, R.L., A.C. Hoffman, G.D. Sandrock and R.L. Dreshfield. 1968. Development of a cobalt-tungsten ferromagnetic, high-temperature, structural alloy. NASA technical note TN D-4338.

Araki, H., T. Mimura, P. Chalermkarnnon, M. Mizuno and Y. Shirai. 2002. Positron lifetime study of defect structures in $B2$ ordered Co-Al alloys. Mater. Trans. 43: 1498–1501.

Baker, I., E.M. Schulson, J.A. Horton. 1987. In-situ straining of Ni_3Al in a transmission electron microscope. Acta Metall., 35: 1533–1541.

Bauer, A., S. Neumeier, F. Pyczak and M. Goëken. 2010. Microstructure and creep strength of different γ/γ'-strengthened Co-base superalloy variants. Scripta Materialia. 63:1197–1200.

Bester, G., B. Meyer and M. Fahnle. 1999. Atomic defects in the ordered compound $B2$ CoAl: A combination of ab initio electron theory and statistical mechanics. Phys. Rev. B: Condens. Matter Mater. Phys. 60:14492–14495.

Bocchini, P.J., E.A. Lass, Kil-Won Moon, M.E. Williams, C.E. Campbell, U.R. Kattner, D.C. Dunand and D.N. Seidman. 2013. Atom-probe tomographic study of γ/γ' interfaces and compositions in an aged Co-Al-W superalloy. Scr. Mater. 68: 563–566.

Booth-Morrison, C., E.D. Noebe and D.N. Seidman. 2008. Effects of a tantalum addition on the morphological and compositional evolution of a model Ni–Al–Cr superalloy. pp. 73–79. *In:* Proc. Conf. "Superalloys 200: TMS. TMS - The Minerals, Metals & Materials Society.

Bozzolo, G., R.D. Noebe and F. Honecy. 1998. Modeling of substitutional site preference in ordered intermetallic alloys. pp. 1–37. *In:* Proc. "Interstitial and Substitutional Effects in Intermetallics" Fall Meeting, Rosemont, Illinois.

Breuer, J. and E.J. Mittemeijer. 2003. Thermodynamics of ordered intermetallic compounds containing point defects. Z. Metallkd. 94: 954–961.

Bushneev, L.S., Yu.R. Kolobov and M.M. Mishlaev. 1990. Basis of the electron microscopy. Tomsk: Tomsk university.

Buschow, K.H.J. and F.R. de Boer. 2003. Physics of Magnetism and Magnetic Materials. Kluwer Academic Publishers: New York, Boston, Dordrecht, London, Moscow.

Cantor, Ch. and P. Schimmel. 1980. Biophysical Chemistry. Part 2. Techniques for the Study of Biological Structure and Function. New York: W. H. Freeman and Co.

Carvalho, P.A., P.M. Bronsveld, B.J. Kooi and J.Th.M. De Hosson. 2002. On the fcc-D0$_{19}$ transformation in Co-W alloys. Acta Materialia. 50: 4511–4526.

Chakrabortty, S.B. and E.A. Starke. 1975. Deformation twinning of Cu$_3$Au. Acta Metal. 23(1): 63–71.

Chang, H., G. Xu, X.-G. Lu, L. Zhou, K. Ishida and Y. Cui. 2015. Experimental and Phenomenological Investigations of Diffusion in Co-Al-W alloys. Scr. Mater. 106: 13–16.

Chen, Q. and B. Sundman. 2001. Calculation of Debye temperature for crystalline structures — a case study on Ti, Zr, and Hf. Acta mater. 49: 947–961.

Christian, J.W. 1951. Theory of the transformations in pure cobalt. Proceedings of the Royal Society of London. Series A, Mathematical and Physical Sciences. 206: 51–64.

Cooper, M.J. 1963. An investigation of the ordering of the phases CoAl and NiAl. Philos. Mag. 89: 805–810.

Cui, Y.F., X. Zhang, G.L. Xu, W.J. Zhu, H.S. Liu and Z.P. Jin. 2011. Thermodynamic Assessment of Co-Al-W System and Solidification of Co-Enriched Ternary Alloys. J. Mater. Sci. 46: 2611-2621.

Davidov, D.I., N.N. Stepanova, N.V. Kazantseva, M.B. Rigmant, D.A. Shishkin. 2015. Study of the magnetic properties, structure, and phase transformation in the alloys of the Co-Al-W system. AIP Conference Proceedings. 1683: 020035-1–020035-4.

Dimitrov, O., A.V. Korznikov, G.F. Korznikova and G. Tram. 2000. Nanostructure formation and phase transformation in intermetallic compounds during severe plastic deformation. J. Phys. IV France. 10: Pr6-33 – Pr6-37.

Diologent, F. and P. Caron. 2004. On the creep behavior at 1033K of new generation single-crystal superalloys. Materials Science and Engineering A. 385(1–2): 245–257.

Dmitrieva, G., T.S. Cherepova and A.K. Shurin. 2005. Phase equilibriums in the alloys of Co- CoAl-W system (P.1). Metallznavstvo ta obraoitka metali. 3–6.

Dmitrieva, G., T.S. Cherepova and A.K. Shurin. 2006. Phase equilibriums in the alloys of Co-CoAl-W system (P.2). Metallznavstvo ta obraoitka metali. 22–25.

Dmitrieva, G., V. Vasilenko and I. Melnik. 2008. Al-Co-W fusion diagram in the Co-CoAl-W part. Chem. Met. Alloys. 338–342.

Doi, M., T. Miyazaki and T. Wakatsuki. 1984. The Effect of Elastic Interaction Energy on the Morphology of γ' Precipitates in Nickel-based Alloys. Materials Science and Engineering. 67: 247–253.

Dutkiewicz, J. and G. Kostorz. 1991. Structure of martensite in Co-W alloys. Materials Science and Engineering. A132: 267–272.

Fatmi, M., M.A. Ghebouli, B. Ghebouli, T. Chihi, S. Boucetta and Z.K. Heiba. 2011. Study of structural, elastic, electronic,optical and thermal properties of Ni$_3$Al. Rom. Journ. Phys. 56(7–8): 935–951.

Feng, Q., F. Xue, H. Zhou, W. Li, William Yi Wang and Zi-Kui Liu. 2016. Creep behaviors and microstructural stabilities of Co-Al-W-Ta-Ti-based Superalloys. ECI Symposium Series.

Fitzgerald, S.P. and Z. Yao. 2009. Shape of prismatic dislocation loops in Anisotropic α-Fe. Phil. Mag. Lett. 89: 581–588.

Geller, Yu.A. 1983. Material Science. M: Metallurgy.

Gol'dshtein, M.I., V.S. Litvinov and B.K. Bronshtein. 1986. Metal Physics of High-Strength Alloys (in Russian). Moscow: Metallurgiya.

Greenberg, B.A., Yu.N. Gornostyrev, L.E. Kar'kina and L.I. Yakovenkova. 1976. On the problem of formation of a perfect twin by a pole source. Fiz. Met. Metalloved. 41(4): 714 – 723.

Greenberg, B.A. and M.A. Ivanov. 2002. Ni$_3$Al and TiAl intermetallic compounds. Ekaterinburg: UrD RAS. (in Russian).

Greeshenko, O.P., and Yu.N. Koval. 2000. Study of the effect of shape memory in Co-5%Al alloy. Metallophysics and new technology. 22(8): 60–63

Guillermet, A.F. 1989. Thermodynamic properties of the Co-W-C system. Metallurgical and Material Transaction A. 20A: 935–956.

Herzer, G. 1996. In: K.H.J. Buschow (Ed.). Magnetic materials, Amsterdam: North Holland Publ. Co.

Hirth, J.P. and J. Lothe. 1972. Theory of Dislocations, New York: McGraw-Hill Book Company.

Huber, B., A. Kodentsov and K.W. Richter. 2011. The Al-Co-Si phase diagram. Intermetallics. 19: 307–320.

Inamura, T., Y. Takahashi, H. Hosoda, K. Wakashima, T. Nagase, T. Nakano, Y. Umakoshi and S. Miyazaki. 2006. Martensitic Transformation Behavior and Shape Memory Properties of Ti-Ni-Pt Melt-Spun Ribbons. Materials Transactions B. 47: 540–545.

International Tables for Crystallography. 2006. Vol. C. P. 263-271.

Ishida, K. 2008. Recent progress on Co-base alloys—Phase diagram and application. Arch. Metall. Mater. 53: 1075–1088.

Ishida, K., et al. 2009. Intermetallic Compounds in Co-Base Alloys - Phase Stability and Application to Superalloys. pp. 1128-U06-06. In: Material Research Society Symposium, Boston, Massachusetts, November 3- December 4, 2009.

Joshi, S.R., K.V. Vamsi and S. Karthikeyan. 2014. First principles study of structural stability and site preference in Co3(W,X). MATEC Web of Conferences 14. 18001.

Kazantseva, N.V., B.A. Greenberg, E.V. Shorokhov, A.N. Pirogov and Y.A. Dorofeev. 2005. Phase transformations in nickel superalloy after shock-wave loading. Physics of Metals and Metallography. 99: 1–10.

Kazantseva, N.V., N.I. Vinogradova and N.N. Stepanova. 2010. Electron microscope study of planar defects in Ni$_3$Al single crystal after high-temperature deformation at 1200-1250°C. Deformation and Razrushenie. Mater. 9: 1–6. (Russian).

Kazantseva, N.V., N.N. Stepanova, M.B. Rigmant., D.I. Davidov, D.A. Shishkin and E.P. Romanov. 2015a. Effect of casting conditions on the structure and magnetic properties of Co-19 at.% Al-6 at.% W alloy. pp. 385-393. In: J.S. Carpenter, C. Bai, J.P. Escobedo-Diaz, J-Y. Hwang, S. Ikhmayies, B. Li, J. Li, S.N. Monteiro, Z. Peng and M. Zhang (eds). Characterization of Minerals, Metals and Materials-2015. New Jersey: John Willey & Sons, Inc., Hoboken.

Kazantseva, N.V., N.N. Stepanova, M.B. Rigmant, D.I. Davydov, D.A. Shishkin, E.P. Romanov, S.L. Demakov and M.A. Ryzhkov. 2015b. Study of magnetic properties and structural and phase transformations in the Co-19 at% Al-6 at% W alloy. Physics of Metals and Metallography. 116: 531–537.

Kazantseva, N.V., S.L. Demakov, A.S. Yurovskih, N.N. Stepanova, N.I. Vinogradova, D.I. Davydov and S.V. Lepikhin. 2016a. Study of the Co-Al-W phase diagram. Structure and Phase Transformations near concentration range of existence of intermetallic compounds $Co_3(Al,W)$. Physics of Metals and Metallography. 117(7): 701–709.

Kazantseva, N.V., N.N. Stepanova, N.I. Vinogradova, D.I. Davydov, D.A. Shishkin, M.B. Rigmant, E.P. Romanov, S.L. Demakov and A.S. Yurovskikh. 2016b. Study of the martensitic transformation in the Co-9 at% Al alloy. Physics of Metals and Metallography. 117: 42–48.

Kazantseva, N.V., D.I. Davydov, P.B. Terent'ev, D.A. Shishkin, S.L. Demakov, A.S. Yurovskikh and E.P. Romanov. 2017a. Mechanical and magnetic properties of alloys near the concentration range of the existence of $Co_3(Al,W)$ intermetallic compound. Physics of Metals and Metallography. 118(5): 432–438.

Kazantseva, N.V., I.V. Ezhov, D.I. Davydov, N.I. Vinogradova and P.B. Terent'ev. 2017b. Investigation of the Intermetallic $\beta'(B2)$-Phase in the Co-Al-Si System. Physics of Metals and Metallography. 118(3): 249–255.

Kazantseva, N.V., D. Davidov, D. Shishkin, P. Terent'ev, N. Vinogradova, S.L. Demakov and Yu. Kabanov. 2017c. $Co_3(Al,W)$ intermetallic compound: from soft to hard ferromagnets. International Journal of Applied Engineering Research. 12(23): 13767–13772.

Kear B.H., J.M. Oblak and A.F. Giamei. 1970. Stacking faults in gamma prime Ni3(Al,Ti) precipitation hardened nickel-base alloys. Metallurgical Transactions. 1: 2477.

Kobayashi, S. 2009. Determination of phase equilibria in the Co-rich Co-Al-W ternary system with a diffusion-couple technique. Intermetallics. 17: 1085–1089.

Kosevich, V.M. 1976 Electron transmission images of dislocation and stacking faults. Moscow: Nauka (Russian).

Kulikov, N.I., A.V. Postnikov, G. Borstel and J. Braun. 1999. Onset of magnetism in $B2$ transition-metal aluminides. Phys. Rev. B: Condens. Matter Mater. Phys. 59: 6824–6833.

Lall, C., S. Chin and D.P. Pope. 1979. The orientation and temperature dependence of the yield stress of $Ni_3(Al, Nb)$ single crystals. Met. Trans. A. 10(9): 1323–1332.

Lass, E.A., M.E. Williams, C.E. Campbell, Kil-Won Moon and U.R. Kattner. 2014. γ' Phase Stability and Phase Equilibrium in TernaryCo-Al-W at 900°C. Journal of Phase Equilibria and Diffusion. 35: 711–723.

Liu, Y., Q. Xing, W.E. Straszheim, J. Marshman, P. Pedersen, R. McLaughlin and T.A. Lograsso. 2016. Formation mechanism of superconducting phase and its three-dimensional architecture in pseudo-single-crystal $K_xFe_{2-y}Se_2$. Phys. Rev. B. 93: 064509.

Luna Ramírez, A., J. Porcayo-Calderon, Z. Mazur, V.M. Salinas-Bravo and L. Martinez-Gomez. 2016. Microstructural Changes during High Temperature Service of a Cobalt-Based Superalloy First Stage Nozzle. Advances in Materials Science and Engineering. 1–7.

Lyakishev, N.P. (Ed.) 1996. State Diagrams of Binary Metallic Systems, Moscow: Mashinostroenie.

Marshall, G.W. and J.O. Brittain. 1976. Hot stage TEM investigations of dislocation climb in NiAl. Metall.Trans. A. 7: 1013–1020.

Marcinkowski, M.J., N. Brown and R.M. Fisher. 1961. Dislocation configurations in AuCu3 and AuCu type superlattices. Acta Metallurgica. 9(2): 129-137.

Markstom, A., B. Sundman and K. Frisk. 2005. A revised thermodynamic description of the Co-W-C system. JPEDAV. 26: 152–160.

McAllster, A.J. 1989. The Al-Co (Aluminum-Cobalt) system. Bull. Alloy Phase Diagrams. 10: 646–649.

Meher, S., H.-Y. Yan, S. Nag, D. Dye and R. Banerjee. 2012. Solute partitioning and site preference in γ/γ' cobalt-base alloys. Scripta Materialia. 67: 850–853.

Mishin, D.D. 1981. Magnetic materials. Moscow: Vishaya skola.

Mishin, Y. 2004. Atomistic modeling of the γ and γ'-phases of the Ni-Al system. Acta Materialia. 52(6): 1451–1467.

Miura, S., K. Ohkubo and T. Mohri. 2007. Mechanical Properties of Co-Based $L1_2$ Intermetallic Compound $Co_3(Al,W)$. Materials Transactions A. 48 (9): 2403–2408.

Moniruzzaman, Md., H. Fukaya, Y. Murata, K.Tanaka and H. Inui. 2012. Diffusion of Al and Al Substituting Elements in Ni_3Al at Elevated Temperatures. Materials Transactions. 53(12): 2111–2118.

Mottura, A., A. Janotti and T.M. Pollock. 2012. Alloying effects in the γ'-phase of Co-based superalloys. pp. 685-693. *In*: Eric S. Huron, Roger S. Reed, Michael M. Mils, Rick E. Montero, Pedro D. Portella, Jack Telesman. (eds). Superalloys 2012. TMS, Warrendale, PA

Nakamura, R., K. Yoshimi and S. Tsurekawa. 2007. Supersaturated vacancies and vacancy complexes in rapidly solidified $B2$ aluminide ribbons. Mater. Sci. Eng., A. 449–451: 1036–1040.

Neumeier, S., C.H. Zenk, L.P. Freund and M. Göken. 2016. γ/γ' Co-base superalloys – new high temperature materials beyond Ni-base Superalloys? ECI Symposium Series. 1–17.

Nikitenko, V.I., V.S. Gornakov, Yu.P. Kabanov, A.J. Shapiro, R.D. Shull, C.L. Chien, J.S. Jiang and S.D. Bader. 2003. Magneto-optical indicator film study of the hybrid exchange spring formation and evolution processes. Journal of Magnetism and Magnetic Materials. 258-259(1-2): 19–24.

Nikolin, B.I. and N.N. Shevchenko. 1981. Formation of new multilayered martensite phases in Co-Al alloys — demonstration of polytypism in metallic alloys. Fiz. Met. Metalloved. 51: 316–325.

Novak, P., I. Marek, J. Kubasek, J. Serak and D. Vojtech. 2011. Effect of silicon on the formation of transition metal aluminides. Proc. Metal. 18: 1–6.

Okamoto, H. 2008. Co-W (Cobalt-Tungsten). Journal of Phase Equilibria and Diffusion. 29: 119.

Okamoto, H., T. Oohashi, H. Adachi, K. Kishida, H. Inui and P. Veyssiere. 2011. Plastic deformation of polycrystals of $Co_3(Al,W)$ with the $L1_2$ structure. Phil. Mag. 91(28): 3667–3684.

Omori, T., Y. Sutou, K. Oikawa, R. Kainuma and K. Ishida. 2003. Shape memory effect associated with fcc–hcp martensitic transformation in Co–Al alloys. Mater. Trans. 44: 2732–2735.

Ooshima, M., K. Tanaka, N.L. Okamoto, K. Kishida and H. Inui. 2010. Effects of quaternary alloying elements on the γ' solvus temperature of Co-Al-W based alloys with fcc/$L1_2$ two-phase microstructures. J. Alloys Compd. 508: 71–78.

Petrushin, N.V., E.S. Elyutin, E.V. Filonova and R.M. Nazarkin. 2015. Segregation of alloying elements during the solidification with a planar front of the γ'-strengthened Co-Al-W-Ta-base superalloy. Vestn. Ross. Fond. Fundam. Issl. 1: 11–17.

Pettinari, F., J. Douin, G. Saada, P. Caron, A. Coujou and N. Clément. 2002. Stacking fault energy in short-range ordered γ-phases of Ni-based superalloys. Mater. Sci. Eng. A 325: 511–519.

Pollock, T.M., J. Dibbern, M. Tsunekane, J. Zhu and A. Suzuki. 2010. New Co-Based γ - γ' High-Temperature Alloys. JOM. 62: 58–63.

Polukhin, P.I., S.S. Gorelik and V.K. Vorontsov. 1982. Physical base of plastic eformation. Moskow: Metallurgy (in Russian).

Povstugar, I., Choi Pyuck-Pa, S. Neumeier, A. Bauer, C.H. Zenk, M. Goeken and D. Raabe. 2014. Elemental partitioning and mechanical properties of Ti- and Ta containing Co-Al-W-base superalloys studied by atom probe tomography and nanoindentation. Acta Mater. 78: 78–85.

Pugacheva, N.B. 2015. Current trends in the development of heat-resistant coatings based on iron, nickel, and cobalt aluminides. Diagnostics, Resour. Mechan. Mater. Struct. 3: 51–82.

Pyczak, F., A. Bauer, M. Goken, U. Lorenz, S. Neumeier, M. Oehring, J. Paul, N. Schell, A. Schreyer, A. Stark and F. Symanzik. 2015. The effect of tungsten content on the properties of $L1_2$-hardened Co-Al-W alloys. J. Alloys Compd. 632: 110–115.

Rama, R. 1986. The Co-W (Cobalt-Tungsten) system. J. Alloy Phase Diagrams. 2(1): 43–52.

Richter, K.W. and D.T. Gutierrez 2005. Phase equilibria in the system Al-Co-Si. Intermetallics. 13: 848–856.

Rinkevich, A.B., N.N. Stepanova and D.P. Rodionov. 2008. Velocities of elastic waves and the elasticity moduli of nickel-base superalloys and of the 60N21 alloy. Physics of Metal and Metallography. 105: 509–516.

Romanov, E.P., N.V. Kazantseva, N.N. Stepanova, S.L. Demakov, D.I. Davydov and D.A. Shishkin. 2017. Heat-Resistant Alloys Based on Intermetallic Co3(Al, W). Doklady Chemistry. 473(2): 88–91.

Sato, J., K. Oikawa, R. Kainuma and K. Ishida. 2005. Experimental verification and magnetically induced phase separation in αCo phase and thermodynamic calculations of phase equilibria in the Co-W system. Materials Transactions. 46: 1199–1207.

Sato, J., T. Omori, O. Kikawa, I. Ohnuma, R. Kainuma and K. Ishida. 2006. Cobalt-base high-temperature alloys. Science, 7: 312.90–91.

Savin, O.V., B.A. Baum, N.N. Stepanova, Yu.N. Akshentsev, V.A. Sazonova, Yu.E.Turkhan. 1999. Structure and properties of Ni3Al alloyed with a third element: I. Effect of alloying on phase equilibria. Physics of Metals and Metallography. 88(4): 377–383.

Shtremel, M.A. 1982. Strength of alloys, Moscow: Metallurgy.

Skinner, A.J., J.V. Lill and J.Q. Broughton. 1995. Free energy calculation of extended defects through simulated alchemy: application to Ni_3Al antiphase boundaries. Model. Simul. Mater. Sci. Eng. 3: 359–370.

Stepanova, N.N., D.I. Davydov, A.P. Nichipuruk, M.B. Rigmant, N.V. Kazantseva, N.I. Vinogradova, A.N. Pirogov and E.P. Romanov. 2011. Structure and magnetic properties of nickel heat-resistant alloy after high-temperature deformation. Physics of metal and metallography. 112: 328–336.

Suzuki, T. and Y. Oya. 1981. The temperature dependence of the strength of pseudo-binary platinum-based $L1_2$ alloys with B-subgroup elements. J. of Mater. Sci. **16**: 2737–2744.

Suzuki, A., C. Garret DeNolf and T.M. Pollock. 2007. Flow stress anomalies in γ/γ' two-phase Co-Al-W-base alloys. Scripta Materialia. 56: 385–388.

Suzuki, A. and T.M. Pollock. 2008. High-temperature strength and deformation of γ/γ' two-phase Co-Al-W-base alloys. Acta Materialia, 56(6): 1288–1297.

Tamminga, Y. 1973. Magnetic Properties of Some Intermetallic Compounds with the CsCl Structure. Amsterdam: Academisch Proefschrift.

Tan, X.P., D. Mangelinck, C. Perrin-Pellegrino, L. Rougier, Ch.-A. Gandin, A. Jacot, D. Ponsen and V. Jaquet. 2014. Spinodal decomposition mechanism of the precipitation in a single crystal Ni-based superalloy. Metall. Trans. A. 45: 4725–4729.

Tanaka, K., T. Ohashi, K. Kishida and H. Inui. 2007. Single-crystal elastic constants of $Co_3(Al,W)$ with the $L1_2$ structure. Applied physics letters. 91: 181907-1–181907-3.

Titus, M.S., A. Suzuki and T.M. Pollock. 2012. High Temperature Creep of New $L1_2$-containing Cobalt-base Superalloys. p. 693. *In*: E. Huron, M. Hardy, M. Mills, R. Montero, P. Portella, J. Telesman and R.C. Reed (eds.). Superalloys 2012, Seven Springs, PA, USA, 2012. The Minerals, Metals and Materials Society, Warrendale, PA, USA.

Travina, N.T., and A.A. Nikitin. 1975. Strengthening and the dislocation structure of single crystals of two-phase Ni3Al-based alloys. Physics of Metals and Metallography. 40 (1): 160–165.

Tsukamoto, Y., S. Kobayashi and T. Takasugi. 2010. The Stability of γ'-$Co_3(Al,W)$ Phase in Co-Al-W Ternary System. Materials Science Forum. 654–656: 448–451.

Vamsi, K.V. and S. Karthikeyan. 2017. Yield anomaly in L12 Co3AlxW1−x vis-à-vis Ni3Al. Scripta Materialiya. 130: 269-273.

Vasil'ev, N., V.D. Buchel'nikov, T. Takagi, V.V. Khovailo and E.I. Estrin. 2003. Shape memory ferromagnets. Phys Usp. 46: 559–588.

Vinogradova, N.I., N.V. Kazantseva, N.N. Stepanova, E.P. Romanov and A.N. Pirogov. 2008. Twinning and phase transformations in the high temperaturestrength alloy EP-800 under dynamical loading. Metallography and Heat treatment. 639(9): 28–32.

Wachtel, E., V. Linse and V. Gerold. 1973. Defect structure and magnetic moments in the β-phases of CoAl and CoGa. J. Phys. Chem. Solids. 34: 1461–1466.

Westbrook, J.H. 1957. Temperature dependence of the hardness of secondary phases common in turbine bucket alloys. JOM. 898–904.

Wu, X., N. Tao, Y. Hong, J. Lu and K. Lu. 2005. α-ε martensite transformation and twinning deformation in fcc cobalt during surface mechanical attrition treatment. Scripta materialia. 52: 547–551.

Xue, F., M.L. Wang and Q. Feng. 2011. Phase equilibriain Co-rich Co-Al-W alloys at 1300°C and 900°C. Materials Science Forum. 686: 388–91.

Xue, F., M. Wang and Q. Feng. 2012. Alloying effects on heat-treated microstructure in Co-Al-W-base superalloys at 1300°C and 900°C. p. 813–821. *In*: E.S. Huron, R.S. Reed, M.M. Mils, R.E. Montero, P.D. Portella, J. Telesman (eds.) Proc. 12th International symposium on superalloys. "Superalloys" 2012. (Ed.). TMS - The Minerals, Metals and Materials Society, Warrendale, PA, USA.

Xue, F., H.J. Zhou, X.F. Ding, M.L. Wang and Q. Feng. 2013. Improved High Temperature γ' Stability of Co-Al-W-Base Alloys Containing Ti and Ta. Mater. Lett. 112: 215–218.

Yan, H.-Yu., V.A. Vorontsov and D. Dye. 2014a. Alloying effects in polycrystalline γ' strengthened Co–Al–W base alloys. Intermetallics 48: 44–53.

Yan, H.-Y., J. Coakley, V.A. Vorontsov, N.G. Jones, H.J. Stone and D. Dye. 2014b. Alloying and the micromechanics of Co-Al-W-X quaternary alloys. Mater. Sci. Eng. A. 613: 201–208.

Yan, M., G. Leaf, H. Kaper, V. Novosad, P. Vavassori, R.E. Camley and M. Grimsditch. 2006. Formation of Stripe Domains in Cobalt Bars via a Magnetic Soft Mode Instability. EEE Transactions on Magnetics. TMAG-06-03-0167: 1–3.

Yan, M., G. Leaf, H. Kaper, V. Novosad, P. Vavassori, R.E. Camley and M. Grimsditch. 2007. Dynamic origin of stripe domains in cobalt bars. Journal of Magnetism and Magnetic Materials. 310: 1596–1598.

Yao, Q., H. Xing and J. Sun. 2006. Structural stability and elastic property of the $L1_2$ ordered $Co_3(Al,W)$ precipitate. Applied Physics Letters. 89: 161906-1–161906-3.

Zhao, Ji-Cheng. 1999. The fcc/hcp phase equilibria and phase transformation in cobalt-based binary systems. Z.Metallkd. 90: 223–232.

Zhao, K.-H., Yu.-H. Ma, L.H. Lou and Z.Q.H. Mu. 2005. Phase in a Nickel Base Directionally Solidified Alloy. Materials transactions. 46(1): 54–58.

Chapter 3

American National Standard ANSI/AWS A4.2-86. Standard procedure for calibrating magnetic instruments to measure the delta-ferrite content of austenitic stainless steel weld metal.

Apaev, B.A. 1976. Phase magnetic analysis of alloys. Moscow: Metallurgy.

Aßmus, K. and W. Hübner. 2004. Methods for investigating austenite stability during tribological stressing of FeCrNi alloys. p. 855. *In*: DGZfP-Proceedings BB 90-CD.

ASME Code Section III. Rules for Construction of Nuclear Facility Components - Division 1 –Appendices ISO 8249:2000. Welding – Determination of Ferrite Number (FN) in austenitic and duplex ferritic-austenitic Cr-Ni stainless ssteel weld metals.

ASTME562-08. Standard Test Method for Determining Volume Fraction by Systematic Manual Point.

AWSA4.2M:2006. Standard Procedures for Calibraiting Magnetic Instruments to Measure the Delta Ferrite Content of Austenitic and Duplex Ferritic-Austenitic Stainless Steel Weld Metal.

Belenkova, M.M. and M.N. Miheev. 1967. About of the magnetic method for determination tendency of the austenitic steel to inter-crystalline corrosion. Russian journal of Non-destructuve testing. 5: 65–75.

Bida, G.V., A.P. Nichipuruk and T.P. Tsarkova. 2001. Magnetic properties of steels after quenching and adging. III High chromium steels. Russian journal of Non-destructuve testing. 2: 43–56.

Bosort, R. 1956. Ferromagnetism. IL. Moscow (Russian translation).

Danilenko, V.M., S.Yu. Mironov, A.N. Beliyakov and A.P. Zilaev. 2012. Application of EBSD analysis in physical materials science. - Factory laboratory. Diagnosis of materials. 78(3): 28–46.

Deriagin, A.I., V.A. Zavalishin, V.V. Sagaradze and B.M. Efros. 2007. Formation of nanoscale ferromagnetic phases during plastic deformation and subsequent annealing of stable austenitic steels, Russian journal of non-destructive testing. 7: 8–21.

Elmer, J. and T. Eagar. 1990. Measuring Determination the residual ferrite content of rapidly solidified stainless steel alloy. Welding Research Saplement. 4: 141–150.

Femenia, M., C. Canalias, J. Pana and C. Leygraf. 2003. Scanning Kelvin Probe force microscopy and magnetic force microscopy for characterization of duplex stainless steels J. Electrochem. Soc. 150(6): B274–B281.

Filipov, M.A., V.S. Litvinov and Yu.R. Nemirovski. 1988. Steels with metastable austenite. Moscow: Metallurgy. (in Russian).

Gerasimov, V.G., A.D. Pokrovski and V.V. Suhorukov. 1992. Non-destructive testing. Book 3. Electromagnetic monitoring. Moscow: High School. (in Russian).

Gol'dstein, M.I., S.V. Grachev and Yu.G. Veksler. 1999. Spesal steels. Moscow: MISIS

GOST 22838-77. Heat resistant alloys. Methods of control and evaluation of macrostructure.

GOST RF 11878. Austenitic steel. Methods for determining the content of the ferritic phase in bars.

GOST SU 26364-90. Ferritometers for austenitic steels. General specifications.

GOST RF 8.518-2010. Ferritometers for austenitic steels. Verification procedure.

GOST RF 2246-70 52. Welding steel wire. Technical conditions.

GOST RF 9466-75. Electrodes coated with metal for manual arc welding of steels and surfacing. Classification and general specifications.

GOST RF 53686-2009 (ISO 8249:2009). Welding. Determination of the content of the ferrite phase in the weld metal of austenitic and two-phase ferritic-austenitic chromium-nickel corrosion-resistant steels.

Gorkunov, E.S., V.M. Somova, T.P. Tsarkova and I.A. Kuznetsov. 1998. Express analysis of the chemical and phase composition of steels by the thermoelectric method. Russian journal of non-destructive testing. 3: 3–16.

Gorkunov, E.S., S.M. Zadvorkin, L.S. Goruliova and A.B. Buhvalov. 2012. On the effectiveness of the use of magnetic and electrical parameters of non-destructive control of microarrays of the crystal lattice of carbon steels after heat treatment. Russian journal of non-destructive testing. 3: 27–39.

Gulyaev, A.P. 1977. Metallography. Moscow: Metalurgy (in Russian).

Himchenko, N.V. and V.A. Bobrov. 1978. Nondestructive testing in chemical and petroleum engineering. Moscow: Mechanical Engineering.

International Standard ISO 8249-1985. Welding. Determination of ferrite number in austenitic weld metal deposited by covered Cr-Ni steel electrodes.

Jacobs, H.O., P. Leuchtmann, O.J. Homan and A. Stemmer. 1998. Resolution and contrast in Kelvin probe force microscopy- Journal of Applied Physics. 84(3): 1168–1173.

Kazantseva, N.V., V.P. Piliyugin, V.A. Zavalishin, M.B. Rigmant, D.I. Davidov and N.N. Stepanova. 2013. Effect of deformation on the magnetic properties of Ni3Al-based alloys. Material Letters. 13: 16–19.

Kershenbaum, V.Ia. (Ed.) 1995. Non-destructive testing. 3: 68–128.

Kliyuev, V.V. (Ed.). 2004. Non destructive testing. Moscow: Mechanical engineering.

Korkh, M.K., D.I. Davidov, J.V. Korkh, M.B. Rigmant, A.P. Nichipuruk and N.V. Kazantseva. 2015a. Phase control of austenitic chrome-nickel steel. AIP Conference Proceedings. 1683: 20097–20100.

Korkh, M.K., M.B. Rigmant, D.I. Davydov, D.A. Shishkin, A.P. Nichipuruk and Y.V. Korkh. 2015b. Determination of the phase composition of three-phase chromium-nickel steels from their magnetic properties. Russian journal of non-destructive testing. 51(12): 727–737.

Korkh, M.K., Yu.V. Korkh, M.B. Rigmant, N.V. Kazantseva and N.I. Vinogradov. 2016. Using Kelvin probe force microscopy for controlling the phase composition of austenite–martensite chromium–nickel steel. Russian journal of nondestructive testing. 52: 664 - 672.

Kurdiumov, V.G., L.M. Utevski and R.I. Entin 1977. Transformation in iron and steel. Moscow: Science.

Kuznetsov, I.A. and V.M. Okunev. 1993. Thermoelectric properties of steels and a device for monitoring chemical and phase compositions. Russian journal of non-destructive testing. 8: 78–84.

Litovchenko, I.Yu., N.V. Shevchenko, A.N. Tumentsev and E.P. Naiden. 2006. Phase composition and defective substructure of austenitic steel 02X17H14M2 after deformation by rolling at room temperature. Physical mesomechanics. 9: 137–140.

Maslenkov, S.B. 1983. Heat resistant steel and alloys. Moscow: Metallurgy (In Russian).

Melitz, W., J. Shen, A.C. Kummel and S. Lee. 2011. Kelvin probe force microscopy and its application. Surface Science Reports. 66: 1–27.

Merinov, P.E., S.D. Entin, B.I. Beketov and A.E. Runov. 1978. The magnetic testing of the ferrite content of austenitic stainless steel weld metal. NDT International. 11(1): 9–14.

Merinov, P.E. and A.G. Mazepa. 1997. Determination of martensite deformation in steels of austenitic class by magnetic method. Factory laboratory. 3: 47–49.

Mikheev, M.N. and E.S. Gorkunov. 1985. Magnetic methods for nondestructive testing of the structural state and strength characteristics of heat-treated steels. Russian journal of non-destructive testing. 3: 3–21.

Mirkin, L.I. 1978. X-ray control of engineering materials. Moscow: MGU.

Mironov, V.L. 2004. Fundamentals of scanning probe microscopy. Nizhny Novgorod: Russian Academy of Sciences, Institute of Physics of Microstructures.

Moore, P.O., R.K. Miller and R.K. Hill (Eds.) 2005. Nondestructive Testing Handbook, 3th edition. 6. INC: Acoustic Emission Testing, American Society for Nondestructive Testing.

Netesov, V.M. and A.A. Yaes. 1987. Effect of $\gamma \to \alpha$ transition on the electrical resistance and structure of X18H10T steel. Metalli. 3: 104–106.

Nonnenmacher, M., M.P. O'Boyle and H.K. Wickramasinghe. 1991. Kelvin probe force microscopy. Applied Physics Letters. 58(25): 2921–2923.

Örnek, C. and D.L. Engelberg. 2015. SKPFM measured Volta potential correlated with strain localisation in microstructure to understand corrosion susceptibility of cold-rolled grade 2205 duplex stainless steel. Corrosion Science. 99: 164–171.

Pustovoit, V.N. and Yu.V. Dolgachev. 2007. Features of the martensitiv transformation in steel under quenching in magnetic field. Vestnik of Donetsk State technical university. 7(4): 459–465.

Rigmant, M.B. and E.S. Gorkunov 1996. Meter for ferritic phase content – ferritometer FM-3 IMP. Russian journal of non-destructive testing. 5: 78–83.

Rigmant, M.B., S.V. Gladkovski, E.S. Gorkunov, P.T. Matafonov and S.V. Smirnov. 2000. On the possibility of magnetic non-destructive testing of elastoplastic deformations in steels with metastable austenite. Control. Diagnostics. 9(27): 62–63.

Rigmant, M.B., A.P. Nichipuruk, B.A. Khudyakov, V.S. Ponomarev, N.A. Tereshchenko and M.K. Korkh. 2005. Devices for magnetic phase analysis of products made of austenitic corrosion-resistant steels. Russian journal of non-destructive testing. 11: 3–15.

Rigmant, M.B., A.P. Nichipuruk and M.K. Korkh. 2012. The Possibility of Separate Measurements of the Amounts of Ferrite and Deformation Martensite in Three-Phase Austenitic-Class Steels Using the Magnetic Method. Russian journal of non-destructive testing. 48: 511–522.

Rigmant, M.B., M.K. Korkh, D.I. Davydov, D.A. Shishkin, Y.V. Korkh, A.P. Nichipuruk and N.V. Kazantseva 2015. Methods for revealing deformation martensite in austenitic–ferritic steels. Russian journal of non-destructive testing. 51(11): 680–691.

Runov, A.E. 1959. Control and correcting of quantity of the ferritic phase in phase in the welded and base metal of welded joints of austenitic steels. Welding. 6: 16–17.

Sandovski, V.A., A.I. Uvarov and T.P. Vasechkina. 2000. Influence of the structure on the eddy current transducer signal in aging metastable steels. Russian journal of non-destructive testing. 11: 43–57.

Sandovski, V.A., A.I. Uvarov, T.P. Vasechkina and E.I. Anufrieva. 2001. Study of discontinuous decay in austenitic alloys using an eddy-current transducer. Russian journal of non-destructive testing. 2: 57–66.

Sathirachinda, N., R. Pettersson and J. Pan. 2009. Depletion effects at phase boundaries in in 2205 duplex stainless steel characterized with SKPFM and TEM/EDS. Corrosion Science. 51: 1850–1860.

Schwartz, A. and R. Wiesendanger. 2008. Magnetic sensitive force microscopy. Nanotoday. 2(1-2): 28-39.

Shcherbinin, V.E. and E.S. Gorkunov 1996. Magnetic control of the quality of the metal. Ekaterinburg: UrD.

Stalmasek, E. 1986. Measurements of ferrite content in austenitic stainless steel weld metal giving internationally reproducible results. Welding Research Council Bulletin. 318: 22–97.

Stepanova, N.N., D.I. Davydov, A.P. Nichipuruk, M.B. Rigmant, N.V. Kazantseva, N.I. Vinogradova, A.N. Pirogov and E.P. Romanov. 2011. Structure and magnetic properties of a nickel heat-resistant alloy after a high-temperature treatments. Physics of metal and metallography. 112: 328–336.

Ulianin, E.A. 1994. Corrosion resistance steel and alloys. Moscow: Metallurgy.

Utevski, L.M. 1973. Diffraction electron microscopy. Moscow: Metallurgy.

Vedeniov, M.A., V.S. Ponomariov, V.G., Kuleyov, M.B. Rigmant, N.P. Kolomeez, G.S. Chernova, A.G. Lavrentiev and I.V. Tretyakov. 1993. Device for monitoring the changes in the magnetic state of sheets of weakly magnetic austenitic steels is the F-01 ferritometer. Russian journal of non-destructive testing. 3: 3–9.

Index

///////////////////////////

Acoustic nondestructive testing 98,
    199
Activation energy 14-17, 26, 165
Antiphase boundaries (APB) 71, 94,
    155
Alloying elements:
    for Ni$_3$Al 1-2, 4, 11, 18, 31, 34-35,
        39, 58-60, 66, 70, 73, 82, 86-89,
        127, 160-161
    for Co-base superalloys 135, 169-
        170, 191
    for Iron superalloys 195, 201, 230
Austenitic steels 193-196, 198-209,
    212, 224, 227, 230, 237-239

Blade vibrations 103, 115, 117, 124-
    126
Bridgman method 9, 11-12, 32, 39, 60,
    65-66, 70, 99, 132, 156, 173, 180,
    185

Chemical composition of the superalloys:
    ZhS36-VI 45
    EI-437B 36
    CNK-8MP 36
    ZhS-36 36
    ZhS-26 39
    ZhS-32 39
    VKNA-1V 60
    VKNA-4U 60
    EP-800 74

Rene 80 80
ChS-70V 86
EI-437B 98
TsNK-8MP 99
CMSX-4 32
Chemical composition:
    of the carbides 50, 74
    of the alloys used for coating 57
    of the main grades of corrosion
        resistant steels 193-194, 206
CoAl intermetallic compound (B2)
    150-155, 166-167, 178-181
Co$_3$W intermetallic compound (D0$_{19}$)
    131, 141-145, 148-149, 166, 173,
    178, 186-188
Coefficients of thermal expansion 3
Coercive force 179-183, 189-190, 208,
    218-219, 233
Curie temperature 78, 85, 91-92, 151,
    177-179, 181-183, 229

Deformation behavior:
    of Ni$_3$Al 155-156,
    of Co$_3$(Al,W) 162-164
Degree of long-range order 2, 4, 92,
    146, 152, 182
Debye temperature 13-14, 18, 37, 65,
    172
Deformation of Ni$_3$Al 27, 71, 91, 96,
    155
Diffusion coefficients 15-16

Superalloys—Analysis and control of failure process

Dislocation density 61-62, 70, 72, 90-92, 97,155-157, 162, 182
Disordering process 5, 18
Domain structure 76
Elastic moduli 12-13, 18, 36, 44, 175-176

Electrical resistivity 6-8, 11, 17, 38-40, 45, 193, 198-199, 237, 239
Electrical conductivity 7
Elastic constants 101, 110-111, 171-172

Ferritometry 200-203
Ferrite content 200, 204-205, 208, 217, 219, 222-226, 233
Fermi level 7, 27
Fermi energy 25
Ferromagnetic inclusions 196, 219, 224, 227-229

Gorsky effect 89

Hardness 175-177, 193, 195, 234-235
High-temperature treatment of melt (HTMT) 38-44, 46-47, 49, 53
Hysteresis loops 84, 182, 212, 217-219, 233

Interatomic interactions 5, 12, 24,
Ion-plasma coatings 56-57

Lattice parameter:
  for Ni₃Al-based alloys 2-3, 17, 21, 25, 34, 43-44, 50-51, 54, 68-70, 88, 92-93
  for carbides 75, 77
  for Al-Co-W alloys 133, 143, 146, 151-153, 169-170
  for Iron superalloys 196
Long-period structures 71, 75-77, 94-95, 136
Long-term strength 33-41, 44, 47-49, 52-53, 56, 58, 62, 127

Magnetic Control 203-204
Magnetic properties:
  of the nickel superalloys 85
  of the Co₃(Al,W)-based Alloys 177-180

of the phases in Co₃(Al,W)-based alloys 180-183
of Iron superalloys 192, 200, 205, 210-212, 217-221, 233, 235
Magnetic susceptibility:
  of nickel superalloys 79-81, 83-87, 89-94, 97
  of cobalt superalloys 179
  of Co₃(Al,W) alloys 180
  of Iron superalloys 208, 220, 227-230
Magnetic transitions 98, 173, 181
Magnetic saturation 200-202, 204
Martensitic transformation 132, 136-138, 199, 206-207, 231
Martensite 134,-140, 198-213, 217-227, 229-240
Mechanical properties:
  of Ni₃Al-based alloys 30, 41, 49-50, 59-60, 63-65, 112, 127
  of Co₃(Al,W)-based alloys 131, 171, 175-176, 191
  of austenitic steels 193
Melt-spun ribbon 138-139, 149
Metastable state 9, 12, 66
Metastable phase 9-10, 65-68, 70, 72
Microhardness 57, 163, 176
Miscibility gap 173, 179
Mismatch of the lattice parameters 34, 38, 43, 168-170
Modulated structure 73-77, 94
μ-phase 35-36, 141, 144, 148-150, 167, 170

Nano-scale structure 94, 183, 189
Non-destructive magnetic methods 77

Orientation relationships 67, 69, 139, 145

Pair interaction energy 24-25
Phase diagram 9, 11, 36, 66, 131, 133, 137, 140-141, 143, 148, 151, 167, 177, 179, 186
Phase transitions:
  in Ni₃Al-based alloys 12
  in nickel superalloys 45, 69
  in carbides 76

in Co-Al alloys 132-136
in Co-W alloys 136-140
in Co-Al-W alloys 140, 182
in Iron superalloys 231, 235
Positron annihilation 15, 18-22
Prismatic loops 153-154

Raft structure 35, 47, 62

Segregation 28, 46, 129, 191, 210
Self-diffusion 14, 16
Severe plastic deformation 68, 72-73,
    76, 79, 91, 96
Short-range order 8
Single-Crystal Blades 103, 117, 123-126
Solvus temperature 130, 141, 148, 167,
    170-174, 185, 190
Specific electrical resistance 198, 230-
    231, 237-238
Spinodal decomposition 66, 174, 187
Stacking faults 71, 83, 94, 96, 138,
    140, 155-156, 158, 162
Strain-induced magnetism 80

Stripe magnetic domain structure
    180-184
Superplasticity 29

Temperature distribution 108-110
Thermal stress of the blade 98, 104,
    110, 124
Thermal X-Ray Diffraction (XRD)
    analysis 132
Topologically Close-Packed phases
    (TCP-phases) 35-36
Turbine blades 32, 38, 56, 59, 78-81,
    85, 89-91, 94, 98, 126, 229-230
Twining 31, 138

Ultra-Dispersed Powder of titanium
    carbonitride (UDP) 48-54

X-ray studies 2, 8, 13, 43,

Yield strength 27-29, 191, 193, 207
Yield anomaly in Co3(Al,W) 164
Young's modulus 175-177

Authors' Biography

Nataliya V. Kazantseva is chief researcher, professor, Doctor of Sciences in physics and mathematics. She is a specialist in the TEM study of the structure and phase transformations in the alloys and intermetallic compounds. She has a Best Publication Diploma from Pleiades Publishing, Inc. and Russian Academy of Sciences, 1996. In 2010 and 2013, she took part in the international exchange Program for the best scientists of USA and Russian Federation. She is the author of over 80 scientific works. E-mail: kazantseva@imp.uran.ru

Natalia N. Stepanova is leading researcher, professor, Doctor of Sciences in physics and mathematics. Stepanova N. N. is a specialist in the structural and phase transformations in Ni_3Al-base alloys and nickel super alloys. She also works in collaboration with industry and studies the structure and phase stability of the modern nickel super alloys. Natalia N. Stepanova is the author of over 130 scientific works, including two collective monographs. E-mail: snn@imp.uran.ru

Mikhail B. Rigmant is senior researcher, Professor, PhD in physics and mathematics. He is a member of the Russian Society for Non-Destructive Testing and Technical Diagnostics and of the Bulgarian Society for Non-Destructive Testing (BG S NDT). He has specialized in the magnetic non-destructive testing of the phase composition and structure of the austenitic steel and alloys. He is an author of 7 RF Patents and of over 70 scientific works. E-mail: rigmant@imp.uran.ru

Yurii S. Vorobiov is leading researcher, Professor, Doctor of Technical Sciences. He is a member of the Ukrainian National Committee on Theoretical and Applied Mechanics, Member of Technical Committee Rotor Dynamics of IFFToMM, Academic of Ukraine Universities Academy, Academic of Ukraine Engineering Academy. In 1984 he received the State Prize of Ukraine in the field of science and technology. He was awarded the Order "Badge of Honor" and a number of medals. In 2016 he was awarded the Yaroslav Mudry Award of the University Academy of Ukraine. He is the author of over 550 scientific works. E-mail: vorobiev@ipmach.kharkov.ua

Printed and bound by CPI Group (UK) Ltd, Croydon, CR0 4YY

24/10/2024

01778304-0004